TOWARD BETTER ENVIRONMENTAL DECISION-MAKING

Committee on Assessing and Valuing the Services of Aquatic and Related Terrestrial Ecosystems

Water Science and Technology Board

Division on Earth and Life Studies

NATIONAL RESEARCH COUNCIL
OF THE NATIONAL ACADEMIES

THE NATIONAL ACADEMIES PRESS
Washington, D.C.
www.nap.edu

THE NATIONAL ACADEMIES PRESS 500 Fifth Street, N.W. Washington, DC 20001

NOTICE: The project that is the subject of this report was approved by the Governing Board of the National Research Council, whose members are drawn from the councils of the National Academy of Sciences, the National Academy of Engineering, and the Institute of Medicine. The members of the committee responsible for the report were chosen for their special competences and with regard for appropriate balance.

Support for this project was provided by the U.S. Environmental Protection Agency under Award No. X-82872401; U.S. Army Corps of Engineers Award No. DACW72-01-P-0076; U.S. Department of Agriculture, Cooperative State Research, Education, and Extension Service under Award No. 2001-38832-11510; U.S. Department of Agriculture-Research, Education, and Economics, Agricultural Research Service, Administrative and Financial Management, Extramural Agreements Division under Award No. 59-0790-1-136. Any opinions, findings, conclusions, or recommendations expressed in this publication are those of the author(s) and do not necessarily reflect the views of the organizations or agencies that provided support for the project.

International Standard Book Number 0-309-09318-X (Book)
International Standard Book Number 0-309-54586-2 (PDF)

Library of Congress Control Number 2005924663

Additional copies of this report are available from the National Academies Press, 500 Fifth Street, N.W., Lockbox 285, Washington, DC 20055; (800) 624-6242 or (202) 334-3313 (in the Washington metropolitan area); Internet, http://www.nap.edu.

Cover design by Van Nguyen, National Academies Press. Cover photograph by Lauren Alexander, Staff Officer with the Water Science and Technology Board, National Research Council.

Printed in the United States of America.

THE NATIONAL ACADEMIES

Advisers to the Nation on Science, Engineering, and Medicine

The **National Academy of Sciences** is a private, nonprofit, self-perpetuating society of distinguished scholars engaged in scientific and engineering research, dedicated to the furtherance of science and technology and to their use for the general welfare. Upon the authority of the charter granted to it by the Congress in 1863, the Academy has a mandate that requires it to advise the federal government on scientific and technical matters. Dr. Bruce M. Alberts is president of the National Academy of Sciences.

The **National Academy of Engineering** was established in 1964, under the charter of the National Academy of Sciences, as a parallel organization of outstanding engineers. It is autonomous in its administration and in the selection of its members, sharing with the National Academy of Sciences the responsibility for advising the federal government. The National Academy of Engineering also sponsors engineering programs aimed at meeting national needs, encourages education and research, and recognizes the superior achievement of engineers. Dr. Wm. A. Wulf is president of the National Academy of Engineering.

The **Institute of Medicine** was established in 1970 by the National Academy of Sciences to secure the services of eminent members of appropriate professions in the examination of policy matters pertaining to the health of the public. The Institute acts under the responsibility given to the National Academy of Sciences by its congressional charter to be an adviser to the federal government and, upon its own initiative, to identify issues of medical care, research, and education. Dr. Harvey V. Fineberg is president of the Institute of Medicine.

The **National Research Council** was organized by the National Academy of Sciences in 1916 to associate the broad community of science and technology with the Academy's purposes of furthering knowledge and advising the federal government. Functioning in accordance with general policies determined by the Academy, the Council has become the principal operating agency of both the National Academy of Sciences and the National Academy of Engineering in providing services to the government, the public, and the scientific and engineering communities. The Council is administered jointly by both Academies and the Institute of Medicine. Dr. Bruce M. Alberts and Dr. Wm. A. Wulf are chair and vice-chair, respectively, of the National Research Council.

www.national-academies.org

COMMITTEE ON ASSESSING AND VALUING THE SERVICES OF AQUATIC AND RELATED TERRESTRIAL ECOSYSTEMS

GEOFFREY M. HEAL, *Chair*, Columbia University, New York
EDWARD B. BARBIER, University of Wyoming, Laramie
KEVIN J. BOYLE, University of Maine, Orono
ALAN P. COVICH, University of Georgia, Athens
STEVEN P. GLOSS, Southwest Biological Science Center, U.S. Geological Survey, Tucson, AZ
CARLTON H. HERSHNER, Virginia Institute of Marine Science, Gloucester Point
JOHN P. HOEHN, Michigan State University, East Lansing
CATHERINE M. PRINGLE, University of Georgia, Athens
STEPHEN POLASKY, University of Minnesota, St. Paul
KATHLEEN SEGERSON, University of Connecticut, Storrs
KRISTIN SHRADER-FRECHETTE, University of Notre Dame, Notre Dame, Indiana

National Research Council Staff

MARK C. GIBSON, Study Director
ELLEN A. DE GUZMAN, Research Associate

Preface

The development of the ecosystem services paradigm has enhanced our understanding of how the natural environment matters to human societies. We now think of the natural environment, and the ecosystems of which it consists, as natural capital—a form of capital asset that, along with physical, human, social, and intellectual capital, is one of society's important assets. As President Theodore Roosevelt presciently said in 1907,

> *The nation behaves well if it treats the natural resources as assets which it must turn over to the next generation increased and not impaired in value.*[1]

Economists normally value assets by the value of services that they provide: Can we apply this approach to ecological assets by valuing the services provided by ecosystems?

An ecosystem is generally accepted to be an interacting system of biota and its associated physical environment. Aquatic and related terrestrial ecosystems are among the most important ecosystems in the United States, and Congress through the Clean Water Act has recognized the importance of the services they provide and has shown a concern that these services be restored and maintained. Such systems intuitively include streams, rivers, ponds, lakes, estuaries, and oceans. However, most ecologists and environmental regulators include vegetated wetlands as aquatic ecosystems, and many also think of underlying groundwater aquifers as potential members of the set. Thus, the inclusion of "related terrestrial ecosystems" for consideration in this study is a reflection of the state of the science that recognizes the multitude of processes linking terrestrial and aquatic systems.

Many of the policies implemented by various federal, state, and local regulatory agencies can profoundly affect the nation's aquatic and related terrestrial ecosystems, and in consequence, these bodies have an interest in better understanding the nature of their services, how their own actions may affect them, and what value society places on their services. The need for this study was recognized in 1997 at a strategic planning session of Water Science and Technology Board (WSTB) of the National Research Council (NRC). The Committee on Assessing and Valuing the Services of Aquatic and Related Terrestrial Ecosystems was established by the NRC in early 2002 with support from the U.S. Environmental Protection Agency (EPA), U.S. Army Corps of Engineers

[1] Inscribed on the wall of the entrance hall of the American Museum of Natural History, Washington, D.C.

(USACE), and U.S. Department of Agriculture (USDA). Its members are drawn from the ranks of economists, ecologists, and philosophers who have professional expertise relating to aquatic ecosystems and the valuation of ecosystem services.

In drafting this report the committee members have sought to understand and integrate the disciplines, primarily ecology and economics, that cover the field of ecosystem service valuation. In fact, the committee quickly discovered that this is not an established field—ecologists have only recently begun to think in terms of ecosystem services and their determinants, while economists have likewise only very recently begun to incorporate the factors affecting ecosystem services into their valuations of these services. If we as a society are to understand properly the value of our natural capital, which is a prerequisite for sensible conservation decisions, then this growing field must be developed further and this report provides detailed recommendations for facilitating that development. Although the field is relatively new, a great deal is understood, and consequently the committee makes many positive conclusions and recommendations concerning the methods that can be applied in valuing the services of aquatic and related terrestrial ecosystems. Furthermore, because the principles and practices of valuing ecosystem services are rarely sensitive to whether the underlying ecosystem is aquatic or terrestrial, the report's various conclusions and recommendations are likely to be directly, or at least indirectly applicable to valuation of the goods and services provided by any ecosystem.

The study benefited greatly from the knowledge and expertise of those who made presentations at our meetings, including Richard Carson, University of California, San Diego; Harry Kitch, USACE; John McShane, EPA; Angela Nugent, EPA; Michael O'Neill, USDA; Mahesh Podar, EPA (retired); John Powers, EPA; Stephen Schneider, Stanford University; and Eugene Stakhiv, USACE Institute for Water Resources. The success of the report also depended on the support of the NRC staff working with the committee, and it is a particular pleasure to acknowledge the immense assistance of study director Mark Gibson and WSTB research associate Ellen de Guzman. Finally, of course, the committee members worked extraordinarily hard and with great dedication, expertise, and good humor in pulling together what was initially a rather disparate set of issues and methods into the coherent whole that follows.

This report was reviewed in draft form by individuals chosen for their diverse perspectives and technical expertise in accordance with the procedures approved by the NRC's Report Review Committee. The purpose of this independent review is to provide candid and critical comments that will assist the institution in making its published report as sound as possible and to ensure that the report meets institutional standards for objectivity, evidence, and responsiveness to the study charge. The review comments and draft manuscript remain confidential to protect the integrity of the deliberative process. We wish to thank the following individuals for their review of this report: Mark Brinson, East Carolina University, Greenville, North Carolina; J. Baird Callicott, University of North Texas, Denton; Nancy Grimm, Arizona State University, Tempe;

Michael Hanemann, University of California, Berkeley; Peter Kareiva, The Nature Conservancy, Seattle, Washington; Raymond Knopp, Resources for the Future, Washington, D.C.; Sandra Postel, Global Water Policy Project, Amherst, Massachusetts; and Robert Stavins, Harvard University, Cambridge.

Although the reviewers listed above have provided many constructive comments and suggestions, they were not asked to endorse the conclusions or recommendations, nor did they see the final draft of the report before its release. The review of this report was overseen by John Boland, Johns Hopkins University, Baltimore. Appointed by the National Research Council, he was responsible for making certain that an independent examination of the report was carefully carried out in accordance with institutional procedures and that all review comments were carefully considered. Responsibility for the final content of this report rests entirely with the authoring committee and the NRC.

Geoffrey M. Heal, *Chair*

Contents

Executive Summary

OVERVIEW

Ecosystems provide a wide variety of marketable goods, fish and lumber being two familiar examples. However, society is increasingly recognizing the myriad functions—the observable manifestations of ecosystem processes such as nutrient recycling, regulation of climate, and maintenance of biodiversity—that they provide, without which human civilizations could not thrive. Derived from the physical, biological, and chemical processes at work in natural ecosystems, these functions are seldom experienced directly by users of the resource. Rather, it is the services provided by ecosystems, such as flood risk reduction and water supply, together with ecosystem goods, that create value for human users and are the subject of this report.[1]

Aquatic ecosystems include freshwater, marine, and estuarine surface waterbodies. These incorporate lakes, rivers, streams, coastal waters, estuaries, and wetlands, together with their associated flora and fauna. Each of these entities is connected to a greater ecological and hydrological landscape that includes adjacent riparian areas, upland terrestrial ecosystems, and underlying groundwater aquifers. Thus, the term "aquatic ecosystems" in this report includes these related terrestrial ecosystems and underlying aquifers. Aquatic ecosystems perform numerous interrelated environmental functions and provide a wide range of important goods and services. Many aquatic ecosystems enhance the economic livelihood of local communities by supporting commercial fishing and agriculture and by serving the recreational sector. The continuance or growth of these types of economic activities is directly related to the extent and health of these natural ecosystems.

However, human activities, rapid population growth, and industrial, commercial, and residential development have all led to increased pollution, adverse modification, and destruction of remaining (especially pristine) aquatic ecosys-

---

[1] *Ecosystem structure* refers to both the composition of the ecosystem (i.e., its various parts) and the physical and biological organization defining how those parts are organized. A leopard frog or a marsh plant such as a cattail, for example, would be considered a component of an aquatic ecosystem and hence part of its structure. *Ecosystem function* describes a process that takes place in an ecosystem as a result of the interactions of the plants, animals, and other organisms in the ecosystem with each other or their environment. Primary production (the process of converting inorganic compounds into organic compounds by plants, algae, and chemoautotrophs) is an example of an ecosystem function. Ecosystem structure and function provide various *ecosystem goods* and *services* of value to humans such as fish for recreational or commercial use, clean water to swim in or drink, and various esthetic qualities (e.g., pristine mountain streams or wilderness areas) (see Box 3-1 for further information).

tems—despite an increase in federal, state, and local regulations intended to protect, conserve, and restore these natural resources. Increased human demand for water has simultaneously reduced the amount available to support these ecosystems. Notwithstanding the large losses and changes in these systems, aquatic ecosystems remain broadly and heterogeneously distributed across the nation. For example, there are almost 4 million miles of rivers and streams, 59,000 miles of ocean shoreline waters, and 5,500 miles of Great Lakes shoreline in the United States; there are 87,000 square miles of estuaries, while lakes, reservoirs, and ponds account for more than 40 million acres.

Despite growing recognition of the importance of ecosystem functions and services, they are often taken for granted and overlooked in environmental decision-making. Thus, choices between the conservation and restoration of some ecosystems and the continuation and expansion of human activities in others have to be made with an enhanced recognition of this potential for conflict and of the value of ecosystem services. In making these choices, the economic values of the ecosystem goods and services must be known so that they can be compared with the economic values of activities that may compromise them and so that improvements to one ecosystem can be compared to those in another.

This report was prepared by the National Research Council (NRC) Committee on Assessing and Valuing the Services of Aquatic and Related Terrestrial Ecosystems, overseen by the NRC's Water Science and Technology Board, and supported by the U.S. Army Corps of Engineers, U.S. Environmental Protection Agency, and the U.S. Department of Agriculture (see Box ES-1). The committee consisted of 11 volunteer experts drawn from the fields of ecology, economics, and philosophy who have professional expertise relating to aquatic ecosystems and to the valuation of ecosystem services. This report's contents, conclusions, and recommendations are based on a review of relevant technical literature, information gathered at five committee meetings, and the collective expertise of committee members. Because of space limitations, this Executive Summary includes only the major conclusions and related recommendations of the committee in the general order of their appearance in the report. More detailed conclusions and recommendations can be found throughout the report.

Valuing ecosystem services requires the successful integration of ecology and economics and presents several challenges that are discussed throughout this report. The fundamental challenge of valuing ecosystem services lies in providing an explicit description and adequate assessment of the links between the structures and functions of natural systems, the benefits (i.e., goods and services) derived by humanity, and their subsequent values (see Figure ES-1).

Ecosystems are complex however, making the translation from ecosystem structure and function to ecosystem goods and services (i.e., the ecological production function) is even more difficult. Similarly, in many cases the lack of markets and market prices and of other direct behavioral links to underlying values makes the translation from quantities of goods and services to value (and the direct translation from ecosystem structure to value) quite difficult, though

BOX ES-1
Statement of Task

The committee will evaluate methods for assessing services and the associated economic values of aquatic and related terrestrial ecosystems. The committee's work will focus on identifying and assessing existing economic methods to quantitatively determine the intrinsic value of these ecosystems in support of improved environmental decision-making, including situations where ecosystem services can be only partially valued. The committee will also address several key questions, including:

- What is the relationship between ecosystem services and the more widely studied ecosystem functions?
- For a broad array of ecosystem types, what services can be defined, how can they be measured, and is the knowledge of these services sufficient to support an assessment of their value to society?
- What lessons can be learned from a comparative review of past attempts to value ecosystem services—particularly, are there significant differences between eastern and western U.S. perspectives on these issues?
- What kinds of research or syntheses would most rapidly advance the ability of natural resource managers and decision makers to recognize, measure, and value ecosystem services?
- Considering existing limitations, error, and bias in the understanding and measurement of ecosystem values, how can available information best be used to improve the quality of natural resource planning, management, and regulation?

both are given by an economic valuation function. Probably the greatest challenge for successful valuation of ecosystem services is to integrate studies of the ecological production function with studies of the economic valuation function. To do this, the definitions of ecosystem goods and services must match across studies. Failure to do so means that the results of ecological studies cannot be carried over into economic valuation studies. Attempts to value ecosystem services without this key link will either fail to have ecological underpinnings or fail to be relevant as valuation studies.

Where an ecosystem's services and goods can be identified and measured, it will often be possible to assign values to them by employing existing economic valuation methods. The emerging desire to measure the environmental costs of human activities, or to assess the benefits of environmental protection and restoration, has challenged the state of the art in environmental evaluation in both the ecological and the social sciences. Some ecosystem goods and services cannot be valued because they are not quantifiable or because available methods are not

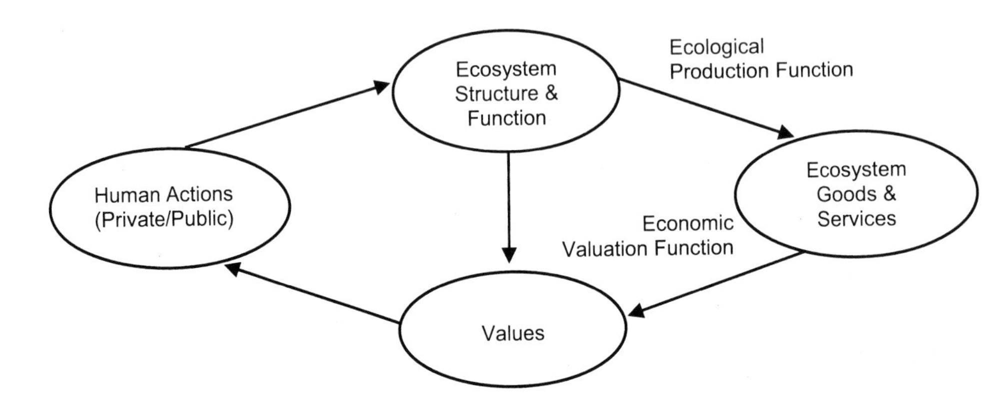

FIGURE ES-1 Components of ecosystem valuation: ecosystem structure and function, goods and services, human actions, and values. (See Figure 7-1 for an expanded version of this figure.)

appropriate or reliable. Economic valuation methods can be complex and demanding, and the results of applying these methods may be subject to judgment, uncertainty, and bias. However, based on an assessment of a very large literature on the development and application of various economic valuation methods, the committee concludes that they are mature and capable of providing useful information in support of improved environmental decision-making.

From an ecological perspective, the challenge is to interpret basic research on ecosystem functions so that service-level information can be communicated to economists. For economic and related social sciences, the challenge is to identify the values of both tangible and intangible goods and services associated with ecosystems and to address the problem of decision-making in the presence of partial valuation. The combined challenge is to develop and apply methods to assess the values of human-induced changes in ecosystem functions and services.

Finally, this report concerns valuing the goods and services that ecosystems provide to human societies, with principal focus on those provided by aquatic and related terrestrial ecosystems. However, because the principles and practices of valuing ecosystem goods and services are rarely sensitive to whether the underlying ecosystem is strictly aquatic or terrestrial, many of the report's conclusions and recommendations are likely to be directly or at least indirectly applicable to the valuation of goods and services provided by any ecosystem.

THE MEANING OF VALUE AND USE OF ECONOMIC VALUATION IN THE ENVIRONMENTAL POLICY DECISION-MAKING PROCESS

In order to develop a perspective on valuing aquatic ecosystems, it is necessary to first provide a clear discussion and statement of what it means to value something and of the role of "valuation" in environmental policymaking. In this regard, environmental issues and ecosystems have been at the core of many recent philosophical discussions regarding value (see Chapter 2). Fundamentally, these debates about the value of ecosystems derive from two points of view. The first is that the values of ecosystems and their services are non-anthropocentric and that nonhuman species have moral interests or rights unto themselves. The other, which includes the economic approach to valuation, is that all values are anthropocentric. This report focuses on the sources of value that can be captured through economic valuation.[2] However, the committee recognizes that all forms of value may ultimately contribute to decisions regarding ecosystem use, preservation, or restoration.

Although economic valuation does not capture all sources or types of value (e.g., intrinsic values on which the notion of rights is founded), it is much broader than usually presumed. It recognizes that economic value can stem from the use of an environmental resource (use values), including both commercial and noncommercial uses, or from its existence even in the absence of use (nonuse value). The broad array of values included under this approach is captured by using the total economic value (TEV) framework to identify potential sources of this value. Use of the TEV framework helps to provide a checklist of potential impacts and effects that need to be considered in valuing ecosystem services as comprehensively as possible. By its nature, economic valuation involves the quantification of values based on a common metric, normally a monetary metric. The use of a dollar metric for quantifying values is based on the assumption that individuals are willing to trade the ecological service being valued for more of other goods and services represented by the metric (more dollars). Use of a monetary metric allows measurement of the costs or benefits associated with changes in ecosystem services.

The role of economic valuation in environmental decision-making depends on the specific criteria used to choose among policy alternatives. If policy choices are based primarily on intrinsic values, there is little need for the quantification of values through economic valuation. However, if policymakers consider trade-offs and benefits and costs when making policy decisions, then quantification of the value of ecosystem services is essential. Failure to include some measure of the value of ecosystem services in benefit-cost calculations will implicitly assign them a value of zero. The committee believes that considering

---

[2] Unless otherwise noted, use of the terms "value," "valuing," or "valuation" refers to economic valuation, more specifically, the economic valuation of ecosystem goods and services.

the best available and most reliable information about the benefits of improvements in ecosystem services or the costs of ecosystem degradation will lead to improved environmental decision-making. The committee recognizes, however, that this information is likely to be only one of many possible considerations that influence policy choice.

The benefit and cost estimates that emerge from an economic valuation exercise will be influenced by the way in which the valuation question is framed. In particular, the estimates will depend on the delineation of changes in ecosystem goods or services to be valued, the scope of the analysis (in terms of both the geographical boundaries and the inclusion of relevant stakeholders), and the temporal scale. In addition, the valuation question can be framed in terms of two alternative measures of value, willingness to pay (WTP) and willingness to accept (compensation) (WTA). These two approaches imply different presumptions about the distribution of property rights and can differ substantially, depending on the availability of substitutes and income limitations. In many contexts, methodological limitations necessitate the use of WTP rather than WTA.

Finally, because ecosystem changes are likely to have long-term impacts, some accounting of the timing of impacts is necessary. This can be done through discounting future costs and benefits. It is essential, however, to recognize that consumption discounting is distinct from the discounting of utility, which reflects the weights put on the well-being of different generations.

Based on these conclusions, the committee makes the following recommendations (Chapter 2):

- Policymakers should use economic valuation as a means of evaluating the trade-offs involved in environmental policy choices; that is, an assessment of benefits and costs should be part of the information set available to policymakers in choosing among alternatives.
- If the benefits and costs of a policy are evaluated, the benefits and costs associated with changes in ecosystem services should be included along with other impacts to ensure that ecosystem effects are adequately considered in policy evaluation.
- Economic valuation of changes in ecosystem services should be based on the comprehensive definition embodied in the TEV framework; both use and nonuse values should be included.
- The valuation exercise should be framed properly. In particular, it should value the *changes* in ecosystem good or services attributable to a policy change.
- In the aggregation of benefits and/or costs over time, the consumption discount rate, reflecting changes in scarcity over time, should be used instead of the utility discount rate.

AQUATIC AND RELATED TERRESTRIAL ECOSYSTEMS

An ecosystem is generally accepted to be an interacting system of biota and its associated physical environment; ecologists tend to think of these systems as identifiable at many different scales with boundaries selected to highlight internal and external interactions. The phrase "aquatic and related terrestrial ecosystems" recognizes the impossibility of analyzing aquatic systems absent consideration of the linkages to adjacent terrestrial environments. For many of the ecosystem functions and derived services considered in this report, it is not possible, necessary, or appropriate to delineate clear spatial boundaries between aquatic and related terrestrial systems (see also Box 3-1). Indeed, to the extent there is an identifiable boundary, it is often dynamic in both space and time.

The conceptual challenges of valuing ecosystem services are explicit description and adequate assessment of the link between the structure and function of natural systems and the goods or services derived by humanity (see Figure ES-1). Describing structure is a relatively straightforward process, even in highly diverse ecosystems. However, ecosystem functions are often difficult to infer from observed structure in natural systems. Furthermore, the relationship between structure and function, as well as how these attributes respond to disturbance, are not often well understood. Without comprehensive understanding of the behavior of aquatic systems, it is clearly difficult to describe thoroughly all of the services these systems provide society. Although valuing ecosystem services that are not completely understood is possible (see more below), when valuation becomes an important input in environmental decision-making, there is the risk that it may be incomplete.

There have only been a few attempts to develop explicit maps of the linkage between aquatic ecosystem structure/function and value. There are, however, a multitude of efforts to separately identify ecosystem functions, goods, services, values, and/or other elements in the linkage, without developing a comprehensive argument. One consequence of this disconnect is a diverse literature that suffers somewhat from indistinct terminology, highly variable perspectives, and considerable, divergent convictions. However, the development of an interdisciplinary terminology and a universally applicable protocol for valuing aquatic ecosystems was ultimately identified by the committee as unnecessary. From an ecological perspective, the value of specific ecosystem functions/services is entirely relative. The spatial and temporal scales of analysis are critical determinants of potential value. Ecologists have described the structure and function of most types of aquatic ecosystems qualitatively, and general concepts regarding the linkages between ecosystem function and services have been developed. Although precise quantification of these relationships remains elusive, the general concepts seem to offer sufficient guidance for valuation to proceed with careful attention to the limitations of any ecosystem assessment. Further integration of economics and ecology at both intellectual and practical scales will improve ecologists' ability to provide useful information for assessing and valuing aquatic ecosystems.

There remains a need for a significant amount of research in the ongoing effort to codify the linkage between ecosystem structure and function and the provision of goods and services for subsequent valuation. The complexity, variability, and dynamic nature of aquatic ecosystems make it likely that a comprehensive identification of all functions and derived services may never be achieved. Nevertheless, comprehensive information is not generally necessary to inform management decisions. Despite this unresolved state, future ecosystem valuation efforts can be improved through use of several general guidelines and by research in the following areas (Chapter 3):

- Aquatic ecosystems generally have some capacity to provide consumable resources, habitat for plants and animals, regulation of the environment, and support for nonconsumptive uses, and considerable work remains to be done in documentation of the potential of various aquatic ecosystems for contribution in each of these broad areas.
- Because delivery of ecosystem goods and services occurs in both space and time, investigation of the spatial and temporal thresholds of significance for various ecosystem services is necessary to inform valuation efforts.
- Natural systems are dynamic and frequently exhibit nonlinear behavior, and caution should be used in extrapolation of measurements in both space and time. Although it is not possible to avoid all mistakes in extrapolation, the uncertainty warrants explicit acknowledgment. Methods are needed to assess and articulate this uncertainty as part of system valuations.

METHODS OF NONMARKET VALUATION

In response to the committee's statement of task (see Box ES-1), this report outlines the major nonmarket methods currently available for estimating monetary values of aquatic and related terrestrial ecosystem services. This includes a review of the economic approach to valuation, which is based on the aforementioned TEV framework. In addition to presenting valuation approaches, the applicability of each method to valuing ecosystem services is discussed. All of this is provided within the context of the committees' implicit objective of assessing the literature in order to facilitate original studies that will develop a closer link between aquatic ecosystem functions, services, and value estimates. It is important to note however, that the report does not provide instructions on how to apply each of the methods, but rather provides a rich listing of references that can be used to develop a greater understanding of any of the methods.

There is a variety of nonmarket valuation approaches that are currently available to be applied in valuing aquatic and related terrestrial ecosystem services. Revealed-preference methods (e.g., averting behavior, travel cost, hedonics) can be applied only to a limited number of ecosystem services. However, both the range and the number of services that can potentially be valued are increasing with the development of new methods, such as dynamic production function approaches, general equilibrium modeling of integrated ecological-economic systems, and combined revealed- and stated-preference approaches.

Stated-preference methods, including contingent valuation and conjoint analysis, can be more widely applied, and certain values can be estimated only through the application of such techniques. On the other hand, the credibility of estimated values for ecosystem services derived from stated-preference methods has often been criticized. For example, contingent valuation methods have come under such scrutiny that it led to National Oceanic and Atmospheric Administration guidelines of "good practice" for these methods in the early 1990s.

Benefit transfers and replacement cost and cost of treatment methods are increasingly being used in environmental valuation, although their application to aquatic ecosystem services is still limited. Economists generally consider benefit transfers as to be a "second-best" valuation method and have devised guidelines governing their use. In contrast, replacement cost and cost of treatment methods should be used with great caution if at all. Although economists have attempted to design strict guidelines for using replacement cost as a last resort "proxy" valuation estimation for an ecological service, in practice estimates employing the replacement cost or cost of treatment approach rarely conform to the conditions outlined by such guidelines.

At least three basic questions arise for any method that is chosen to value aquatic ecosystem services. First, are the services that have been valued those that are the most important for supporting environmental decision-making and policy analyses involving benefit-cost analysis, regulatory impact analysis, legal judgments, and so on? Second, can the services of the aquatic ecosystem that are valued be linked in some substantial way to changes in the functioning of the system? Last, are there important services provided by aquatic ecosystems that have not yet been valued so that they are not being given full consideration in policy decisions that affect the quantity and quality of these systems? In many ways, the answers to these questions are the most important criteria for judging the overall validity of the valuation method chosen.

Only a limited number of ecosystem services have been valued to date, and effective treatment of aquatic ecosystem services in benefit-cost analyses requires that more services be valued. Nonuse values require special consideration; these may be the largest component of total economic value for aquatic ecosystem services. Unfortunately, nonuse values can be estimated only with stated-preference methods, and this is the application in which these methods have been soundly criticized.

Although a variety of valuation methods are currently available, no single method can be considered best at all times and for all types of aquatic ecosystem applications. In each application it is necessary to consider what method(s) is the most appropriate. Based on its assessment of the current literature and the preceding conclusions, the committee makes the following recommendations (Chapter 4):

- Specific attention should be given to funding research at the "cutting edge" of the valuation field, such as dynamic production function approaches, general equilibrium modeling of integrated ecological-economic systems, conjoint analysis, and combined stated-preference and revealed-preference methods.
- Specific attention should be given to funding research on improved valuation study designs and validity tests for stated-preference methods applied to determine the nonuse values associated with aquatic and related terrestrial ecosystem services.
- Benefit transfers should be considered a "second-best" method of ecosystem services valuation and should be used with caution and only if appropriate guidelines are followed.
- The replacement cost method and estimates of the cost of treatment are not valid approaches to determining benefits and should not be employed to value aquatic ecosystem services. In the absence of any information on benefits, and under strict guidelines, treatment costs could help determine cost-effective policy action.

TRANSLATING ECOSYSTEM FUNCTIONS TO THE VALUE OF ECOSYSTEM SERVICES: CASE STUDIES AND LESSONS LEARNED

Although there has been great progress in ecology in understanding ecosystem processes and functions, and in economics in developing and applying nonmarket valuation techniques for their subsequent valuation, at present there often remains a gap between the two. There has been mutual recognition among at least some ecologists and economists that addressing issues such as conserving ecosystems and biodiversity requires the input of both disciplines to be successful. Yet there are few examples of studies that have successfully translated knowledge of ecosystems into a form in which economic valuation can be applied in a meaningful way. Several factors contribute to this ongoing lack of integration. First, ecology and economics are separate disciplines—one in the natural sciences, the other in the social sciences. Traditionally, academic organization and the reward structures for scientists make collaboration across disciplinary boundaries difficult even when the desire to do so exists. Second, the concept of ecosystem services and attempts to value them are still relatively recent; building the necessary working relationships and integrating methods across disciplines will take time.

Nevertheless, some useful integrated studies on the value of aquatic and related terrestrial ecosystem goods and services are starting to emerge. Chapter 5 of this report provides a series of case studies of the integration of ecology and economics necessary for valuing the services of aquatic and related terrestrial ecosystems (including those from both the eastern and the western United States; see Box ES-1). More specifically, this review begins with situations in which the focus is on valuing a single ecosystem service. Typically these are cases in which the service is well defined, there is reasonably good ecological understanding of how the service is produced, and there is reasonably good economic understanding of how to value it. Even when valuing a single ecosystem service however, there can be significant uncertainty either about the production of the ecosystem service, the value of the ecosystem service, or both. Next, attempts to value multiple ecosystem services are reviewed. Since ecosystems produce a range of services, and these services are frequently closely connected, it is often hard to discuss valuation of a single service in isolation. However, valuing multiple ecosystem services typically multiplies the difficulty of evaluation. Last to be reviewed are analyses that attempt to encompass all services produced by an ecosystem. Such cases can arise with natural resource damage assessment, where a dollar value estimate of total damages is required, or with ecosystem restoration efforts, and will typically face large gaps in understanding and information in both ecology and economics.

Proceeding from single services to entire ecosystems illustrates the range of circumstances and methods for valuing ecosystem goods and services. In some cases, it may be possible to generate relatively precise estimates of value. In other cases, all that may be possible is a rough categorization (e.g., "a lot" versus "a little"). Whether there is sufficient information for the valuation of ecosystem services to be of use in environmental decision-making depends on the circumstances and the policy question or decision at hand (see Chapters 2 and 6 for further information). In a few instances, a rough estimate may be sufficient to decide that one option is preferable to another. Tougher decisions will typically require more refined understanding of the issues at stake. This progression from situations with relatively complete to relatively incomplete information also demonstrates what gaps in knowledge may exist and the consequences of those gaps. Of course, part of the value of going through an ecosystem services evaluation is to identify the gaps in existing information to show what types of research are needed.

Chapter 5 includes an extensive discussion of various implications and lessons learned from the case studies that are reviewed. These examples show that the ability to generate useful information about the value of ecosystem services varies widely across cases and circumstances. For some policy questions, enough is known about ecosystem service valuation to help in decision-making. As other examples make clear, knowledge and information may not yet be sufficient to estimate the value of ecosystem services with enough precision to answer policy-relevant questions. In general, the inability to generate relatively precise and reliable estimates of ecosystem values may arise from any combination of the following three reasons: (1) insufficient ecological knowledge or information to estimate the quantity of ecosystem services produced or to estimate how ecosystem service production would change under alternative scenarios, (2) an inability of existing economic methods to generate precise estimates of value for the provision of various levels of ecosystem services, and (3) a lack of integration of ecological and economic analysis.

Studies that focus on valuing a single ecosystem service show promise of delivering results that can inform important policy decisions. In no instance, however, should the value of a single ecosystem service be confused with the value of the entire ecosystem. Unless it is clearly understood that valuing a single ecosystem service represents only a partial valuation of the natural processes in an ecosystem, such single service valuation exercises may provide a false signal of total value. Even when the goal of a valuation exercise is focused on a single ecosystem service, a workable understanding of the functioning of large parts or possibly the entire ecosystem may be required. Although the valuation of multiple ecosystem services is more difficult than the valuation of a single service, interconnections among services may make it necessary to expand the scope of the analysis. As noted previously, ecosystem processes are often spatially linked, especially in aquatic ecosystems. Full accounting of the consequences of actions on the value of ecosystem services requires understanding these spatial links and undertaking integrated studies at suitably large spatial scales to fully cover important effects. In generating estimates of the value of ecosystem services across larger spatial scales, extrapolation may be unavoidable, but it should be applied with careful scrutiny. Lastly, the value of ecosystem services depends upon underlying conditions. Ecosystem valuation studies should clearly present assumptions about underlying ecosystem and market conditions and how estimates of value could change with changes in these underlying conditions.

Building on the implications and lessons learned and on these preceding conclusions, the committee provides the following recommendations (Chapter 5):

- There is no perfect answer to questions about the proper scale and scope of analysis in ecosystem services valuation. One way to accomplish the integration of ecology and economics to value ecosystem services is to design the study to answer a particular policy question. The policy question then serves as the unifying frame that directs both ecological and economic analysis.
- Estimates of ecosystem value need to be placed in context. Assumptions about conditions in ecosystems outside the target ecosystem and assumptions about human behavior and institutions should be clearly specified.
- Concerted efforts should be made to overcome existing institutional barriers that prevent ready and effective collaboration among ecologists and economists regarding the valuation of ecosystem services. Furthermore, existing and future interdisciplinary programs aimed at integrated environmental analysis should be encouraged and supported.

JUDGMENT, UNCERTAINTY, AND VALUATION

The valuation of aquatic and related terrestrial ecosystem services inevitably involves investigator judgments and some amount of uncertainty. Although unavoidable, uncertainty and the need to exercise professional judgment are not debilitating to ecosystem valuation. However, when such judgments are made it is important to explain why they are needed and to indicate the alternative ways in which judgment could have been exercised. It is also important that the sources of uncertainty be acknowledged, minimized, and accounted for in ways that ensure that a study's results and related decisions regarding ecosystem valuation are not systematically biased and do not convey a false sense of precision.

There are several cases in which investigators must use professional judgment in ecosystem valuation regarding how to frame a valuation study, how to address the methodological judgments that must be made during the study, and how to use peer review to identify and evaluate these judgments. Of these, perhaps the most important choice in any ecosystem valuation study is the selection of the question to be asked and addressed (i.e., "framing" the study). The case studies discussed in Chapter 6 illustrate the fact that the policy context unavoidably affects the framing of an ecosystem valuation study and therefore the type and level of analysis needed to answer it. Framing also affects the way in which people respond to any given issue. Analysts need to be aware of this and sensitive to the different ways of presenting data and issues, and should make a serious attempt to address all perspectives in their presentations because failure to do so could undermine the legitimacy of an ecosystem valuation study.

In most ecosystem valuation studies, an analyst will be called on to make various methodological judgments about how the study should be designed and conducted. Typically, these judgments will address issues such as whether, and at what rate, future benefits and costs should be discounted; whether to value goods and services by what people are willing to pay or what they would be willing to accept if these goods and services were reduced or lost; and how to account for and present distributional issues arising from possible policy measures. In many cases, different choices regarding some of these issues will make a substantial difference in the final valuation. The unavoidable need to make professional judgments in ecosystem valuation through choices of framing and methods suggests that there is a strong case for peer review to provide input on these issues before study design is complete and relatively unchangeable.

There are several major sources of uncertainty in the valuation of aquatic ecosystem services and several options for policymakers and analysts to respond. Model uncertainty arises for the obvious reason that in many cases the relationships between certain key variables are not known with certainty (i.e., the "true model" will not be known). Parameter uncertainty is one level below model uncertainty in the logical hierarchy of uncertainty in the valuation of ecosystem services. The almost inevitable uncertainty facing analysts involved in ecosystem valuation can be more or less severe depending on the availability of good probabilistic information or lack thereof (i.e., the amount of ambiguity). A favorable case would be one in which although there is uncertainty about some key magnitudes of various parameters, the analyst nevertheless has good probabilistic information. An alternative and common scenario in ecosystem valuation is one in which there is really no good probabilistic information about the likely magnitude of some variables, and what is available is based only on expert judgment. However, just as there are different types of uncertainty in ecosystem valuation, there are also different ways and decision criteria that an analyst can use to allow for uncertainty in the support of environmental decision-making; these are reviewed in Chapters 2 and 6. One of these is the use of Monte Carlo simulations as a method of estimating the range of possible outcomes and the parameters of its probability distribution. The outcome of an environmental policy choice under uncertainty is necessarily unpredictable, and risk aversion is a measure of what a person is willing to pay to avoid an uncertain outcome. In a heterogeneous population, the analyst will have to make an assumption about the level of risk aversion that is appropriate for the group as a whole.

Although considerable uncertainty exists about the value of ecosystem services, there is often the possibility of reducing this uncertainty over time through passive and/or active learning. Regardless of its source, the possibility of reducing uncertainty in the future through learning can affect current decisions, particularly when the impacts of those decisions are (effectively) irreversible (e.g., the construction or removal of a dam). With learning, there is an "option value" that needs to be incorporated into the analysis as part of the expected net benefits that reflects the value of the additional flexibility. This flexibility allows future decisions to respond to new information as it becomes available. It follows that in a cost-benefit analysis, measurement of the benefits of ecosystem protection through ecosystem valuation should consider the possibility of learning (i.e., should incorporate the option value). At present, only a limited amount of empirical work has been done on estimating the magnitude of option value. A natural extension of the observation that better decisions can be made if one waits for additional information is through the use of adaptive management. Adaptive management is a relatively new but increasingly used paradigm for confronting the inevitable uncertainty arising among management policy alternatives for large complex ecosystems or ecosystems in which functional relationships are poorly known. It provides a mechanism for learning systematically about the links between human societies and ecosystems, although it is not a tool for ecosystem valuation or a method of valuation per se.

Based on these conclusions, the committee makes the following recommendations regarding judgment and uncertainty in ecosystem valuation activities and methods and approaches to effectively and proactively respond to them (Chapter 6):

- Analysts must be aware of the importance of framing in designing and conducting ecosystem valuation studies so that the study is tailored to address the major questions at issue. Analysts should also be sensitive to the different ways of presenting study data, issues, and results and make a concerted attempt to address all relevant perspectives in their presentations.
- The decision to use WTP or WTA as a measure of the value of an eco-

system good or service is a choice about how an issue is framed. If the good or service being valued is unique and not easily substitutable with other goods or services, then these two measures are likely to result in very different valuation estimates. In such cases, the committee cannot reasonably recommend that the analyst report both sets of estimates in a form of sensitivity analysis because this may effectively double the work. Rather, the analyst should document carefully the ultimate choice made and clearly state that the answer would probably have been higher or lower had the alternative measure been selected and used.

- Because even small differences in a discount rate for a long-term environmental restoration project can result in order-of-magnitude differences in the present value of net benefits, in such cases the analyst should present figures on the sensitivity of the results to alternative choices for discount rates.
- Ecosystem valuation studies should undergo external review by peers and stakeholders early in their development when there remains a legitimate opportunity for revision of the study's key judgments.
- Analysts should establish a range for the major sources of uncertainty in an ecosystem valuation study whenever possible.
- Analysts will often have to make an assumption about the level of risk aversion that is appropriate for use in an ecosystem valuation study. In such cases, the best solution is to state clearly that the assumption about risk aversion will affect the outcome and to conduct sensitivity analyses to indicate how this assumption impacts the outcome of the study.
- There is a need for further research about the relative importance of and estimating the magnitude of option values in ecosystem valuation.
- Under conditions of uncertainty, irreversibility, and learning, there should be a clear preference for environmental policy measures that are flexible and minimize the commitment of fixed capital or that can be implemented on a small scale on a pilot or trial basis.

ECOSYSTEM VALUATION: SYNTHESIS AND FUTURE DIRECTIONS

The final chapter of this report seeks to synthesize the current knowledge regarding ecosystem valuation in a way that will be useful to resource managers and policymakers as they incorporate the value of ecosystem services into their decisions. A synthesis of the report's general premises and major conclusions regarding ecosystem valuation suggests that a number of issues or factors enter into the appropriate design of a study of the value of aquatic ecosystem services. The context of the study and the way in which the resulting values will be used play a key role in determining the type of value estimate that is needed. In addition, the type of information that is required to answer the valuation question and the amount of information that is available about key economic and ecological relationships are important considerations. This strongly suggests that the valuation exercise will be very context specific and that a single, "one-size-fits-all" or "cookbook" approach cannot be used. Instead, the resource manager or decision maker who is conducting a study or evaluating the results of a valuation study must assess how well the study is designed in the context of the specific problem it seeks to address. In this regard, Chapter 7 provides a checklist to aid in this assessment that identifies questions that should be openly discussed and satisfactorily resolved in the course of the valuation exercise.

Finally, Chapter 7 identifies what the committee feels are the most pressing recommendations for improving the estimation of ecosystem values and their use in decisions regarding ecosystem protection, preservation, or restoration. These overarching recommendations are based on, and in some cases build on, the more specific recommendations presented at the ends of the previous chapters; they include (1) overarching recommendations for conducting ecosystem valuation and (2) overarching research needs, which imply recommendations regarding future research funding.

1
Introduction

The biota and physical structures of ecosystems provide a wide variety of marketable goods—fish and lumber being two familiar examples. Moreover, society is increasingly recognizing the myriad life support functions, the observable manifestations of ecosystem processes that ecosystems provide and without which human civilizations could not thrive (Daily, 1997; Naeem et al., 1999). These include water purification, recharging of groundwater, nutrient recycling, decomposition of wastes, regulation of climate, and maintenance of biodiversity. Derived from the physical, biological, and chemical processes at work in natural ecosystems, these functions are seldom experienced directly by users of the resource. Rather, it is the services provided by the ecosystems—services that create value for human users, such as flood risk reduction and water supply—together with the ecosystem goods, that are the subject of this report.

Despite the importance of ecosystem functions and services, they are often overlooked or taken for granted and their value implicitly set at zero in decisions concerning conservation or restoration (Bingham et al., 1995; Heal, 2000; Postel and Carpenter, 1997). Choices between the conservation and restoration of ecosystems and the continuation and expansion of human activities have to be made however in the recognition of conflicts between the expansion of certain human activities and the continued provision of valued ecosystem goods and services. In making these choices, the economic values of ecosystem goods and services should be assessed and compared with the economic values of activities that may compromise them. Although factors other than economic values may ultimately enter into the choices, these values are important inputs to the environmental policy decision-making process.

Aquatic ecosystems include freshwater, marine, and estuarine surface waterbodies. These incorporate lakes, rivers, streams, coastal waters, estuaries, and wetlands, together with their associated flora and fauna. Each of these entities is connected to a greater ecological and hydrological landscape that includes adjacent riparian areas, upland terrestrial ecosystems, and underlying groundwater aquifers. As discussed in detail in Chapter 3, the term "aquatic ecosystems" used in this report includes related terrestrial ecosystems and underlying aquifers.

Historically, the United States had an abundance of aquatic ecosystems. However, many of these systems have been lost altogether, or the species of plants and animals they support have been diminished in kind and number. For example, between the time of European settlement and about 1950, it is estimated that more than half of the nation's wetlands were converted for agricultural or other land uses (Heinz Center, 2002; NRC, 2001). An additional 10 percent of the wetlands remaining in 1950 have since been converted to another use (see also Table 1-1). In addition, less than 2 percent of the nation's 3.1 million miles of rivers and stream remain free flowing for longer than 125 miles and include more than 75,000 dams larger than 6 feet and 2.5 million smaller dams (TNC, 1998). Within the United States, more than 60 percent of freshwater mussels and crayfish are considered rare or imperiled and 35 percent or more of fish and aquatic amphibian species are at some risk of extinction (Abell et al., 2000). Thus, the number and amount of intact functional aquatic ecosystems have been substantially reduced in recent decades. This relative scarceness has called increasing attention to the need to better understand the functionality and value of the remaining ecosystems to society.

Despite the large losses and changes in these systems, aquatic ecosystems remain broadly and heterogeneously distributed across the nation. At a glance, there are almost 4 million miles of rivers and streams, 59,000 miles of ocean shoreline waters, and 5,500 miles of Great Lakes shoreline in the United States (EPA, 2002). There are 87,000 square miles of estuaries, while lakes, reservoirs, and ponds account for more than 40 million acres. As of 1997, the lower 48 states contained about 165,000 square miles (105.5 million acres) of wetlands of all types—an area about the size of California (Dahl, 2000). Figure 1-1 shows major rivers and streams. Figure 1-2 shows major aquifers in the United States classified by major features that affect the occurrence and availability of groundwater. A variety of federal programs report on the extent, status, and related trends of aquatic ecosystems located throughout the United States. Although it is beyond the scope of this report to review systematically or even summarize all such programs, a few of the largest and most important programs are described briefly in Chapter 3.

TABLE 1-1 Recent Wetland Losses in the United States

Period	Losses Due to Agriculture	Losses Due to Non-Agriculture[a]	Total Acreage Lost[b] (Annual Average Loss)
Mid-1970s to mid-1980s (10 years)	137,540 acres per year (54% of loss)	117,230 acres per year (46% of loss)	2,547,700 acres (254,770 acres per year)
1986 to 1997 (11 years)	15,222 acres per year (26% of loss)	43,324 acres per year (76% of loss)	644,000 acres (58,545 acres per year)

SOURCE: Adapted from Dahl (2000); Dahl and Johnson (1991); NRC (2001).
[a] Non-agricultural losses include those from silviculture, urban, and rural development uses.
[b] Total acreage lost was determined by multiplying the annual acreage loss by the total number of years in that time period.

FIGURE 1-1 Major rivers and streams of the conterminous United States. SOURCE: Generated from the National Atlas of the United States (available on-line at *http://www.nationalatlas.gov*).

As noted above, aquatic ecosystems collectively perform numerous interrelated functions and provide a wide range of services. In addition, many aquatic ecosystems support the economic livelihood of local communities through commercial fishing and by serving the recreational sector. To illustrate the importance of these activities, recreational fishing alone generated an estimated $116 billion in total economic output the United States in 2001 (American Sportsfishing Association, 2002). The continuance or growth of these types of economic activities is directly related to the extent and health of these natural ecosystems. However, human activities and rapid population growth (often preferentially in or near aquatic ecosystems), along with historical and ongoing industrial, commercial, and residential development, have led to increased pollution, adverse modification, and destruction of remaining (especially pristine) aquatic ecosystems (Baron et al., 2003; Carpenter et al., 1998; Howarth et al., 2000; NRC, 1992). At the same time, increased human demand for water has reduced the amount available to support these ecosystems (Heinz Center, 2002; Jackson et al., 2001).

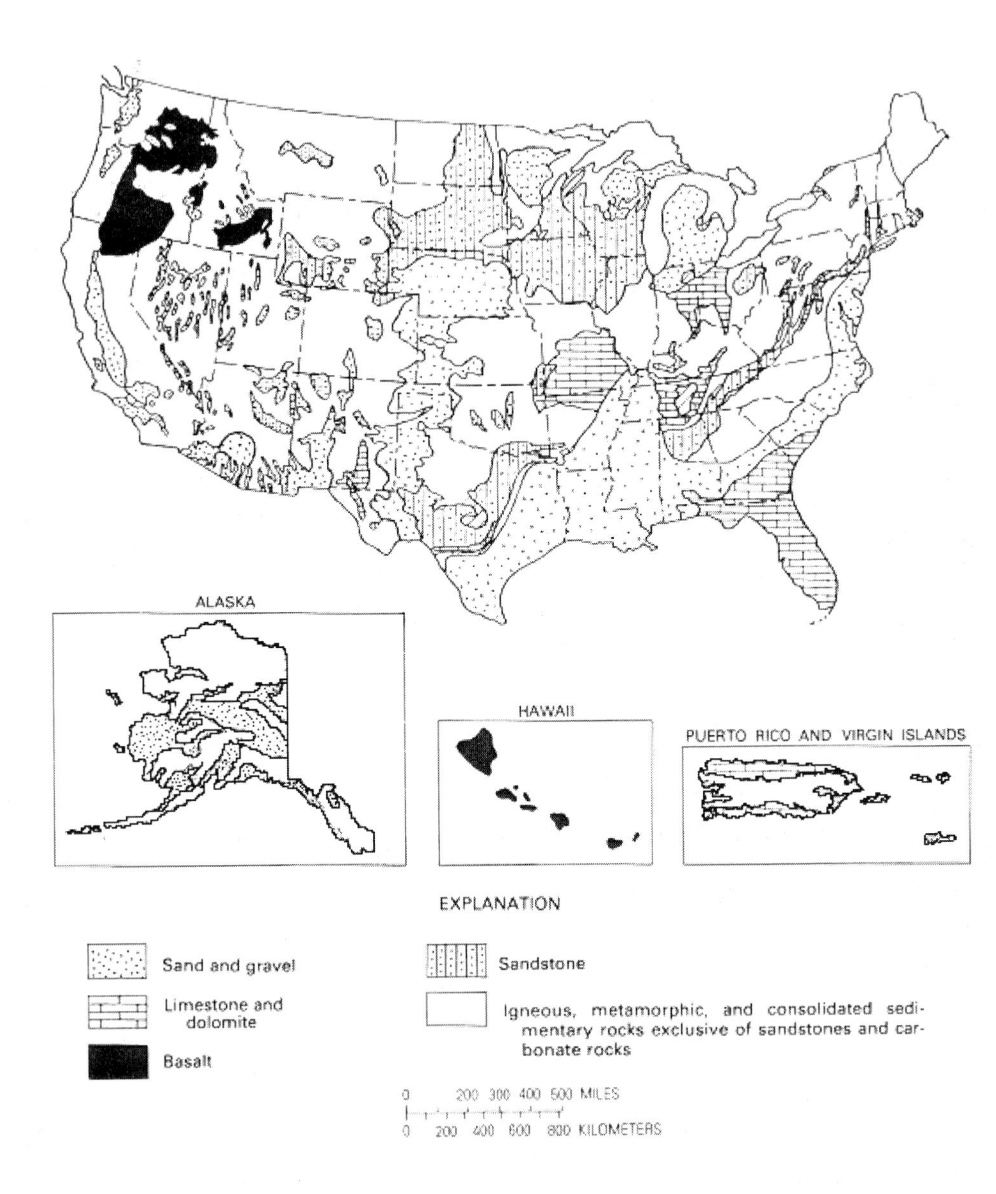

FIGURE 1-2 Groundwater regions in the United States. Note: Shading refers to principal types of water-bearing rocks. SOURCE: Heath (1984).

In the case of commercial and recreational fishing, pollution of aquatic ecosystems has adversely affected annual fish catch. For example, coastal areas and estuaries provide important nurseries for many species of commercially valuable fish and shellfish and have been adversely affected by nutrient pollution and habitat loss (Beck et al., 2001, 2003). Moreover, increasing demand for the services of aquatic ecosystems has resulted in a huge increase in the raising of fish (aquaculture) worldwide, which itself is having substantive effects on natural aquatic ecosystems (Naylor, 2001). This has occurred despite an increase in federal, state, and local regulations intended to restore and protect these natural resources. In this regard, many of the regulatory efforts to control pollution stem from the Clean Water Act (CWA),[1] which originally focused on controlling point source pollution and limiting the destruction of wetlands.

Initially, certain large point sources of pollution were exempted from this federal act, such as concentrated or confined animal feeding operations (CAFOs), which have been responsible for pollution of a number of important aquatic ecosystems. However, CAFOs have recently been required to meet tighter discharge standards (EPA, 2003a) under the CWA. At present, nonpoint source (NPS) pollution is widely considered the leading remaining cause of water quality problems throughout much of the United States. The sources of NPS pollution to aquatic ecosystems are varied and range from runoff of fertilizers and pesticides applied to farm fields to atmospheric deposition of rainfall polluted from automobile emissions (Carpenter et al., 1998; Howarth et al., 2002).

This chapter serves as an introduction to the extent and importance of aquatic and related terrestrial ecosystems throughout the United States. It provides a statement of the problem of attempting to assess and value the services of aquatic and related ecosystems, summarizes the origin and scope of the study, and describes the perspective of the committee and this report. Chapter 2 provides an overview of the different sources and meanings of "value" in the policy process with a focus on economic valuation and the role it can play in improving environmental decision-making. Chapter 3 reviews some existing definitions of aquatic and related terrestrial ecosystems; describes their associated structures and functions; and introduces their translation to ecosystem goods and services. Chapter 4 provides a review of key existing methods of nonmarket valuation for aquatic ecosystems and issues related to their development and successful application. Chapter 5 focuses on translating ecosystem functions into services using an extensive series of case studies that compare and contrast such efforts in or-

[1] Growing public awareness of and concern for controlling water pollution nationwide led to enactment of the Federal Water Pollution Control Act (FWPCA; enacted in 1948) Amendments of 1972. The Clean Water Act, as it became known, arose from 1977 amendments to the FWPCA and is a comprehensive statute intended to restore and maintain the chemical, physical, and biological integrity of the nation's waters. To accomplish this national objective, the CWA seeks to attain a level of water quality that "provides for the protection and propagation of fish, shellfish, and wildlife, and provides for recreation in and on the water." Primary authority for implementation and enforcement of the CWA—which has been amended almost yearly since its inception—rests with the U.S. Environmental Protection Agency.

der to develop "lessons learned" that can be applied in future ecosystem valuation activities. Chapter 6 assesses judgment and uncertainty associated with ecosystem valuation and suggests how analysts and decision-makers can and should respond. Lastly, Chapter 7 synthesizes the current knowledge regarding ecosystem services valuation and builds on the preceding chapters in order to provide guidelines for policymakers and planners concerned with the management, protection, and restoration of aquatic ecosystems. It also identifies what the committee feels are overarching recommendations for improving the valuation of ecosystem services and related research needs.

STATEMENT OF THE PROBLEM

Some believe that environmental amenities and services lie outside the scope of economic analyses, arguing that the need to protect environmental assets is self-evident and not properly the subject of economic analyses (see Chapter 2 for further discussion). However, wherever there is scarcity and the need to choose between alternatives, the question of relative values is unavoidable. It may be costly to protect, conserve, and restore aquatic ecosystems, and the costs are borne by giving up benefits in other parts of the economy, now or in the future. When ecosystem protection projects and policies are proposed, it is appropriate to ask whether they achieve the stated goals in a cost-effective and efficient manner, whether the costs are commensurate with the benefits received, what society's costs are if protection is not provided, and whether costs and benefits are properly allocated across the present population and across generations.

Economic valuation requires that ecosystems be described in terms of the goods and services they provide to humans or other beneficiaries. Goods and services, in turn, must be quantified and measured on a common (though not necessarily monetary) scale if improvements to one ecosystem are to be compared to improvements to another. Although the issues that this raises apply to all types of ecosystems, the use of such information has started to come into particularly sharp focus for aquatic ecosystems and especially for wetlands (NRC, 2001).

Studying ecosystem services presents several challenges that are discussed throughout this report. The most fundamental challenge lies in providing an explicit description of the links between the structure and function of natural systems and the benefits (i.e., goods and services) derived by humanity. This problem is complicated by the fact that humans are an integral part of the system; by incomplete knowledge of how ecosystems function; and by the fact that ecosystem services tend to be specific to locations and situations, thus making it difficult to develop generic principles or identify generic characteristics.

The challenges to both ecologists and economists implicit in valuing ecosystem services are summarized in Figure 1-3. Human actions affect the structure, functions, and goods and services of ecosystems. Ecosystem conditions are also affected by various biophysical parameters (not shown in figure). The translation from ecosystem structure and functions to ecosystem goods and services is given by an ecological production function, and the translation from ecosystem goods and services to value is given by an economic valuation function. There may be occasions in which the structure of the ecosystem is valued directly by humans, without the intermediation of functions, goods, or services. For example, people may value the existence of redwood forests in their own right rather than because of any functions, goods, or services that they might provide; a possibility indicated in Figure 1-3 by the direct connection from ecosystem structure to values (also given by an economic valuation function). Estimating the value of ecosystem services requires uncovering both the ecological production function and the economic valuation function. As Chapters 3, 4, and 5 illustrate, uncovering each of these functions is difficult. Furthermore, because aquatic ecosystems are complex, the production of goods and services can be complicated and indirect; this in turn makes the translation from ecosystem structure and function to ecosystem goods and services difficult. The lack of markets and market prices and of other direct behavioral links to underlying values makes the translation from quantities of goods and services to value difficult as well.

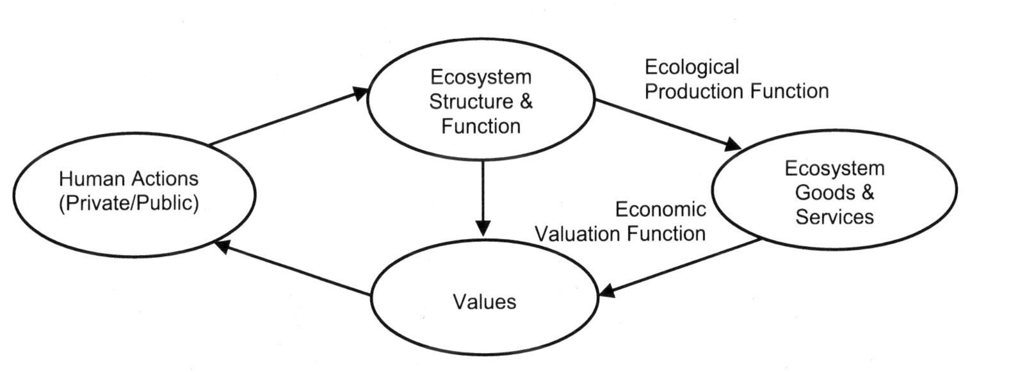

FIGURE 1-3 Components of ecosystem valuation: ecosystem structure and function, goods and services, human actions (policies), and values (see Figure 7-1 for an expanded version of this figure).

Although valuing ecosystem services does not require knowledge of the function that maps human actions into ecosystem conditions, evaluating whether certain actions are in society's best interest does require this knowledge. For example, knowing whether to allow housing development in a watershed or timber harvesting in a forest patch requires predictions of how these actions will perturb ecosystems. This perturbation will change the production and value of ecosystem goods and services, and can then be compared to the direct economic value generated by the action (e.g., housing values, value of timber harvest) to see whether or not the action generates positive net benefits.

Where an ecosystem's goods and services can be identified and measured, it will often be possible to assign values to them by employing existing economic valuation methods. Chapter 4 provides a summary of key existing nonmarket valuation methods for (primarily aquatic) ecosystem services. Some ecosystem goods and services cannot be valued because they are not quantifiable or because available methods are not appropriate or reliable. In other cases, the cost of valuing a particular service may rule out the use of a formal method. Available economic valuation methods are complex and demanding. The results of applying these methods may be subject to judgment and uncertainty and must be interpreted with caution. Still, the general sense of a very large literature on the development and application of various methods is that they are relatively well evolved and capable of providing useful information in support of improved ecosystem valuation. There is little to be gained from a comprehensive National Academies review of these valuation methods. Indeed, the literature contains numerous authoritative reviews and critiques, and some federal agencies have published their own assessments and guidelines, which are cited and discussed briefly in Chapter 4. Thus, an important question for this committee was not how to use any particular valuation method, but how to address ecosystem services for which no existing valuation method has been identified, and how to integrate economic and ecological analysis to obtain economic values of ecosystem conservation. Similarly, while not repeating existing reviews or assessments of valuation methods, this report addresses the decision-making consequences of judgment and uncertainty, including the implications for the selection of methods in specific applications.

Probably the greatest challenge for successful valuation of ecosystem services is to integrate studies of the ecological production function with studies of the economic valuation function. After all, an understanding of the goods and services provided by a particular ecological resource, the interactions among them, and their sustainable levels can come only from ecological research and models. To integrate economic and ecological studies, the definitions of ecosystem goods and services must match across studies. In other words, the quantities of goods and services must be defined in a similar manner for both ecological studies and economic valuation studies. Failure to do so means that the results of ecological studies cannot be carried over into economic valuation studies. Attempts to value ecosystem services without this key link will either fail to have ecological underpinnings or fail to be relevant as valuation studies.

Although there has been great progress in ecology in improving our understanding of aquatic ecosystem structure and function and in economics in developing and applying nonmarket valuation techniques, there remains a gap between the two. There are few examples of studies that have successfully translated knowledge about ecosystems into a form where economic valuation can be applied in a meaningful way. Several factors contribute to this continued lack of integration. First, some ecologists and economists hold vastly different views on the current "state of the world" and the direction in which it is headed. More recently, however, there has been mutual recognition among at least some ecologists and economists that addressing issues such as conserving ecosystems and biodiversity requires the input of both disciplines to be successful. A second reason for the lack of integration is that ecology and economics are separate disciplines, one in natural science and the other in social science. The traditional academic organization and the reward structure for scientists often make collaboration across disciplinary boundaries difficult even when the desire to do so exists (e.g., Bingham et al., 1995). Third, the ecosystem services paradigm is relatively new, as are attempts to value ecosystem services. Building the necessary working relationships and integrating methods across disciplines will take time.

Integrated studies of the value of ecosystem goods and services are now emerging. Chapter 5 reviews several such studies, beginning with situations in which the focus is on valuing a single ecosystem service, progressing to attempts to value multiple ecosystem services, and ending by reviewing analyses that attempt to encompass all services produced by an ecosystem. In some cases, it may be possible to generate relatively precise estimates of value; in other cases, all that may be possible is a rough categorization ("a lot" versus "a little"). Whether this is sufficient information depends on the circumstances. In some instances, a rough estimate may be sufficient to decide that one option is preferable to another, whereas tougher decisions will require more refined information. This progression from situations with good to poor information also demonstrates what types of information will often be lacking and the consequences of those gaps. Indeed, part of the value of going through an ecosystem services evaluation is to point out the gaps in existing information and show what research is needed to fill these gaps.

It is clear that more categories of human endeavor will in the future be evaluated to some extent in terms of environmental effects and impacts on quality of life. The emerging desire to measure the environmental costs of human activities, or to assess the benefits of environmental protection and restoration, has challenged the state of the art in environmental evaluation in both the ecological and the social sciences. From an ecological perspective, the challenge is to interpret basic research on ecosystem functions so that service-level information can be communicated to economists. For economics and related social sciences, the challenge is to identify the values of both tangible and intangible goods and services associated with ecosystems and to address the problem of decision-making in the presence of partial valuation. The combined challenge is to develop and apply methods to assess the values of human-induced changes in ecosystem functions and services.

STUDY ORIGIN AND SCOPE

This study was conceived in 1997 at a strategic planning session of the Water Science and Technology Board (WSTB) of the National Research Council (NRC). Initially, the NRC organized and hosted a planning workshop to assess the feasibility of and need for an NRC study of the functions and associated economic values of aquatic ecosystems. Fourteen key experts involved or interested in the management, protection, and restoration of aquatic ecosystems—including representatives of the study sponsors, the U.S. Army Corps of Engineers (USACE), U.S. Environmental Protection Agency (EPA), and U.S. Department of Agriculture (USDA)—participated in the workshop that was held early in November 1999 in Washington, D.C. All participants agreed that an NRC study of valuation methods used to assess aquatic ecosystem services, rather than functions, was feasible and timely and would make a significant contribution toward advancing the understanding and appropriate use of economic valuation methods in environmental decision-making. However, it is important to note that the NRC has released several reports in the last decade that are somewhat related to this study. These are listed and briefly summarized in ascending chronological order in Appendix A. Furthermore, there has been a general increase in interest in the area of economic valuation of ecosystem services and its role in environmental policy and decision-making since the committee was formed in early 2002 (discussed below). For example, the EPA's Science Advisory Board (SAB) recently established a panel to review EPA's draft Environmental Economics Research Strategy (EPA, 2003b).[2]

The WSTB developed a full study proposal and while several minor changes were made to the proposal in response to the sponsoring (and nonsponsoring) agencies, one significant change was made. As a compromise to the USACE's desire to expand the scope of the study to include all ecosystems, it was decided and subsequently agreed by the NRC and all study sponsors to expand the study proposal to include "related terrestrial ecosystems." The original basis for this change in language and study focus was the key 1983 water resources planning report *Economic and Environmental Principles and Guidelines for Water and Related Land Resources Implementation Studies* (WRC, 1983). The implications of linking "related terrestrial ecosystems" to aquatic ecosystems are discussed more fully in Chapter 3.

The committee's statement of task (see Box ES-1) was to evaluate methods

---

[2] The panel consists of members of the existing SAB Environmental Economics Advisory Committee to which several experts were added (including several members of this NRC committee) to form the Advisory Panel on the Environmental Economics Research Strategy (see *http://www.epa.gov/sab/pdf/apeers_bios_for-web.pdf* and *http://es.epa.gov/ncer/events/news/2003/06_23_03a.html* for further information).

for assessing the economic value of the goods and services provided by aquatic and related terrestrial ecosystems. More specifically, it asks "What lessons can be learned from a comparative review of past attempts to value ecosystem services—particularly, are there significant differences between eastern and western U.S. perspectives on these issues?" As is evident throughout this report, the committee made extensive use of case studies in ecosystem services valuation (especially in Chapter 5) to help develop many of its conclusions and recommendations and respond to this and other elements of the statement of task. Although the case studies are drawn primarily from throughout the United States, including eastern and western areas, the committee decided early in its deliberations that it would not make geographic distinctions in developing implications and lessons learned from the case studies.

This report is about placing values on the goods and services that ecosystems provide to human societies, with its principal focus on the goods and services provided by aquatic and related terrestrial ecosystems. Furthermore, the report focuses on freshwater and estuarine systems, eschewing extensive consideration of marine and groundwater systems. This reflects an intentional effort to focus on management and valuation issues confronting state and federal agencies for these ecosystems. However, because the principles and practices of valuing ecosystem goods and services are rarely sensitive to whether the underlying ecosystem is aquatic or terrestrial, the report's various conclusions and recommendations are likely to be directly or at least indirectly applicable to the valuation of the goods and services provided by any ecosystem.

PERSPECTIVE OF THIS REPORT

Several elements are fundamental to the perspective taken by the committee as it developed this report. The first is that ecosystems provide goods and services, sometimes very important ones, to society (see for example, Daily, 1997; de Groot et al., 2002; Ewel, 2002; Peterson and Lubchenco, 2002; Postel and Carpenter, 1997). The second element is that in many cases these goods and services can be quantified and an economic value can be placed on them. In large part, the remaining chapters discuss how to do this. A third element is that economic valuation can often be useful in support of environmental policy decision-making. Although the economic value of an ecosystem may not capture all of the reasons it is valued and conserved, economic valuation captures some of these reasons—perhaps most of them under certain circumstances. This valuation, in turn, becomes a necessary input to decisions about environmental conservation, particularly in situations where there is an apparent conflict between conservation or restoration and a conventional idea of economic progress, as indicated by gross national or state product measured at market prices.

In many cases, some reviewed in the following chapters, careful valuation shows that conservation is economically beneficial, whereas the destruction or modification of natural systems is economically harmful. Finally, the concept of economic value is very inclusive, much more so than is recognized and appreciated outside the economics profession. Consequently, many of what noneconomists typically consider to be noneconomic values are in fact captured (at least to some extent) by economists' estimates of value—especially by what is called "existence value."

The reason economic valuation is more comprehensive than generally recognized is that economists recognize two basic types of value, use and nonuse values (see Chapters 2 and 4 for a more complete discussion). In brief, use values are those that derive from using a good or service provided by an ecosystem, such as using a lake for fishing or swimming, lake water for drinking or irrigation, or an estuary for boating. On the other hand, an example of a type of nonuse value is an existence value; a person may value the existence of a species even though he or she will never make any *use* of this species or of any of its members. Existence values, although often difficult and controversial to measure, are legitimate and indeed important economic values since people are willing to pay (see more below) for the continued existence of species or landscapes. Existence values also affect the way people behave, and anything that changes behavior has economic consequences. For example, even if people are not able to pay directly for the preservation of a species, the value they place on it might affect other aspects of their behavior, such as how they vote or their choice of products in the market. Values that lead to behavior changes are therefore economic values, even though their origins may lie in ethical, aesthetic, or religious beliefs (see Chapter 2 for further information). However, there could be occasions on which people value ecosystems, but this value is not reflected in any change in their behavior and is never revealed. For example, they might for some reason wish to keep their valuation secret. In such a case, economic methods of measuring values would fail to reflect a person's valuation.

Valuation studies may be conducted in many different contexts, and the context can affect some aspects of the study. A study may be conducted as part of a policy analysis, as in the case of the restoration of the New York Catskills watershed, or in the context of environmental litigation related to the *Exxon Valdez* oil spill (see Chapter 5). Alternatively, a valuation study may be conducted in the context of a NRDA (natural resource damage assessment) required by the federal Comprehensive Environmental Response, Compensation, and Liability Act (CERCLA).[3] As can be seen in the case studies developed in later chapters, the context can have an impact on the way a valuation study is framed (see Chapters 2 and 6) and on the way it is developed.

---

[3] In response to growing public concern over health and environmental risks posed by hazardous waste sites, Congress enacted CERCLA, commonly known as the Superfund program, in 1980 to identify and clean up such sites. Superfund is administered by EPA in cooperation with individual sites throughout the United States; further information can be found at *http://www.epa.gov/superfund/action/law/cercla.htm*.

SUMMARY AND CONCLUSIONS

Aquatic and related terrestrial ecosystems are broadly distributed across the nation, perform numerous interrelated functions, and provide a wide range of important goods and services. In addition, many aquatic ecosystems enhance the economic livelihood of local communities by supporting commercial fishing, supporting agriculture, and serving the recreational sector. The continuance or growth of these types of economic activities is directly related to the extent and health of these natural ecosystems. However, human activities, rapid population growth, and industrial, commercial, and residential development have all led to increased pollution, adverse modification, and destruction of remaining aquatic ecosystems—despite an increase in federal, state, and local regulations intended to protect, conserve, and restore these natural resources. Increased human demand for water has simultaneously reduced the amount available to support these ecosystems.

Despite growing recognition of the importance of ecosystem functions and services, they are often taken for granted and overlooked in environmental decision-making. Thus, choices between the conservation and restoration of some ecosystems and the continuation and expansion of human activities in others have to be made with an enhanced recognition of this potential for conflict. In making these choices, the economic values of these ecosystem goods and services to society have to be known, so that they can be compared with the economic values of activities that may compromise them and improvements to one ecosystem can be compared to those in another.

The fundamental challenge of valuing ecosystem services lies in providing an explicit description and adequate assessment of links between the structures and functions of natural systems and the benefits (i.e., goods and services) derived by humanity and is summarized in Figure 1-3. Ecosystems are complex however, making the translation from ecosystem function to ecosystem goods and services (i.e., the ecological production function) difficult. Similarly, the lack of markets and market prices and of other direct behavioral links to underlying values makes the translation from quantities of goods and services to value (i.e., the economic valuation function) quite difficult.

Probably the greatest challenge for successful valuation of ecosystem services is to integrate studies of the ecological production function with studies of the economic valuation function. To do this, the definitions of ecosystem goods and services must match across studies. Failure to do this means that the results of ecological studies cannot be carried over into economic valuation studies. Attempts to value ecosystem services without this key link will either fail to have ecological underpinnings or fail to make be relevant as valuation studies.

Where an ecosystem's services and goods can be identified and measured, it will often be possible to assign values to them by employing existing economic (primarily nonmarket) valuation methods. Some ecosystem goods and services cannot be valued because they are not quantifiable or because available methods are not appropriate or reliable; in other cases, the cost of valuing a particular service may rule out the use of a formal method. Economic valuation methods are complex and demanding, and the results of applying these methods may be subject to judgment, uncertainty, and bias and must be interpreted with caution. However, based on an assessment of a very large literature on the development and application of various economic valuation methods, the committee concludes that they are relatively mature and capable of providing useful information in support of improved environmental decision-making.

Although there has been great progress in ecology in better understanding ecosystem structure and functions, and in economics in developing and applying nonmarket valuation techniques, there remains a gap between the two. The challenge from an ecological perspective is to interpret basic research on ecosystem functions so that service-level information can be communicated to economists. The challenge for economics and related social sciences is to identify the values of both tangible and intangible goods and services associated with ecosystems while addressing the problem of decision-making in the presence of partial valuation. The combined challenge is to develop and apply methods to assess the values of human-induced changes in ecosystem functions and services.

Lastly, this report is primarily concerned with valuing the goods and services that aquatic and related terrestrial ecosystems provide to human societies. However, because the principles and practices of valuing ecosystem goods and services are rarely sensitive to whether the underlying ecosystem is strictly aquatic or terrestrial, many of its conclusions and recommendations are likely to be directly or at least indirectly applicable to the valuation of goods and services provided by any ecosystem.

REFERENCES

Abell, R.A., D.M. Olson, D.M. Dinerstein, P.T. Hurley, J.T. Diggs, W. Eichbaum, S.Walters, W. Wettengel, T. Allnutt, C.J. Loucks, and P. Hedao. 2000. Freshwater Ecoregions of North America: A Conservation Assessment. Washington, D.C.: Island Press.

ASA (American Sportsfishing Association). 2002. Sportsfishing in America: Values of Our Traditional Pastime. Alexandria, VA: American Sportsfishing Association.

Baron, J.S., N.L. Poff, P.L. Angermeier, C.N. Dahm, P.H. Glecik, N.G. Hairston, Jr., R.B. Jackson, C.A. Johnston, B.D. Richter, and A.D. Steinman. 2003. Sustaining healthy freshwater ecosystems. Issues in Ecology No. 10. Washington, D.C.: Ecological Society of America.

Beck, M.W., K.L. Heck, Jr. , K. W. Able, D.L. Childers, D.B. Eggleston, B.M. Gillanders, B. Halpern, C. G. Hays, K. Hoshino, T.J. Minello, R.J. Orth, P.F. Sheridan, and M. M. Weinstein. 2001. The identification, conservation and management of estuarine and marine nurseries for fish and invertebrates. Bioscience 51:633-641.

Beck, M.W., K.L. Heck, Jr. , K.W. Able, D.L. Childers, D.B. Eggleston, B.M. Gillanders, B. Halpern, C.G. Hays, K. Hoshino, T.J. Minello, R.J. Orth, P.F. Sheridan, and M.P. Weinstein. 2003. The role of nearshore ecosystems as fish and shellfish nurseries. Issues in Ecology 11. Washington, D.C.: Ecological Society of America.

Bingham, G., R. Bishop, M. Brody, D. Bromley, E. Clark, W. Cooper, R. Costanza, T. Hale, G. Hayden, S. Kellert, R. Norgaard, B. Norton, J. Payne, C. Russell, and G. Suter. 1995. Issues in ecosystem valuation. Ecological Economics 14(2):67-125.

Carpenter, S., N. Caraco, D.L. Correll, R.W. Howarth, A.N. Sharpley, and V.H. Smith. 1998. Nonpoint pollution of surface waters with phosphorus and nitrogen. Ecological Applications 8:559-568.

Dahl, T.E. 2000. Status and Trends of Wetlands in the Conterminous United States 1986 to 1997. Washington, D.C.: U.S. Department of the Interior, Fish and Wildlife Service.

Dahl, T.E. and C.E. Johnson. 1991. Wetlands, Status and Trends in the Conterminous United States, mid 1970's to mid-1980's. Washington, D.C.: U.S. Department of the Interior, Fish and Wildlife Service.

Daily, G.C. 1997. Introduction: What are ecosystem services? Pp. 1-10 in G.C. Daily (ed.) Nature's Services: Societal Dependence on Natural Ecosystems. Washington, D.C.: Island Press.

De Groot, R.S., M.A. Wilson, and R.M.J. Boumans. 2002. A typology for the classification, description and valuation of ecosystem functions, goods and services. Ecological Economics 41:393-408.

EPA (U.S. Environmental Protection Agency). 2001. National Coastal Condition Report. EPA-620/R-01/005. Washington, DC: U.S. EPA, Office of Research and Development/Office of Water. Also available on-line at *http://www.epa/gov/owow/oceans/NCCR/index*. Accessed October 2002.

EPA. 2002. 2000 National Water Quality Inventory. EPA-841-R-2-001. Washington, D.C.: Office of Water.

EPA. 2003a. National Pollutant Discharge Elimination System Permit. Available online at *http://www.mass.gov/czm/envpermitnpdes.htm*. Accessed October 11, 2004.

EPA. 2003b. Science Advisory Board, Economics Advisory Committee, Advisory Panel on the Environmental Economics Research Strategy; Request for Nominations. Federal Register 68(120):37151-37152

Ewel, K.C. 2002. Water quality improvement by wetlands. Pp. 329-344 in G.C. Daily (ed.) Nature's Services: Societal Dependence on Natural Ecosystems. Washington, D.C.: Island Press.

Heal, G. 2000. Nature and the Marketplace. Washington D.C.: Island Press.

Heath, R.C. 1984. Ground-water Regions of the United States. Geological Survey Water-Supply Paper 2242. Washington, D.C.: U.S. Government Printing Office.

Heinz Center (The H. John Heinz III Center for Science, Economics and the Environment). 2002. The State of the Nation's Ecosystems: Measuring the Lands, Waters, and Living Resources of the United States. Cambridge, UK: Cambridge University Press.s

Howarth, R., D. Anderson, J. Cloern, C. Elfring, C. Hopkinson, B. LaPointe, T. Malone, N. Marcus, K. McGlathery, A. Sharpley, and D. Walker. 2002. Nutrient pollution of coastal rivers, bays, and seas. Issues in Ecology No. 7. Washington, D.C.: Ecological Society of America.

Jackson, R., S. Carpenter, C. Dahm, D. McKnight, R. Naiman, S. Postel, and S. Running. 2001. Water in a changing world. Issues in Ecology No. 9. Washington, D.C.: Ecological Society of America.

Naeem, S., F.S. Chapin III, R. Costanza, P.R. Ehrlich, F.B. Golley, D.U. Hooper, J.H. Lawton, R.V. O'Neill, H.A. Mooney, O.E. Sala, A.J. Symstad, and D. Tilman. 1999. Biodiversity and ecosystem functioning: Maintaining natural life support processes. Issues in Ecology No. 4. Washington, D.C.: Ecological Society of America.

Naylor, R., R. Goldburg, J. Primavera, N. Kautsky, M. Beveridge, J. Clay, C. Folke, J. Lubchenco, H. Mooney, and M. Troell. 2001. Effects of aquaculture on world fish supplies. Issues in Ecology No. 8. Washington, D.C.: Ecological Society of America.

NRC (National Research Council). 1992. Restoration of Aquatic Ecosystems: Science, Technology, and Public Policy. Washington, D.C.: National Academy Press.

NRC. 2001. Compensating for Wetland Losses Under the Clean Water Act. Washington, D.C.: National Academy Press.

Peterson, C.H., and J. Lubchenco. 2002. Marine ecosystem services. Pp. 177-194 in G.C. Daily (ed.) Nature's Services: Societal Dependence on Natural Ecosystems. Washington, D.C.: Island Press.

Postel, S.L., and S. Carpenter. 1997. Freshwater ecosystem service. Pp. 195-214 in G.C. Daily (ed.) Nature's Services: Societal Dependence on Natural Ecosystems. Washington, D.C.: Island Press.

TNC (The Nature Conservancy). 1998. Rivers of Life: Critical Watersheds for Protecting Freshwater Biodiversity. L.L. Master, S.R. Flack, and B.A. Stein (eds.). Arlington, Va.: The Nature Conservancy.

WRC (Water Resources Council). 1983. Economic and Environmental Principles and Guidelines for Water and Related Land Resources Implementation Studies. Washington, D.C.: Government Printing Office.

2

The Meaning of Value and Use of Economic Valuation in the Environmental Policy Decision-Making Process

INTRODUCTION

In developing a perspective and providing expert advice on valuing aquatic and related terrestrial ecosystems, it is necessary to begin with a clear discussion and statement of what it means to value something and of the role of "valuation" in environmental policy decision-making. Environmental issues and ecosystems have been at the core of many recent philosophical discussions regarding value (e.g., Goulder and Kennedy, 1997; Sagoff, 1997; Turner, 1999). Fundamentally, these debates about the value of ecosystems derive from two points of view. One view is that some values of ecosystems and their services are non-anthropocentric—that nonhuman species have moral interests or value in themselves. The other view, which includes the economic approach to valuation, is that all values are anthropocentric.

While acknowledging the potential validity of the first point of view, the committee was charged (see Chapter 1 and Box ES-1) specifically with assessing methods of valuing aquatic and related terrestrial ecosystems using economic methods, an approach that views values as inherently anthropocentric. For that reason, this report focuses on the sources of ecological value that can be captured through economic valuation.[1] However, the committee recognizes that all kinds of value may ultimately contribute to decisions regarding ecosystem use, preservation, or restoration. The committee's approach is consistent with the approach taken in the international Millennium Ecosystem Assessment,[2] which focuses on contributions of ecosystems to human well-being while at the same time recognizing that potential for non-anthropocentric sources of value.

Although this report focuses on the subset of values that can be captured through economic valuation, it is important to emphasize that this subset of values is quite broad; indeed, it is much broader than is often presumed. There are many misconceptions about the term "economic valuation." For example, many believe that the term refers simply to an assessment of the commercial value of

---

[1] Unless otherwise noted, use of the terms "value," "valuing," or "valuation" in this report refers to economic valuation; more specifically, the economic valuation of ecosystem goods and services.

[2] The Millennium Ecosystem Assessment was launched in June 2001 to help meet the needs of decision-makers and the public for scientific information concerning the consequences of ecosystem change for human well-being and options for responding to such changes (see Chapter 3 and *http://www.millenniumassessment.org/en/index.aspx* for further information).

something. In fact, the economic view of value actually includes many components that have no commercial or market basis (Freeman, 1993a; Krutilla, 1967), such as the value that individuals place on the beauty of a natural landscape or the existence of a species that has no commercial value. Thus, although economic valuation does not include all sources of value that have been identified or that are potentially important, it encompasses a very broad array of values. In addition, it provides a systematic way in which those values can be factored into environmental policy choices. This chapter provides an overview of economic valuation and the role it can play in improving environmental decision-making. The purpose is first to identify the values that are, and those that are not, captured by the economic approach to valuation and then to discuss how a quantification of these values can contribute to better environmental decision-making.

The chapter is divided into two main sections. The first discusses the role of economic valuation in the policy process and addresses the different meanings and sources of value in this context. The role and importance of quantifying values are discussed next, followed by a discussion of how information about values can be used in policy decisions. Finally, the importance of "framing" the valuation question appropriately is discussed, since the way in which a valuation exercise is defined can have a significant impact on the results that emerge from it.

Given this overview, the following section provides a more detailed examination of economic valuation. The section begins with a description of the "total economic value" framework, from which it is clear that economic valuation includes a wide array of values—many (in some cases most) of which are unrelated to any market or commercial value. This is followed by a discussion of quantifying value using a monetary metric. Two monetary metrics are described, willingness to pay (WTP) and willingness to accept (WTA), and the implications of using one versus the other are discussed. Finally, a discussion of discounting follows because many environmental policy impacts extend over long durations and it is important to incorporate the timing of these impacts into any valuation analysis. Discounting is the approach most commonly used in economic analysis to capture the timing of benefits and costs. The important distinction between discounting as a means of weighing the utility of future generations differently from that of present generations (utility discounting) and discounting as a means of weighing consumption (through benefits and costs) differently at different times (consumption discounting) is highlighted. The chapter closes with a summary of its conclusions and recommendations.

The broad overview of economic valuation provided in this chapter is followed in subsequent chapters by more detailed discussions of the types of ecosystem services that can be valued, the economic methods that can currently be used to quantify those values, and the role of professional judgment and uncertainty in ecosystem valuation.

ROLE OF ECONOMIC VALUATION

Different Sources and Meanings of Value

Given the crucial role that ecosystems and their services play in supporting human, animal, plant, and microbial populations, there is now widespread agreement that ecosystems are "valuable" and that decision-makers ranging from individuals to governments should consider the "value" of these ecosystems and the services they provide to society (Daily, 1997). However, there are different views on what this means and on the sources of that value. The literature on environmental philosophy and ethics distinguishes between (1) *instrumental* and *intrinsic* values, (2) *anthropocentric* and *biocentric* (or *ecocentric*) values, and (3) *utilitarian* and *deontological* values (Callicott, 2004). In order to place economic valuation in the context of these distinctions, each is discussed briefly below.

The instrumental value of an ecosystem service is a value derived from its role as a means toward an end other than itself. In other words, its value is derived from its usefulness in achieving a goal. In contrast, intrinsic value is the value that exists independently of any such contribution; it reflects the value of something for its own sake. For example, if a fish population provides a source of food for either humans or other species, it has instrumental value. This value stems from its contribution to the goal of sustaining the consuming population. If it continued to have value even if it were no longer "useful" to these populations (e.g., if an alternative, preferred food source were discovered), that remaining value would be its intrinsic value. For example, if the Grand Canyon and the Florida Everglades have intrinsic value, that component of value would be independent of whether humans directly or indirectly use them—either as sites for recreation, study, or even contemplation. Intrinsic value can also stem from heritage or cultural sources, such as the value of culturally important burial grounds. Because intrinsic value is the value of something unrelated to its instrumental use of any kind, it is often termed "noninstrumental" value.

Anthropocentricism assumes that only human beings have intrinsic value and that the value of everything else is instrumental to human goals. To say that all values are anthropocentric, however, assumes that only humans assign value, and thus the value of other organisms stems from their usefulness to humans. Non-anthropocentric or biocentric values assume that certain things have value even if no human being thinks so. Thus, a biocentric approach assigns intrinsic value to all individual organisms, including but not limited to humans. Within this framework, intrinsic value or worth reflects more than humans caring about nonhumans and includes, in addition, the recognition that nonhumans have worth or value that is independent of any human caring or any satisfaction humans might receive from them. For example, a biocentric approach would assign a positive value to an obscure fish population (e.g., the snail darter; see more below) even if no human being feels that it is valuable and thus worth preserving. Clearly, both instrumental value and intrinsic value can be either anthropocentric or non-anthropocentric (see Callicott, 2004; Turner, 1999).

Intrinsic value is related but not identical to what economists call "existence value," which reflects the desire by some individuals to preserve and ensure the continued existence of certain species or environments. Existence value is an anthropocentric and utilitarian concept of value. Utilitarian values stem from the ability to provide "welfare," broadly defined to reflect the overall well-being of an individual or group of individuals. In this sense, utilitarian values are instrumental in that they are viewed as a means toward the end result of increased human welfare as defined by human preferences, without any judgment about whether those preferences are "good" or "bad." Existence values still stem from the fact that continued existence generates welfare for those individuals, rather than from the intrinsic value of nonhuman species. As such, there is the potential for substitution or replacement of this source of welfare with an alternative source (i.e., more of something else). In fact, implicit in the economic definition of existence values is the possibility of a welfare-neutral trade-off between continued existence of the species or environment and other things that also provide utility (see more detailed discussion below). Thus, the utilitarian approach implicitly assumes that existence value is an anthropocentric instrumental value that is potentially substitutable.[3]

In contrast, under the deontological (or duty-generating) approach, intrinsic value implies a set of rights that include a right of existence. Under this approach, something with intrinsic value is irreplaceable, implying that a loss cannot be offset or "compensated" by having more of something else. For example, a human person's own life is of intrinsic value to that person because it cannot be offset or compensated by that person having more of something else. This approach has its roots in the writings of the philosopher Immanual Kant, who wrote extensively about intrinsic value (e.g., Kant translated in 1987). However, Kant used the concept of rationality to determine the realm of beings that have intrinsic value and rights. He argued that human beings were the only beings who were rational and thus that only human beings have intrinsic value and rights. In this sense, Kant's views were strictly anthropocentric. Since Kant's writings, others have suggested alternative criteria for determining the realm for intrinsic value and rights (see footnote 31 in Callicott, 2004) and hence have argued that rights should extend to nonhumans, including animals (either individual animals or species) and in some cases all biological creatures (i.e., all plant and animal life) or the biota collectively. The modern notion of intrinsic value (as used in the context of ecosystem valuation) reflects the notion that rights should be extended beyond human beings (Stone, 1974).[4]

As discussed in more detail below, the economic approach to valuation is an anthropocentric approach based on utilitarian principles. It includes considera-

---

[3] This assumption rules out fixed proportions preferences between the different categories of values.

[4] A good reference regarding the relationship between intrinsic value and legal rights is Christopher Stone's *Should Trees Have Standing? Towards a Theory of Legal Rights for Natural Objects* (Stone, 1974).

tion of all instrumental values, including existence value. Environmental policy and law may also be based on intrinsic value, as exemplified by the Endangered Species Act of 1973. Because it is utilitarian based, economic valuation assumes that the potential for substitutability between the different sources of value that contribute to human welfare. The main categories of value that are not captured by the economic approach are non-anthropocentric values (e.g., biocentric values) and intrinsic values on which the concept of rights is based.

Finally, it is important to keep in mind that economic valuation is based on the notion that the values assigned by an individual reflect that individual's preferences or marginal willingness to trade one good or service for another, and that societal values are the aggregation of individual values. At any point in time, individual preferences can be influenced by a variety of factors, including culture and information, which can change over time. In addition, an individual's willingness to trade one good for another will reflect the amount of the goods and services currently available to him, which will in turn depend at least partially on income. If income changes over time, the economic measure of value for an individual can be expected to change as well. For these reasons, the values measured through economic valuation are inherently time- and context-specific.

Quantifying Values

Recognition that ecosystems or ecosystem services are valuable, possibly in a variety of ways or for a variety of reasons, does not necessarily imply a *quantification* of that value (i.e., its *valuation*).[5] In fact, those people who affirm the intrinsic value of ecosystems object to the very idea of trying to quantify the value of environmental goods and services (see, for example, Dreyfus, 1982; MacLean, 1986; Sagoff, 1993, 1994, 1997). For them, that would be as objectionable as quantifying the value of human life. The quantification of the value of ecosystems is by definition anthropocentric since humans are doing it. In addition, it implies a ranking of values (i.e., a statement of which goods or services are "more valuable," and possibly by how much). Some people object to one or both of these implications of quantification as being analogous to ranking the value of different human beings based, for example, on gender or ethnicity.

However, there are a number of contexts in which quantification of such values may be useful or even necessary, including (1) informing policy decisions in which trade-offs are considered, (2) providing damage estimates for natural resource damage assessment (NRDA) or similar cases, and (3) incorpo-

---

[5]It is important to distinguish between "values," which are an attribute of a good or service, and "valuation," which is the process of quantifying that attribute.

rating environmental assets and services into national income accounts.[6] For example, if an environmental policy decision involves a trade-off in the choice between providing one ecosystem service (such as a particular habitat or an ecological service) and providing another good or service (such as agricultural output), then information about the relative values of these alternative goods or services can lead to better-informed and more defensible choices. This requires a ranking of values, which follows from quantification. A recognition that quantification or valuation may be useful or necessary in informing policy decisions is explicit in the remainder of the committee's statement of task (see Box ES-1). Given the committee's charge, the remainder of this report focuses on the role of valuation in the context of policy decisions and improved environmental decision-making. Although not the focus of this study, the committee believes that quantification is also important (in fact, necessary) in the other two contexts as well. In NRDA cases, a quantification of lost value is necessary to determine the compensation that must be paid by responsible parties.[7] Similarly, in order to incorporate changes in environmental and other natural assets into national income accounts, these changes must be quantified in a manner comparable to the quantification of other components of national income (Heal and Kriström, 2003; NRC, 1999).

If quantification is deemed to be a useful or necessary input for policy decisions, a particular quantification or valuation approach must be selected. As noted above, given the committee's charge, this report focuses on the quantification embodied in the economic approach to valuation. In this approach to valuation, the metric that is used to quantify values in nearly all applications is a monetary metric, such as U.S. dollars.[8] In the context of ecosystem goods or services that are bought and sold in markets, dollars or some other currency provide a natural metric for quantification since such prices, absent any market distortions, reflect the consumer valuation of that good (see further discussion in Chapter 4). Thus, when policies involve trade-offs between market goods (already valued in dollar terms) and ecosystem services that are not traded in markets, quantifying the value of these nonmarket services using the same metric (e.g., a dollar metric) allows a direct assessment of the trade-offs.

However, the use of a dollar metric for quantifying values is based on the assumption that individuals are willing to trade the good being valued for something else that can also be quantified by the dollar metric. It thus assumes that

---

[6] Note that the type of quantification that is necessary can vary across these different contexts. For example, NRDA requires a point estimate of the total damages or lost benefits from an environmental reduction in ecosystem services resulting from some event (e.g., an oil spill). In contrast, in a policy context, quantification of the value of a subset of services may be sufficient (see Chapters 5 and 6 for further discussion).

[7] Quantification of values is not necessary if compensation is measured in physical units (e.g., when based on habitat equivalency). However, a habitat equivalency approach to compensation implicitly assumes that the value of the restored or replaced habitat is equivalent to the value of the degraded one.

[8] Some have advocated the use of energy analysis as an alternative currency or metric for measuring value. See Chapter 3 and Box 3-7 for further information.

the good being valued is in principle substitutable or replaceable with other goods or services that are also of value and that money can buy; this reflects the utilitarian principles that underlie economic valuation.[9]

Role of Valuation in the Policy Process

Although economic valuation requires a quantification of values, the specific design of the valuation exercise should depend on its purpose or the role that it will play in the policy process. One approach is to base policy decisions regarding preservation of environmental resources on moral principles, stemming from a political consensus about what is morally right or wrong. While adherence to moral principles relating to intrinsic value will inevitably involve trade-offs, under this approach these trade-offs are of little or no consequence to the policy choice. If policy choices are to be based on the notion of intrinsic values and rights, then these rights have to be identified, but the values are implied by that identification need not be quantified in order to choose among alternatives (unless the decision to protect one intrinsic value implies a loss of something else with intrinsic value). Thus, with this decision rule, valuation of ecosystem services has no effect on policy choices and hence plays a very limited role (see Goulder and Kennedy, 1997).[10]

Strict utilitarianism, on the other hand, implies that a decision is based solely on economic efficiency, that is, maximization of the net benefits to society (Goulder and Kennedy, 1997). This decision rule is implemented through the use of benefit-cost analysis (BCA). Economic valuation plays a central role in the application of BCA, since BCA requires an estimate of the benefits and costs of each alternative using a common method (economic valuation) and metric (dollars) so that the two can be compared. The comparison of costs and benefits allows an explicit consideration of the trade-offs that are inevitably involved in most environmental policy decisions. It recognizes that achieving a particular objective or goal such as preservation of a particular ecosystem comes at a cost, since the resources that must be devoted to this preservation are not available for use in providing other goods and services. A typical BCA asks whether the benefits of that preservation are "worth" the costs involved. In this

---

[9] Several environmental philosophers argue that while a monetary metric is an appropriate metric for utilitarian values, it is inappropriate for non-utilitarian values such as non-anthropocentric intrinsic values or values based on notions of morals, rights, and duties (deontological values) (e.g.,Callicott, 2004; Sagoff, undated and 1997). This raises the question of what, if any, metric might be used to quantify, or at least rank, these non-utilitiarian values. Callicott (2004) suggests use of a "penalty metric." He argues that the severity of the penalties imposed for violations of certain types of protections that reflect intrinsic value provides a democratically determined measure, or at least ordinal ranking, of those values.

[10] Of course, valuation could be used in this context to determine whether adherence to a moral principle came at a net cost or benefit to society. However, under such an approach, this information would be a "curiosity" rather than a determinant of the policy choice.

sense, it ensures that the limited resources used to provide goods and services to society are used in the most efficient way—that is, to achieve the greatest net benefit.

In addition, a benefit-cost approach provides a means of combining heterogeneous views of what is desirable. Although some may prefer preservation of the environment or a particular ecosystem, others may prefer an alternative (e.g., development of the land). These different views can stem from differences in an individual's net benefits from the alternatives. Those who realize a net gain from preservation would be expected to prefer preservation, whereas those who realize a net gain from the alternative are likely to prefer it. The benefit-cost approach provides a mechanism for combining these disparate views to reach a decision that incorporates both perspectives. Of course, in doing so, it assigns equal weights to the net benefits of all individuals, a property of BCA that may draw criticism (Azar, 1999; Layard, 1999; Potts, 1999).

If BCA is to be used to evaluate environmental policy options, it is imperative that all costs and benefits be considered.[11] In particular, for policy decisions that impact ecosystems, the benefits that the ecosystem generates through the various goods and services it provides must be included in calculating the benefits of preserving the ecosystem or the costs (forgone benefits) of allowing it to be degraded. As noted in Chapter 1, failure to assign a dollar value to these benefits (e.g., on the principle that they cannot be valued accurately or that the values are "incalculable") effectively assigns them a zero value or a zero weight in the calculation of net benefits, implying that changes in those services will not be incorporated into the net benefit calculation (Epstein, 2003).

Political and legal decisions are often made on the basis of information about many sources of value, including intrinsic and moral values, as well as economic values, and some decision rules seek to incorporate different types of values explicitly. For example, decision rules that imply adherence to moral principles or a premise of intrinsic value unless the cost is too high (as under a "safe minimum standard" rule; see Chapter 6 for further information) incorporate concern about both intrinsic value and economic welfare, and implicitly allow some trade-offs between the two. Similar trade-offs are also implied by decision rules that apply a benefit-cost test to environmental policy choices but constrain the decisions to ensure that certain conditions reflecting intrinsic value are not violated. Possible constraints include ensuring (1) that basic notions of justice and fairness are not violated, (2) that populations or levels of critical ecosystem services do not fall below standards necessary to ensure their continuation, and (3) that uncertainties regarding outcomes are not deemed too great. In such cases, information about benefits and costs as determined by economic valuation will be a useful input into the policy decision but will not solely de-

[11] In some cases, the decision implied by a benefit-cost analysis may be clear without a full quantification of all values. For example, if a proposal or project would pass a benefit-cost test with a complete quantification of costs and an incomplete quantification of benefits, then it would also pass with a complete quantification of benefits. In such a case, quantification of the remaining benefits would not change the results of the test.

termine it, since the net benefits from the various alternatives will be only one of the factors considered when making a policy choice.

Examples of different weights put on intrinsic values versus utilitarian welfare can be found throughout environmental policies in the United States. For example, the Clean Air Act requires a periodic assessment of the costs and benefits of the act, although it clearly states that the costs or impacts of any standard or regulation promulgated under the act shall not be a basis for changes that preclude the U.S. Environmental Protection Agency (EPA) from carrying out its central mission to "protect human health and welfare." Thus, information about costs and benefits is intended to inform but not drive policy decisions. In contrast, Executive Order 12291[12] required a strict cost-benefit approach to evaluating regulations. The order stated that "regulatory action shall not be undertaken unless the potential benefits to society for the regulation outweigh the potential costs to society." This order, and a related order (Executive Order 12866), were later replaced by Executive Order 13258, issued in 1996, which replaced the strict benefit-cost criterion for decision-making with a weaker version that instead simply required that the benefits of the regulation *justify* the costs (OMB, 1996; see also Chapter 4). Under this more recent order, BCA is an input into regulatory decisions but not the sole criterion for them.

Other environmental policies appear to reject more explicitly a consideration of benefits and costs in favor of an approach based on intrinsic value and rights. For example, Callicott (2004) has argued that the protection granted to species under the Endangered Species Act (ESA) is based primarily on principles regarding the duty to preserve species because of their intrinsic value. In *Tennessee Valley Authority vs. Hill*, the U.S. Supreme Court found that although "the burden on the public through the loss of millions of unrecoverable dollars would [seem to] greatly outweigh the loss of the snail darter. . ., *neither the Endangered Species Act nor Article III of the Constitution provides federal courts with authority to make such fine utilitarian calculations*" [emphasis added]. On the contrary, the plain language of the act, buttressed by its legislative history, shows clearly that Congress viewed the value of endangered species as "incalculable" (e.g., Telico dam-snail darter case; U.S. Supreme Court, 1978).[13] In response to this finding, Congress immediately amended the ESA to allow at least the possibility of consideration of benefits and costs and to create a committee with authority to grant exceptions to the law's prohibitions under very limited conditions that consider, but do not simply compare, benefits and costs.

It is clear from the preceding overview that in many policy contexts relating to the use and preservation of environmental resources, some consideration is given to the magnitude of benefits and costs, even though this information is likely to be only one of many possible considerations that influence policy choice. To provide this information, those benefits and costs must be measured, and economic valuation provides a means of measuring them. It is the judgment

[12] See Federal Register 46(33), February 19, 1981, for further information.

[13] See Erdheim (1981) for a discussion of this seminal case.

of this committee that having the best available and most reliable information about the economic valuation of ecosystem services will lead to improved environmental decision-making. It will allow policymakers to identify and evaluate trade-offs and, if appropriate, incorporate a consideration of these trade-offs into environmental policy design.

Framing the Valuation Question

In order to be useful in the evaluation of environmental policy options, the valuation exercise should be designed or framed to provide the necessary information to policymakers. A number of dimensions are important in framing the analysis. Some of these dimensions are discussed briefly below (see also Chapter 6).

First, it is important to recognize that policy choices, and the benefits and costs associated with them, imply *changes* in environmental quality or the level of environmental services (e.g., changes in ecosystem goods and services), either positive or negative, and that the valuation exercise is the quantification of the value of those changes.[14] Thus, in a policy context, economic valuation is not concerned with quantifying the value of an entire ecosystem (unless the policy under consideration would effectively destroy the entire ecosystem); rather, it is concerned with translating the physical changes in the ecosystem and the resulting change in ecosystem services into a common metric of associated changes in the welfare (utility or "happiness") of members of the relevant population. Thus, the valuation of ecosystem services should be framed in terms of valuing the changes in those services implied by different policy choices.

A second important dimension of framing is the scope of the analysis. Scope refers to the inclusion or exclusion, by choice or necessity, of certain ecosystem functions or services and/or certain types of value. Thus, a valuation exercise may focus on only a subset of ecosystem services; for example, an exercise might seek to value changes in flood control or water purification services but not changes in the quantity or quality of habitat. Similarly, the valuation exercise may focus (by necessity) on the quantification of certain types or sources of value and may not capture other sources. Although a broader scope provides a more accurate picture of the total impact of the policy change, in some policy contexts a partial approach may be sufficient. For example, if the results of a benefit-cost analysis based on a measure of the partial value of ecosystem preservation imply that the benefits of a particular policy or activity outweigh the costs, then inclusion of additional benefits (by valuing additional services or including additional sources of value) will only reinforce this conclusion (see also footnote 11).

[14] An important consideration is the benchmark used for measuring these changes. Different benchmarks imply different assumptions about property rights and require different valuation measures. The link between valuation measures and property rights is discussed later in this chapter.

The outcome of the valuation exercise will also depend on its spatial or geographic scale (see Chapters 3 and 5 for further information). Spatial scale has two components. The first is definition of the geographic extent of the relevant ecosystem(s). In defining the physical impacts of a given policy, one can restrict consideration to fairly localized impacts or consider spillover impacts on related ecosystems that are not impacted directly but change indirectly through those linkages.[15] Consideration of these indirect impacts will yield a more inclusive analysis, but these indirect effects may be difficult to identify and quantify accurately. In addition, some policies (particularly at the national level) can affect many ecosystems. For example, a categorical exclusion under the National Environmental Policy Act (NEPA) of federal activity in all wetlands 10 acres or less in size will affect the hundreds or thousands of wetlands across the United States. In such cases, the aggregate impact across all affected ecosystems should be valued.

The second component of spatial or geographic scale is definition of the relevant population (i.e., the stakeholders). In estimating the value that individuals place on ecosystem changes, one must identify which individuals (whose values) to include. In other words, what is the relevant population for estimating the benefits and costs of the policy change? For example, in valuing possible damages from a major oil spill, should calculations reflect damages to the local population, to the population within the state, to the population within the nation, or to the world population? Because an oil spill that leads to loss of wildlife may negatively impact those outside the local area who value the existence of the animals, the aggregate measure of damages will generally vary directly with the extent of the population considered (Carson et al., 2001). The appropriate population to include will depend on the perspective of the decision-maker, his or her jurisdiction, and the target population of concern to the decision-maker when assessing the aggregate welfare impacts of the policy change. Thus, local officials may be concerned primarily with the costs and benefits borne by their local constituents, while national policymakers can be expected to take a broader view.

In addition to the spatial or geographical scale, the valuation exercise is also affected by the temporal scale of the analysis (i.e., the period of time over which benefits and costs are distributed). Most policy impacts last for extended periods, and some last (effectively) forever because they lead to irreversible changes. This is particularly likely in the context of ecosystems, where stock effects are important and losses of key ecosystem services may be irreversible. When the benefits and/or costs extend over time, the period of analysis becomes a key factor in determining the results of a valuation exercise. For example, if land conversion for development purposes causes irreversible loss of critical habitat, an analysis that considers only a short time period will not accurately assess the benefits and costs of that conversion. In addition, the analysis should

---

[15] This distinction is comparable to the economic distinction between partial and general equilibrium analysis (see further discussion below).

account for differences in the timing of impacts across alternatives. One approach to this is the use of discounting to weight impacts differently depending on when they occur. The meaning and use of discounting are discussed later in this chapter (see also Chapter 6). At this point, it is sufficient to note that the temporal framing of the valuation exercise—the time period chosen and the method used to reflect differences in the timing of impacts—plays a crucial role in determining its results.

The discussion thus far suggests that the quantification of ecosystem value using the economic approach to valuation can and does play an important role in environmental policy analysis and decision-making. However, the results that emerge from this quantification or the valuation exercise will be influenced significantly by the way in which the valuation question is framed. To provide meaningful input to decision-makers, it is imperative that the valuation exercise seeks to value the *changes* in ecosystem goods or services attributable to the policy change, that the scope considers all relevant impacts and stakeholders, and that the temporal scale of the analysis is consistent with the scale of the impacts. The results will also depend on a number of methodological and data issues. These issues are discussed in detail in Chapter 4 and illustrated through the case studies provided in Chapter 5.

THE ECONOMIC APPROACH TO VALUATION

Having discussed economic valuation and its role in general terms, a more detailed discussion of the economic approach to valuation follows. As noted earlier, the economic concept of value is based on an anthropocentric, utilitarian approach to defining value based on individual preferences. As such, it does not encompass all possible sources of value. However, it is much broader than the narrow concept of commercial or financial value, and includes all values, tangible as well as intangible, that contribute to human satisfaction or welfare. This broad definition is reflected in the "total economic value" framework that underlies economic valuation and is described below.

The Total Economic Value Framework: Use and Nonuse Values

The *total economic value* (TEV) framework is based on the presumption that individuals can hold multiple values for ecosystems. It provides a basis for a taxonomy of these various values or benefits. Although any taxonomy of such values is somewhat arbitrary and may differ from one use to another, the TEV framework is necessary to ensure that all components of value are given recognition in empirical analyses and that "double counting" of values does not occur when multiple valuation methods are employed (Bishop et al., 1987; Randall, 1991). It is important to state that the TEV framework does not imply that the "total value" of an ecosystem should be estimated for each policy of concern. Even a marginal change in ecosystem services can give rise to changes in multiple values that can be held by the same individual, and the TEV framework simply implies that all values that an individual holds for a change should be counted.

In the simplest form, TEV distinguishes between *use* values and *nonuse* values. The former refer to those values associated with current or future (potential) use of an environmental resource by an individual, while nonuse values arise from the continued existence of the resource and are unrelated to use. Typically, use values involve some human "interaction" with the resource whereas nonuse values do not. The distinction between use and nonuse values is similar but not identical to the distinction between instrumental and intrinsic value discussed earlier. Clearly, use values are instrumental and utilitarian, but, as noted above, the concept of existence value is not identical to the notion of intrinsic value, because the latter is deontological and includes non-anthropocentric values while the former does not.

Within the TEV framework an individual can hold both use and nonuse values for the services of an aquatic ecosystem. Consider an oil spill on a popular coastal beach resulting in forgone recreational trips to the beach—this is a lost use value. In addition, the oil spill could damage the ecosystem in ways that would not affect beach use and that beach users would never observe. It might, for example, kill marine mammals that live off the beach and are not seen by beach users, and beach users, as well as those who do not visit the beach, might experience a loss because of this ecosystem damage. The loss by those who do not visit the beach would be a loss of nonuse value, though there could also be a loss of nonuse value on the part of beach users. The TEV framework implies that analysts proceed to investigate the potential loss in use and in nonuse values of beach users and in nonuse values of people who do not visit the beach. It is not necessary to estimate the total value of the coastal ecosystem, only the total loss in value associated with the oil spill.

A number of TEV frameworks have been proposed in recent decades (e.g., Bishop et al., 1987; Freeman, 1993a; Randall, 1991). Although varied in detail and application, the distinction between use and nonuse values is a fundamental theme. The TEV framework, as applied to typical aquatic system services for the purposes of this report, is illustrated in Table 2-1. In the discussion below, distinctions are drawn between the components of TEV, but when people hold both use and nonuse values, the literature cited above argues for estimating peoples' TEV rather than estimating the components and then adding the component estimates to compute a TEV. However, the discussion of valuation methods in Chapter 4 shows that some methods are better able to measure selected components of TEV than others.

TABLE 2-1 Classification and Examples of Total Economic Values for Aquatic Ecosystem Services

Use Values		Nonuse Values
Direct	Indirect	Existence and Bequest Values
Commercial and recreational fishing Aquaculture Transportation Wild resources Potable water Recreation Genetic material Scientific and educational opportunities	Nutrient retention and cycling Flood control Storm protection Habitat function Shoreline and river bank stabilization	Cultural heritage Resources for future generations Existence of charismatic species Existence of wild places

SOURCE: Adapted from Barbier (1994) and Barbier et al. (1997).

Use Values

Use values are generally grouped according to whether they are *direct* or *indirect*. The former refers to both *consumptive* and *nonconsumptive* uses that involve some form of direct physical interaction with the resources and services of the system. Consumptive uses involve extracting a component of the ecosystem for an anthropocentric purpose such as harvesting fish and wild resources. In contrast, nonconsumptive direct uses involve services provided directly by aquatic ecosystems without extraction, such as use of water for transportation and recreational activities such as swimming. Although nonconsumptive uses do not involve extraction and hence diminution in the quantity of the resource available, they can diminish the quality of aquatic ecosystems through pollution and other external effects.

It is also increasingly recognized that the livelihoods of populations in areas near aquatic ecosystems may be affected by certain key *regulatory ecological functions* (e.g., storm or flood protection, water purification, habitat functions) (Daily, 1997). The values derived from these services are considered indirect, since they are derived from the support and protection of activities that have directly measurable values (e.g., property and land values, drinking supplies, commercial fishing). For example, mangrove swamps may provide a "storm protection" function in that they may stop coastal storms from wreaking havoc on valuable coastal properties and infrastructure (Janssen and Padilla, 1999). Activities such as reading a book or magazine article about ecosystems, or watching a nature program, are also thought to provide indirect use values.

Nonuse Values

Many natural environments are thought to have substantial existence values; individuals do not make use of these environments but nevertheless wish to see them preserved "in their own right" (Bishop and Welsh, 1992; Boyle and Bishop, 1987; Freeman, 1993b; Madariaga and McConnell, 1987; Randall, 1991; Smith, 1987). The terms "existence," "nonuse," and "passive" use are generally used synonymously in the literature. For the purposes of this report, nonuse values refer to all values people hold that are not associated with the use of an ecosystem good or service. Use values typically arise from a good or service provided by ecosystems that people find desirable. Nonuse values need not arise from a service provided by an aquatic ecosystem; rather, people may benefit from the knowledge that an ecosystem simply exists unfettered by human activity (e.g., Crater Lake). The latter is what was traditionally known as a "pure" existence value in the literature. Other motivations for nonuse values are bequest and cultural or heritage values. The empirical literature generally does not attempt to measure values for individual aspects of nonuse values, but focuses on the estimation of nonuse values irrespective of the underlying motivations people have for holding this value component.

The economic valuation of the impacts of the *Exxon Valdez* oil spill on the aquatic and related ecosystems of Prince William Sound, Alaska, highlights the importance of nonuse values in natural resource damage assessments and project appraisals (Carson et al., 1992). The *Exxon Valdez* study revealed that many Americans who have not visited Alaska and never intend to do so nevertheless place high values on maintaining the pristine and unique but fragile coastal and aquatic ecosystems of Alaska. In the context of the *Exxon Valdez* study, questions were raised about the accuracy with which nonuse values can be estimated (Hausman, 1993; NOAA, 1993). This issue is discussed in greater detail in Chapter 4.

Measurement Using a Monetary Metric: WTP Versus WTA

Economic valuation is concerned with how to estimate the impact of changes in ecosystem services on the welfare of individuals and is based on the principles of utilitarianism. If ecosystem changes result in individuals' judging that they are worse off, one would like to have some measure of the loss of welfare to these individuals. Alternatively, if the changes make people better off, one would want to estimate the resulting welfare gain.

The basic concept used by economists to measure such welfare gains and losses is rooted in the utilitarian notion that for any individual, the different sources of value that affect the individual's utility are potentially substitutable; that is, the individual is willing to trade a reduction in one source of value for an increase in another in a manner that leaves his or her overall utility unchanged.

The essence of this approach is to value a change by determining what people would be willing to trade (i.e., to receive or to give up) so they would be equally satisfied or happy with or without the change.

Consider, for example, a case in which a freshwater lake can be restored to enhance sportfishing opportunities. An economic measure of the benefit of such an improvement to recreational anglers is the maximum that anglers would be willing to pay for this improvement in fishing if he or she had to pay. Each angler's maximum willingness to pay should represent how much money the angler is prepared to give up in exchange for the increase in individual enjoyment gained from the improved recreational fishing. It represents the reduction in income that would be necessary to offset exactly the gain in angler utility resulting from the restoration, thereby leaving anglers at the same utility level as they were prior to any restoration. Maximum willingness to pay could then be aggregated for all anglers who benefit to determine the total benefits of the project.[16] This aggregation, in turn, would facilitate an assessment of whether public funds should be spent on the project.

An alternative measure of the value of the improvement in recreational fishing from restoration of the lake is based not on anglers' willingness to *pay* for the improvement but rather on the amount they would be willing to *accept* to forgo the improvement. If the improvement is promised, then failure to provide this improvement (i.e., failure to restore the lake) would reduce the utility of anglers relative to the level they would have attained with the restoration. The value of this loss or the forgone benefit from restoration can be measured by the minimum amount of income that the anglers would be willing to accept as compensation for forgoing that benefit. The increase in income (i.e., the compensation) would have to increase the utility of anglers by exactly the same amount as the reduction in utility stemming from the failure to restore the lake, so that the combined effect would be to leave utility unchanged (i.e., leave the anglers just as well off without the restoration as they would have been with it).

The preceding example illustrates the two alternative measures of value that are used in economic valuation: WTP and WTA. Each measure looks at potential trade-offs between money and the good or service being valued that leave utility unchanged from some base level. They differ, however, in the base level of utility that is maintained when the hypothetical trade-off is made. In valuing an improvement in environmental quality or services, WTP considers trade-offs that would leave utility at the level that existed prior to the improvement (the pre-change utility level), whereas WTA considers the utility level that would exist after the improvement (the post-change utility level).

In some cases such as when valuing small price changes, WTP and WTA measures of value can be expected to be quite close, differing only because of the different income levels implied by paying rather than receiving compensation (Willig, 1976). However, for many environmental goods and services, the

---

[16] It is important to note that the concept of willingness to pay does not rely on the individual's actually paying for the change.

two can be substantially different. In particular, Hanemann (1991) has shown that when valuing changes in the quantities of goods or services available for which there are no close substitutes (including many ecosystem services), the two measures of value can yield quite different results. For environmental improvements, the amount an individual is willing to accept to forgo that improvement will normally be greater than the amount he or she would be willing to pay to ensure it (WTA > WTP).

Because WTP and WTA measures of ecosystem services could differ significantly, a key issue in the use of economic valuation in this context is the choice between these two possible measures of value. As noted above, the conceptual difference lies in the base level of utility that each is designed to ensure. This reflects a difference in the assumption regarding the underlying allocation of property rights or, equivalently, the baseline levels of utility that society collectively agrees to ensure to each individual within that society. Consider again the case of lake restoration. If anglers do not have a right to the improved conditions, then society is not collectively prepared to ensure them a level of utility that includes the restoration. If these anglers want restoration, then in theory they would have to "buy" it from the rest of society. In such a case, WTP is the appropriate economic measure of the value of the improvement. Conversely, if anglers have a right to the improved conditions, then if society wants to use the resources for other purposes, in theory it would have to buy the right to do so from the anglers and pay or otherwise compensate them for failure to restore the lake. In such a case, WTA is the appropriate economic measure of the value of the water quality improvement.

Economic theory, and hence economic valuation, provides no basis for choosing between the alternative property rights regimes and therefore no basis for preferring one measure of value over the other. Property rights are determined collectively by society. In addition, virtually all theories of property rights recognize that they are not absolute or strong but represent only "weak" rights, insofar as they are subject to modification and based on community welfare in ways that strong rights (e.g., a right to life) are not. They are weak rather than strong because they are not considered essential to human dignity in the way that rights to life or to equal protection are (Dworkin, 1977).

Although in theory economic valuation can seek to measure either WTP or WTA depending on the underlying assignment of property rights, it is common to use WTP as an empirically reliable measure. The primary reason is that most of the existing economic methods for estimating values capture WTP but not WTA (see Chapter 4 for further information). The use of WTP may be inappropriate in a given case because of the implicit property rights assumption embedded in it. However, even in cases where WTA would be the appropriate measure, WTP may still be a reasonable proxy for WTA. In theory and practice, the absolute value of willingness to accept usually exceeds the absolute value of willingness to pay (Hanemann, 1991; Horowitz and McConnell, 2002). Thus, WTP can be viewed as a lower-bound for WTA and hence as a lower-bound for the value of the improvement. In some contexts, a lower-bound estimate of values will be sufficient to inform policy decisions. For example, if the benefits of an increase in ecosystem services exceed the costs when those benefits are measured using WTP, they would also have exceeded costs if measured using a higher WTA. However, if a WTP measure of benefits was lower than cost in a context in which WTA was the correct measure to use, then it is still possible that benefits would have exceeded costs had WTA been used.

In addition to the difference regarding the implicit assumption with respect to underlying property rights, WTP and WTA also differ in another important aspect, namely, the role of income limitations. Clearly, the amount that an individual is willing to pay for an environmental improvement depends on the amount that he or she is *able* to pay. In other words, WTP is constrained by an individual's income since he or she could never be willing to pay more than the amount available. WTA, on the other hand, is not income constrained. The amount of compensation that would be required to compensate an individual for accepting a lower level of environmental quality can exceed a person's income. This difference has important implications in measures of aggregate net benefits. Income constraints imply that, all else being equal, low-income individuals will have a lower WTP than wealthier individuals simply because of their lower ability to pay. This implies that the preferences of wealthy people will get more weight than those of poorer people in net benefit calculations based on WTP. This feature of WTP should be borne in mind when using this measure of value.

Uncertainty and Valuation

Estimates of the values of ecosystem services are frequently somewhat uncertain for a variety of reasons. Chapter 6 explores the major sources and types of uncertainty, indicates which are most significant, and discusses their consequences in ecosystem services valuation. This discussion includes the problems posed by uncertainties about models and parameters, and how analysts and decision-makers can and should respond. Sensitivity analysis and Monte Carlo simulation are discussed as a possible analyst response to model and parameter uncertainties, while risk aversion, quasi option values, adaptive management, safe minimum standards, and the precautionary principle are discussed in the context of use by decision-makers.

Discounting: Utility Versus Consumption

In many ecosystem valuation contexts, the impacts of a particular policy choice will extend over time, and hence an attempt must be made to estimate the costs and benefits not only for current years but well into the future. Deriving an aggregate measure of costs or benefits that reflects their change over time requires an aggregation method that appropriately incorporates the timing of benefits and costs. The most commonly used approach in economic valuation is discounting, that is, weighting future costs and benefits differently than current costs and benefits when summing over time.

The desirability of discounting future costs and benefits has been the subject of intense debate (Heal, 1998; Portney and Weyant, 1999). The simplest explanation of discounting can be found in the financial context. People generally agree, for example, that accountants are correct to discount future income. If a person will receive an income of $20,000 a year for the next 30 years, most people would agree that it is unreasonable to value that total income at 30 times $20,000. Instead, a more reasonable valuation would be $20,000 for the first year, plus $20,000 discounted by some rate (such as 5 percent) for the second year, plus the amount from the second year, discounted by an additional 5 percent, for the third year, and so on. The rationale for such discounting is the productive power of the economy that converts commodities at one time into a greater quantity of commodities at a later time. If one ignores inflation, then money represents a quantity of purchasing power over economic commodities, and therefore commodities available at an earlier time are worth more than commodities available only at a later time. If the economy remains productive, then (even on a simple level) it is easy to see that money at a later time is worth less than money at the present time because, for example, money this year can be converted into more money in the future by depositing it into a bank to earn interest.

However, the issues raised by the use of discounting in cost-benefit analysis, project evaluation, and ecosystem valuation go far beyond the simple arithmetic of compound interest on bank balances. It is important to realize that two different types of discounting may be practiced—utility discounting and consumption discounting. This distinction is absolutely central, although unfortunately it is not as widely understood. The properties of and justifications for these two rates are quite different, and some of the arguments that apply to one are not relevant in the context of the other (Heal, 2004).

This chapter provides only a brief summary of the underlying issues, which are quite complex and the subject of a massive literature.[17] What is normally referred to as "the discount rate" is in fact the *utility discount rate*, also known as the pure rate of time preference, the social rate of discount, or the social rate of time preference.[18] This is the rate to which Frank Ramsey's famous strictures

---

[17] For a more detailed discussion, see Heal (2004).

[18] This is the rate *r* in the utilitarian maximand $\int_0^\infty U(c)e^{-rt}dt$. In the utilitarian approach a proposed policy is evaluated by the weighted sum of the utilities accruing at different points in time. The weight placed on utility at time *t* is given by e^{-rt}, an exponential function of time. The utility discount rate is the rate at which this weight—the weight placed on utility at time *t*—decreases with time. It is the proportional rate of change of e^{-rt} with *t*, which is of course just *r*. The reason for calling this the utility discount rate is obvious; it is the rate at which one discounts utility.

apply and indeed those of Roy Harrod as well.[19] There is no compelling reason for this discount rate to be positive; the value of the utility discount rate reflects the relative valuations that are placed on present and future generations. If one is convinced that future generations should be valued less than present generations, then a positive utility discount rate should be chosen; otherwise this rate should be zero.

The *consumption discount rate* is conceptually and operationally different from the utility discount rate. The utility discount rate, as emphasized above, is intended to represent the relative weights put on present and future utilities. It expresses society's preferences for distribution between generations, with a zero rate representing equal weights for all generations, and a positive rate implying less weight to future people. In contrast, the consumption discount rate represents the weights placed on increments of consumption at different dates. It answers the question, How does one value an extra dollar's worth of consumption (instead of an extra unit of utility) today relative to an extra dollar's worth of consumption in the future?

Even if future utilities are valued the same as present utilities (i.e., there is a zero utility discount rate), one may still value an increment of consumption 20 years in the future differently from the same increment today. There are several reasons for this. One reflects changes in wealth or the standard of living over time. Suppose, for example, that people 20 years from now are expected to be wealthier than those today. If the extra utility generated by additional consumption diminishes with income, then providing the additional consumption in the future when people are wealthier will yield less of an increase in utility than providing the same additional consumption today. This suggests that future consumption should be discounted. If this were done, however, it would not reflect a judgment about the relative merits of present and future people, which is what the utility discount rate does. Rather, it would reflect a distributional judgment about the relative merits of extra consumption going to richer or poorer people, quite independent of the dates at which they live. If this approach is accepted, it implies a positive consumption discount rate when living standards are rising over time and, conversely, a negative rate when they are falling.

The distinction between utility and consumption discounting is important in the context of environmental issues (Heal, 2004). One might feel that access to aquatic ecosystem services will decrease over time as a result of human pressures on natural habitat and that, consequently, peoples' marginal valuations of these services will increase as they become scarcer. As a result, the value of incremental ecosystem services will rise over time and the consumption discount rate to be applied to these will be negative rather than positive. That is to say,

---

[19] Frank Ramsey was an influential economist and mathematician at Cambridge, United Kingdom, in the 1920s. He remarked that "discounting is ethically indefensible and arises purely from a weakness of the imagination" (Ramsey, 1928). Roy Harrod, an Oxford University economist of the same generation, wrote similarly that discounting is a "polite expression for rapacity and the conquest of reason by passion" (Harrod, 1948).

increments in the future will be worth more than those in the present—not because they are in the future but rather because they are being made available at a later date when they are scarcer. This reflects diminishing marginal utility or valuation rather than the result of futurity.

It follows from this discussion that the consumption discount rate is quite flexible and reflects many different characteristics of the underlying problem. If people are concerned with ecosystem goods and services, which are expected to be scarcer in the future than in the present, then the consumption discount rate may be negative, meaning that a unit of consumption in the future would be valued more than a unit at present. If income levels are rising over time, then future income levels will be higher than those at present, so the marginal valuation of income will decrease over time and the consumption discount rate will be positive (i.e., the future should be discounted).

The preceding discussion highlights the existence of two quite distinct concepts of discounting—utility and consumption discounting. It argues that there is no compelling argument for discounting utility, but that there may be reasons for discounting consumption, although the appropriate rate may be positive or negative. When is it appropriate to use the consumption discount rate in ecosystem valuation and when should the utility discount rate be used instead?

In general, the utility discount rate should be used when the policy under consideration is such as to lead to changes in the overall utility or welfare levels of the economy, or at least a significant subsector of it. In economic terms, the utility discount rate is applicable in the context of general equilibrium analyses. The consumption discount rate, on the other hand, is applicable in the context of partial equilibrium problems. These are problems in which only a small part of the economy is being affected by our decisions, and these decisions have only a small impact on overall consumer welfare. Because all of the environmental valuation problems considered in this report are of a partial equilibrium nature, the relevant discount rate to be considered is the consumption rate, which may have either sign. The committee emphasizes that the consumption discount rate is the rate of change of the value placed on an increment of consumption as its date changes. It is not a number that the analyst chooses a priori but one that emerges from the characteristics of the economy, such as whether consumption of the ecosystem good at issue increases or decreases over time. Given this interpretation, one does not argue about whether to discount consumption or at what rate. Discounting consumption—in the very general sense of applying different marginal valuations to increments of consumption at different dates—is unavoidable in the utilitarian framework, and indeed in most other frameworks. One can however argue about the values of parameters that influence, but do not fully determine, the consumption discount rate and in particular determine whether that rate should be positive or negative—that is, whether future costs and benefits should be weighted less or more heavily than current costs and benefits when those costs and benefits are aggregated over time.

SUMMARY: CONCLUSIONS AND RECOMMENDATIONS

This chapter provides an overview of economic valuation and the role it plays in the policy and environmental decision-making process. Although economic valuation does not capture all sources or types of value (e.g., intrinsic values on which the notion of rights is founded), it is much broader than usually presumed. It recognizes that economic value can stem from use of an environmental resource (use values), including both commercial and noncommercial uses, or from its existence even in the absence of use (nonuse values). The broad array of values included under this approach is captured by using the total economic value framework to identify potential sources of economic value. Use of this framework helps to provide a checklist of potential impacts and effects that must be considered in valuing ecosystem services as comprehensively as possible. It reduces the likelihood of omitting key sources of value, as well as the possibility of double counting values. By its nature, economic valuation involves the quantification of values based on a common metric, normally a monetary metric. The use of a dollar metric for quantifying values is based on the assumption that individuals are willing to trade the ecological service being valued for more of other goods and services represented by the metric (more dollars). Use of a monetary metric allows measurement of the costs or benefits associated with changes in ecosystem services.

The role of economic valuation in environmental decision-making depends on the specific criteria used to choose among policy alternatives. If policy choices are based primarily on intrinsic values, there is little need for the quantification of values through economic valuation. In such cases, the "benefit" of preservation is the protection of the right. In such cases, it may still be important to society to know how much protecting that right (e.g., preserving an intrinsically valuable endangered species) would cost—that is, what is being given up to ensure that protection, but there is no need to quantify the benefit of protection. However, if policymakers consider trade-offs and benefits and costs when making policy decisions, quantification of the value of ecosystem services is essential. Failure to include some measure of the value of ecosystem services in benefit-cost calculations will implicitly assign them a value of zero. The committee believes that considering the best available and most reliable information about the benefits of improvements in ecosystem services or the costs of ecosystem degradation will lead to improved environmental decision-making. The committee recognizes, however, that this information is likely to be only one of many possible considerations that influence policy choice.

The benefit and cost estimates that emerge from an economic valuation exercise will be influenced by the way in which the valuation question is framed. In particular, the estimates will depend on the delineation of the changes in ecosystem goods or services to be valued, the scope of the analysis (in terms of both the geographical boundaries and the inclusion of relevant stakeholders), and the temporal scale. In addition, the valuation question can be framed in terms of two alternative measures of value, willingness to pay and willingness to accept (compensation). These two approaches imply different presumptions about the distribution of property rights and can differ substantially, depending on the availability of substitutes and income limitations. In many contexts, methodological limitations necessitate the use of willingness to pay rather than willingness to accept.

Finally, because ecosystem changes are likely to have long-term impacts, some accounting of the timing of impacts is necessary. This can be done through discounting future costs and benefits. It is essential, however, to recognize that consumption discounting is distinct from the discounting of utility, which reflects the weights put on the well-being of different generations. When the impacts being valued are relatively limited, the discount rate that is used should be the consumption rate rather than the utility rate. The consumption discount rate can be positive or negative, depending on whether consumption is increasing or decreasing. For environmental or ecological services that become scarcer over time, consumption would be decreasing, implying a negative discount rate.

Based on these conclusions, the committee provides the following recommendations:

- Policymakers should use economic valuation as a means of evaluating the trade-offs involved in environmental policy choices; that is, an assessment of benefits and costs should be part of the information set available to policymakers in choosing among alternatives.
- If the benefits and costs of a policy are evaluated, the benefits and costs associated with changes in ecosystem services should be included along with other impacts to ensure that ecosystem effects are adequately considered in policy evaluation.
- Economic valuation of changes in ecosystem services should be based on the comprehensive definition embodied in the total economic value framework; both use and nonuse values should be included.
- The valuation exercise should be framed properly. In particular, it should value the *changes* in ecosystem good or services attributable to a policy change. In addition, the scope should consider all relevant impacts and stakeholders, and the temporal scale of the analysis should be consistent with that of the impacts.
- The valuation exercise should indicate clearly whether (1) WTP or WTA measure of value was used, (2) in that context WTP is likely to differ significantly from WTA, (3) in that context WTP is likely to be strongly influenced by income differentials, and (4) use of the alternative value measure instead would likely have led to different policy prescriptions.
- In the aggregation of benefits and/or costs over time, the consumption discount rate, reflecting changes in scarcity over time, should be used instead of the utility discount rate.

REFERENCES

Azar, C. 1999. Weight factors in cost-benefit analysis of climate change. Environmental and Resource Economics 13(3):249-68.

Barbier, E.B. 1994. Valuing environmental functions: Tropical wetlands. Land Economics 70(2):155-173.

Barbier, E.B., M. Acreman, and D. Knowler. 1997. Economic Valuation of Wetlands: A Guide for Policy Makers and Planners. Geneva: Ramsar Convention Bureau.

Bishop, R.C., and M. P. Welsh. 1992. Existence values in benefit-cost analysis and damage assessment. Land Economics 68(4):405-417.

Bishop, R.C., K. J. Boyle, and M. P. Welsh. 1987. Toward total economic value of Great Lakes fishery resources. Transactions of the American Fisheries Society 116 (3):339-345.

Boyle, K.J., and R.C. Bishop. 1987. Valuing wildlife in benefit-cost analyses: a case study involving endangered species. Water Resources Research 23 (May):943-950.

Callicott, J.B. 2004. Explicit and implicit Values in the ESA. In The Endangered Species Act at Thirty: Retrospect and Prospects, Davies, F., D. Goble, G. Heal, and M. Scott (eds.). Washington, D.C.: Island Press.

Carson, R.T., N.E. Flores, and N.F. Meade. 2001. Contingent valuation: Controversies and evidence. Environmental and Resource Economics 19(2):173-210.

Carson, R.T., R.C. Mitchell, W.M. Hanemann, R.J. Kopp, S. Presser, and P.A. Ruud. 1992. A Contingent Valuation Study of Lost Passive Use Values Resulting from the Exxon Valdez Oil Spill. Report to the Attorney General of the State of Alaska, November. San Diego, Calif.: University of California, San Diego.

Daily, G. C. (ed.). 1997. Nature's Services: Societal Dependence on Natural Ecosystems. Washington, D.C.: Island Press.

Dreyfus, S.E. 1982. Formal models vs. human situational understanding. Technology and People 1:133-165.

Dworkin, R. 1977. Taking Rights Seriously. Cambridge: Harvard University Press.

Epstein, R.A. 2003. The regrettable necessity of contingent valuation. Journal of Cultural Values 27(3-4):259-274.

Erdheim, E. 1981. The wake of the snail darter: Insuring the effectiveness of Section 7 of the Endangered Species Act. Ecology Law Quarterly 9(4):629-682.

Freeman, A.M., III. 1993a. The Measurement of Environmental and Resource Values: Theory and Methods. Washington, D.C.:Resources for the Future.

Freeman, A.M., III. 1993b. Nonuse values in natural resource damage assessment. In Valuing Natural Assets: The Economics of Natural Resource Damage Assessment, Kopp, R. J., and V. K. Smith (eds.). Washington, D.C.: Resources for the Future.

Goulder, L.H., and D. Kennedy. 1997. Valuing ecosystem services: Philosophical bases and empirical methods. Pp. 23-47 in Nature's Services: Societal Dependence on Natural Ecosystems, Daly, G.C. (ed.). Washington, D.C.: Island Press.

Hanemann, W.M. 1991. Willingness to pay and willingness to accept—How much can they differ. American Economic Review 81(3):635-647.

Harrod, R.F. 1948. Towards a Dynamic Economy. London: Macmillan.

Hausman, J.A. 1993. Contingent Valuation: A Critical Assessment. Amsterdam: North Holland.

Heal, G.M. 1998. Valuing the Future. New York: Columbia University Press.

Heal, G.M. 2004. Intertemporal Welfare Economics and the Environment. In Handbook of Environmental Economics, Mäler, K-G and J. Vincent (eds.). Amsterdam: North Holland.

Heal, G.M., and B. Kriström. 2003. National Income in Dynamic Economies. Available on-line at *http://www.sekon.slu.se/~bkr/nat-inc-april5-02.PDF*. Accessed June 3, 2004.

Horowitz, J.K., and K.E. McConnell. 2002. A Review of WTA/WTP studies. Journal of Environmental Economics and Management 44(3):426-47.

Janssen, R., and J.E. Padilla. 1999. Preservation or conversion? Valuation and evaluation of a mangrove forest in the Philippines. Environmental and Resource Economics 14(3):297-331.

Kant, I., (translated by T.K. Abbott). 1987. Fundamental Principles of the Metaphysic of Morals. Buffalo, N.Y.: Prometheus Books.

Krutilla, J.V. 1967. Conservation reconsidered. American Economic Review 57(4):777-786.

Layard, R. 1999. Tackling Inequality. New York: St. Martin's Press.

MacLean, D. 1986. Social values and the distribution of risk. Pp. 75-93 in Values at Risk, MacLean, D. (ed.). Totowa, Md.: Roman and Allenheld.

Madariaga, B., and K.E. McConnell. 1987. Some issues in measuring existence value. Water Resources Research 23:936-942.

NOAA (National Oceanic and Atmospheric Administration) Panel on Contingent Valuation. 1993. Natural Resource Damage Assessment Under the Oil Pollution Act of 1990. Federal Register 58(10):4601-4614.

NRC (National Research Council). 1999. Nature's Numbers. Washington, D.C.: National Academy Press.

OMB (U.S. Office of Management and Budget). 1996. Economic Analysis of Federal Regulations Under Executive Order 12866. Available on-line at *http://www. whitehouse.gov/omb/inforeg /riaguide.html*. Accessed December 2003.

Portney, P.R., and J.P. Weyant. 1999. Discounting and Intergenerational Equity. Washington, D.C.: Resources for the Future.

Potts, D. 1999. Forget the weights, who gets the benefits? How to bring a poverty focus to the economic analysis of projects. Journal of International Development 11 (4):581-95.

Ramsey, F. 1928. A mathematical theory of saving. Economic Journal 38:543-559.

Randall, A. 1991. Total and nonuse values. Chapter 10 in Measuring the Demand for Environmental Quality, Braden, J.B., and C.D. Kolstad (eds.). Amsterdam: North Holland.

Sagoff, M. 1993. Environmental economics: An epitaph. Resources 111:2-7.

Sagoff, M. 1997. Can we put a price on nature's services? Philosophy and Public Policy 17(3):7-12.

Sagoff, M. 1994. Should preferences count? Land Economics 70(2):127-144.

Sagoff, M. undated. On the value of endangered and other species. Available on-line at *http://www.puaf.umd.edu/faculty/papers/sagoff/biodival.pdf.* Accessed June 3,2004.

Smith, V.K. 1987. Nonuse values in benefit cost analysis. Southern Economic Journal 54(1):19-26.

Starr, C. 2003. The precautionary principle versus risk analysis. Risk Analysis 23(1):1-3.

Stone, C. 1974. Should trees have standing? Towards a theory of legal rights for natural objects. Los Altos, Calif.: William Kaufman.

Turner, R.K. 1999. The place of economic values in environmental valuation. In Valuing Environmental Preferences, Bateman, I., and K.G. Willis (eds.). London: Oxford University Press.

U.S. Supreme Court. 1978. Tennessee Valley Authority v. Hill et al. Vol. 437 U.S. 153.
Willig, R.D. 1976. Consumer's surplus without apology. American Economic Review 66:589-597.

3
Aquatic and Related Terrestrial Ecosystems

INTRODUCTION

An ecosystem is generally accepted to be an interacting system of biota and its associated physical environment. Ecologists tend to think of these systems as identifiable at many different scales with boundaries selected to highlight internal and external interactions. In this sense, an aquatic ecosystem might be identified by the dominance of water in the internal structure and functions of an area. Such systems intuitively include streams, rivers, ponds, lakes, estuaries, and oceans. Most ecologists and environmental regulators also include vegetated wetlands as members of the set of aquatic ecosystems, and many think of groundwater aquifer systems as potential members of the set. "Aquatic and related terrestrial ecosystems" is a phrase that recognizes the impossibility of analyzing aquatic systems absent consideration of the linkages to adjacent terrestrial environments.

The inclusion of "related terrestrial ecosystems" for this study is a reflection of the state of the science that recognizes the multitude of processes linking terrestrial and aquatic systems. River ecologists have long understood the important connections between rivers and their floodplains (Junk et al., 1989; Stanford et al., 1996). The inflows of water, nutrients, and sediments from surrounding watersheds are heavily influenced by conditions within the floodplain. Conversely, floodplain plant and animal habitat value and sediment supply and fertility are often determined by river hydrology. This same sort of relationship between terrestrial and aquatic system is now understood to influence many of the functions of wetlands that motivate management efforts (Wetzel, 2001). Wetland ecologists have debated for years about appropriate recognition of capacity and opportunity to perform functions when conducting assessments of wetlands. A classic example of the discussion focuses on two identical wetlands, one in a pristine forested landscape, and the other in an intensely developed landscape. Both are assumed to have equivalent internal capacities to sequester pollutants, modify nutrient loads, and provide habitat, but the surrounding conditions mean that the opportunity for these functions to occur will differ significantly.

For many of the ecosystem functions and derived services considered in this chapter, it is not possible, necessary, or appropriate to delineate clear spatial boundaries between aquatic and related terrestrial systems (see Box 3-1). Indeed, to the extent that there is an identifiable boundary, it is often dynamic in both space and time. Floods, droughts, and seasonal patterns in rainfall are inte-

BOX 3-1
Understanding Ecosystem Terminology

Ecology is a scientific field that studies the relationships between and among (micro)organisms such as plants, animals, and bacteria and their environment. Like most scientists, ecologists use a variety of terms to describe aspects of their discipline. A few of the terms used throughout this report are defined below in the interest of facilitating the readability and understanding of this report.

Ecosystem biodiversity describes a number and kinds of organisms in a specific geographic area that can be distinguished from other areas by its physical boundaries (e.g., lake, forest), though such boundaries can be somewhat arbitrary. In addition to biodiversity, ecosystems have properties such as the amount of plant and animal matter they produce (***primary and secondary production***) and the flow of chemical elements within and through the system (***nutrient cycling***).

Ecosystem structure refers to both the composition of the ecosystem (i.e., its various parts) and the physical and biological organization defining how those parts are organized. A leopard frog or a marsh plant such as a cattail, for example, would be considered a component of an aquatic ecosystem and hence part of its structure. The relationship between primary and secondary production would also be part of the ecosystem structure, because it reflects the organization of the parts.

Ecosystem function describes a process that takes place in an ecosystem as a result of the interactions of plants, animals, and other (micro)organisms in the ecosystem with each other or their environment and that serves some purpose. Primary production (most notably the generation of plant material) is an example of an ecosystem function. The ***net primary production*** in an ecosystem is determined by the number and kinds of plants present; the amounts of sunlight, nutrients, and water available; and the amount of this productivity used internally by the plants themselves.

Ecosystem structure and function provide various ***goods and services*** to humans that have value: for example, rare species of plants or animals, fish for recreational or commercial use, clean water to swim in or drink. The functioning of ecosystems (interaction of organisms and the physical environment) often provides for services such as water purification, recharge of groundwater, flood control, and various aesthetic qualities such as pristine mountain streams or wilderness areas.

gral forcing functions for freshwater systems, just as tides, hurricanes, and sea-level rise constantly revise the boundaries between land and water in coastal systems. For these reasons, and as stated in Chapter 1, "aquatic ecosystems" collectively refers to aquatic and related terrestrial ecosystems unless noted otherwise.

The conceptual challenges of valuing ecosystem services involve explicit description and adequate assessment of the link (i.e., the ecological production function) between the structure and function of natural systems and the goods or services derived by humanity (see Figure 1-3). Describing structure is a relatively straightforward process, even in highly diverse ecosystems. Exceptions sometimes arise at the levels of small invertebrates and microorganisms. However, function is often difficult to infer from observed structure in natural systems. Furthermore, the relationship between ecosystem structure and function as well as how these attributes respond to disturbance are not often well understood. Indeed, ecological investigations of aquatic systems show no signs of running out of questions about how these systems operate. Without comprehensive understanding of the behavior of aquatic systems, it is clearly difficult to describe thoroughly all of the services these systems provide society. Although valuing ecosystem services that are not completely understood is possible (see Chapters 4 and 5 for further information and examples), when valuation becomes an important input in environmental decision-making, there is the risk that the valuation may be incomplete.

There have only been a few attempts to develop explicit maps of the linkage between aquatic ecosystem structure/function and value. There are, however, a multitude of efforts to separately identify ecosystem functions, goods, services, values, and/or other elements in the linkage without developing a comprehensive argument. One consequence of this disconnect is a diverse literature that suffers somewhat from indistinct terminology, highly variable perspectives, and considerable divergent convictions. Despite these shortcomings, the core issue of how to assess and value aquatic ecosystem services is intuitive and important enough to support some synthesis—especially as related to environmental decision-making.

The goal of this chapter is to review and summarize some of the common elements in the published literature concerning the identification of aquatic ecosystem functions and their linkage to goods and services for subsequent economic valuation. It also includes a summary review of the extent and status of aquatic ecosystems in the United States and some of the issues that continue to complicate efforts to value aquatic ecosystem services. The chapter closes with a summary of its conclusions and recommendations.

EXTENT AND STATUS OF AQUATIC AND RELATED TERRESTRIAL ECOSYSTEMS IN THE UNITED STATES

There are impressive examples of almost every kind of aquatic ecosystem within the United States. The country has some of the largest freshwater lakes in the world (see Box 3-2), one of the world's largest river systems (see Box 3-3), one of the world's largest estuaries (see Box 3-4), thousands of miles of coastline, extensive underground aquifers (see Box 3-5), a vast array of tidal and nontidal wetlands (see Box 3-6), and so many small creeks and streams that they are still being mapped. There is a long history of efforts to understand and manage these resources for public and private benefit, and the need to make informed decisions continues to motivate both research and monitoring. These short summaries identify some of the ways that humans have used and benefited from these ecosystems over time and many of the ecosystem services that managers seek to value in efforts to inform decisions. The summaries also identify some of the key management issues that have arisen as a result of evolving and often conflicting interests regarding ecosystem services.

In 2002, U.S. Environmental Protection Agency (EPA) released the *2000 National Water Quality Inventory* (NWQI; EPA, 2002)—the thirteenth installment in a series that began in 1975. These reports are required by Section 305(b) of the Clean Water Act and are considered by EPA to be the primary vehicle for informing Congress and the public about general water quality conditions in the United States. As such, the reports characterize water quality, identify widespread water quality problems of national significance, and describe various programs implemented to restore and protect U.S. waters. Notably, these assessments include streams and rivers, lakes and ponds, coastal resources to include tidal estuaries, shoreline waters (coastal and Great Lakes), and wetlands. Table 3-1 summarizes some of the relevant results and findings from the 2002 NWQI report.[1]

Although EPA, various federal and state partners, and other nongovernmental organizations and scientists have been assessing the condition of estuaries for decades, the *National Coastal Condition Report* (NCCR; EPA, 2001) represents the first comprehensive summary of coastal conditions in the United States and uses data and information collected from 1990 to 2000.[2] The report, a coordinated effort between EPA (lead), the National Oceanic and Atmospheric Administration (NOAA), the U.S. Geological Survey (USGS), and the U.S. Fish and Wildlife Service (USFWS), compiles and summarizes several data sets from

---

[1] The NWQI report includes information about water quality standards, detailed summaries of the results of waterbody assessments by designated uses and states, and a discussion of the data collection and analysis methods used in that report.

[2] Interested readers are directed to the NCCR report (EPA, 2001) for further information and details on the findings as well as data collection and analysis methods used to generate and interpret the regional results. Notably, Chapter 1 of that report includes a comprehensive list of federal programs and initiatives that address coastal issues, many of which are conducted jointly with various coastal states and local organizations.

BOX 3-2
Great Lakes Ecosystem

The Great Lakes ecosystem is the largest freshwater system in the world, comprising Lakes Michigan, Superior, Huron, Erie, and Ontario. Collectively, they cover a land area of 94,000 square miles and contain 5,500 cubic miles of water in the United States and Canada. Rivers and streams running into the lakes drain 201,000 square miles of land. Rain that falls in Chicago or Duluth may eventually leave the ecosystem more than 1,000 water miles to the east at Montreal, although outflows of water and its solutes are small, less than 1 percent by volume per year.

Habitats within the ecosystem are diverse. In the north, forests surrounding Lake Superior support healthy populations of black bears, bald eagles, wolves, and moose. Waterfowl, songbirds, and raptors funnel between Lakes Michigan and Erie during the spring and fall migrations. Lakes, wetlands, and uplands across the basin provide a mixture of habitats for temperate plants and animals of many types. The beaches and dunes of the southern shores are nesting areas for open water birds and wading birds such as the endangered piping plover.

Mining, timbering, agriculture, and industry brought major changes to the ecosystem beginning in the 1800s. Industries of all sorts grew up on the shorelines of lakes and rivers and used these waterbodies to facilitate both waste disposal and shipping. New locks and canals between the lakes allowed access to the Atlantic, while also opening pathways for the introduction of exotic species. For example, saltwater alewives displaced native species and sea lamprey devastated Great Lakes trout populations. Although industry created great wealth and well-being, it also left behind vast quantities of waste, including residues of dichlorodiphenyltrichloroethane and 1,1,1-trichloro-2,2-bis(4-chlorophenyl)ethane (DDT), polychlorinated biphenyls (PCBs), and heavy metals. Sewage and soil erosion turned lake water from clear blue to dark green through eutrophication.

Different trends began in the 1960s. Economic and public policy changes began to stem the flow of pollutants into the system, while aging mines, mills, and refineries closed. Electricity and natural gas replaced coal for heating, and air pollution laws cut power plant and automobile emissions. DDT and PCBs were banned, and the use of heavy metals declined. Treaties with Canada and interstate agreements established ecosystem-wide authorities to identify environmental problems and implement solutions. Marked changes in the former ecosystem followed these economic and regulatory changes. Water quality gradually improved so that the "oligotrophic blue" is reestablished in all the lakes. Between 1974 and 1994, PCB levels in top-of-the-food-web predators dropped by as much as 90 percent. Bald eagles once again breed along lake and river shorelines, and shoreline beaches and dunes are major summer destinations. Boating and recreational fishing are multibillion dollar industries.

Continues

BOX 3-2 Continued

However, history and the daily activities of 33 million people present continuing challenges for the ecosystem. Old harbors and shipping points are still lined by millions of tons of toxic materials and sediments. Although ambient concentrations are low, persistent toxic materials are concentrated by the ecosystem and food web, and levels of metals and PCBs in the blood and tissue of fish, waterfowl, and birds of prey are still high. Fish consumption advisories for recreational anglers remain in effect in across the region, and further reductions in mercury use and emissions remain a regulatory priority.

Restoring habitat and native species is also a priority. Wetland regulations halted the destruction of rare wetland types such as cedar bogs, fens, and salt marshes. Wetland restoration aims at restoring scarce wetland types, especially those along Great Lakes shorelines and bird migration routes. Elk and moose are reestablished in some areas, and significant efforts are under way to strengthen populations of Lake Superior native clams, walleye, brook trout, and sturgeons. Invasive and exotic species such as zebra mussels, lamprey, ruffe, and goby, however, continue to displace and threaten native species.

The Great Lakes region can be viewed a continuing experiment in testing human capability to live and prosper within the bounds of a major aquatic ecosystem, and although the last four decades allow some optimism, major environmental problems remain. During storms, combined sewer and stormwater drainage systems overflow, releasing untreated sewage in otherwise protected waterbodies. Urban and agricultural runoff contribute excessive nutrients into susceptible bays and inlets. Toxic air emissions disperse trace contaminants across the region, feeding the cycle of bioaccumulation. Success in this Great Lakes experiment will not be accidental. Thus, careful choices must be made and subsequent actions taken.

SOURCE: Great Lakes National Program Office (2001, 2002).

federal and state coastal monitoring programs to present a broad baseline picture of the condition of U.S. coastal waters as divided into five discrete regions: Northeast, Southeast, Great Lakes, Gulf of Mexico, and West Coast. The report is intended to serve as a benchmark for assessing the progress of coastal programs in the future and will be followed by subsequent reports on more specialized coastal issues.

It is important to note that the condition of U.S. coastal waters is described primarily in terms of data on estuaries, which are loosely defined in the NCCR as the productive transition areas between freshwater rivers and the ocean. In addition, although the intent of the report is to evaluate the condition of coastal waters (i.e., primarily estuaries) nationwide, the report states that there was insufficient information to completely assess West Coast estuaries and the Great

BOX 3-3
The Missouri River Ecosystem

The Missouri River basin extends over 530,000 square miles and covers approximately one-sixth of the continental United States. The one-hundredth meridian, the widely accepted boundary between the arid western states and the more humid states in the eastern United States, crosses the middle of the basin. The Missouri River's source streams are in the Bitterroot Mountains of northwestern Wyoming and southwestern Montana. The Missouri River begins at Three Forks, Montana, where the Gallatin, Jefferson, and Madison Rivers merge on a low, alluvial plain. From there, the river flows to the east and southeast to its confluence with the Mississippi River just above St. Louis. Near the end of the nineteenth century, the Missouri River's length was measured at 2,546 miles.

Between 1804 and 1806, the famous explorers Meriwether Lewis and William Clark led the first recorded upstream expedition from the river's mouth at St. Louis to the Three Forks of the Missouri, and eventually reached the Pacific coast via the Columbia River. The Missouri River subsequently became a corridor for exploration, settlement, and commerce in the nineteenth and early twentieth centuries, as navigation extended upstream from St. Louis to Fort Benton, Montana. Social values and goals in the Missouri River basin during this period reflected national trends and the preferences of basin inhabitants. Statehood, federalism, and regional demands to develop and control the river produced a physical and institutional setting that generated demands from a wide range of interests.

The Missouri River ecosystem experienced a marked ecological transformation during the twentieth century. At the beginning of the century, the Missouri River was notorious for large floods, a sinuous and meandering river channel that moved freely across its floodplain, and massive sediment transport. However, by the end of the twentieth century, the Missouri River bore little resemblance to the previously wild, free-flowing river. Over time, demands for the benefits associated with the Missouri's control and management resulted in significant and lasting physical and hydrologic modifications of the river. These modifications led to substantial changes in the river and floodplain ecosystem. Numerous reservoirs are scattered across the basin, with seven large dams and reservoirs located on the river's mainstem.

Ecological changes that accompanied changes in hydrology proceeded more slowly but were of a similar magnitude. Large floodplain areas along the upper Missouri were inundated by the reservoirs, and large areas of native vegetation communities in downstream floodplains were converted into farmland. Many native fish and avian species experienced substantial reductions, while nonnative species—especially fish—thrived in some areas. The rich biodiversity of the pre-regulated Missouri River ecosystem was sustained through a regime of natural disturbances that included periodic floods and attendant sediment erosion and deposition. These disturbances, in turn, supported a variety of ecological benefits, including commercial and recreational fishing, timber,

Continues

BOX 3-3 Continued

wild game, trapping and fur production, clean water, soil replenishment processes, and natural recharge of groundwater. Flow regulation and channelization substantially changed the Missouri River's historic hydrologic and geomorphic regimes. The isolation of the Missouri River from its floodplain caused by river regulation structures has in many stretches largely eliminated the flood pulse and its ecological functions and services. As a result of these changes, the production and the diversity of the ecosystem have both markedly declined.

For purposes of comparison, the major benefits of river regulation come from hydropower, water supply, and flood damage reduction, each of which has annual benefits measured in hundreds of millions of dollars. Recreation comes next, with annual benefits measured in tens of millions of dollars. Navigation follows, with annual benefits measured in millions of dollars. The value of ecosystem services that have been forgone in order to achieve other benefits is largely unknown.

Today the Missouri River floodplain ecosystem consists of extensive ecosystems in and around the large reservoirs, open reaches of channel, and riparian floodplains. Some of these systems are recognized producers of recreational opportunities or agriculture. Some traditional ecosysems, particularly those representing the historical habitats of the pre-regulated Missouri, have been less well recognized for the social values provided through ecosystem services. Many ecosystem services, such as fish, game, and aesthetic values, are not monetized and are not traded in markets. They thus tend to be underappreciated and undervalued by the public and by decision-makers.

SOURCE: NRC (2002b).

Lakes, and no assessment was possible for the estuarine systems of Alaska, Hawaii, and other island territories. However, new ecological programs, both newly created and proposed, should permit a comprehensive and consistent assessment of all of the nation's coastal resources by 2005. The NCCR used aggregate scores for a total of seven water quality indicators (water clarity, dissolved oxygen, coastal wetland loss, contaminated sediments, benthos, fish tissue contaminants, and eutrophic condition); 56 percent of assessed estuarine areas (representing more than 70 percent of the estuarine areas of the conterminous United States, excluding Alaska) were found to be in good condition for supporting aquatic life use (plant and animal communities) and human uses (e.g., water supply, recreation, agriculture). In contrast, 44 percent of the nation's estuaries were characterized as impaired for human use (10 percent), aquatic life use (11 percent), or both (23 percent). In general, the nation's coastal areas were rated as poor if the mean conditions for the seven indicators showed that more than 20 percent of the estuarine area in that region was degraded.

BOX 3-4
Chesapeake Bay

The Chesapeake Bay is the largest estuary in the United States and among the largest in the world. The watershed spreads over approximately 64,000 square miles, encompassing major portions of Pennsylvania, Maryland, and Virginia; all of the District of Columbia; and lesser portions of New York, West Virginia, and Delaware. It receives freshwater from six major rivers and has more than 2,000 square miles of relatively protected tidal waters.

The bay has been prized by its human inhabitants for centuries for its ability to provide food, water, navigation, waste disposal, recreation, and aesthetic pleasures. The estuary supports extensive commercial and recreational fisheries for striped bass, menhaden, flounder, perch, and many others. Oyster, crab, and clam harvests have supported local fishermen for generations. In addition, important habitat is provided for sea turtles, sharks, rays, eels, whelks, and an enormous diversity of waterfowl.

Hampton Roads located at the mouth of the bay in Virginia and Baltimore near the head of the bay in Maryland are among the nation's largest ports. Hampton Roads is home to the world's largest naval base, and both ports contain major international shipping terminals. Shipbuilding and repair are major industries in the regional economy. The value of commercial navigation in the bay is rivaled by the tremendous investment in recreational boating that operates from hundreds of marinas and thousands of private docks. The more than 20,000 miles of tidal shoreline in the system also provide highly desired home locations for many of the area's residents.

All of these benefits have led to intensive and continually increasing pressure on the ecosystem as human populations in the region have increased and subsequent use has escalated. One consequence has been emergence of the Chesapeake Bay as one of the most extensively studied estuaries in the world. Interest in the system has been driven by concern for declines in finfish and shellfish populations. These trends are recognized as the result of overharvesting, pollution, habitat destruction, and introduced diseases. The challenge of restoring the system's productivity has motivated investment of millions of dollars of public funds through the Chesapeake Bay Program, a cooperative effort by states and the federal government to reduce impacts and improve conditions in the ecosystem. The extensive and complex array of stakeholder groups, commitments, and programs orchestrated under the umbrella of this program has become a model for similar efforts emerging in other large aquatic ecosystems.

The current focus of the Chesapeake Bay Program is on reduction of nutrient, sediment, and toxic inputs to the system. This is being accomplished through the use of state-of-the-art simulation models, extensive monitoring, outreach and education, and a mix of regulatory and nonregulatory programs to design and implement best management practices throughout the watershed.

Continues

BOX 3-4 Continued

Parallel efforts are under way to restore vital habitats such as wetlands, submerged aquatic vegetation, and oyster reefs; promulgate multispecies and ecosystem management plans; and control the impacts of continuing development.

Estimates of the funding necessary to achieve restoration goals in the Chesapeake Bay extend into the tens of billions of dollars. This amount exceeds currently available resources by several orders of magnitude, creating unavoidable need to prioritize such efforts. To date, the incorporation of economic valuation in bay program management has been informal. Although cost-benefit analyses are implicit in almost every budget decision for Program activities, explicit use of economic assessments is not a characteristic of program management.

SOURCE: Scientific and Technical Advisory Committee (2003).

BOX 3-5
The Edwards Aquifer and Groundwater Recharge in San Antonio, Texas

The Edwards Aquifer of central Texas is a highly permeable karst limestone on the edge of the Chihuahuan Desert. The average annual temperature is 20.5°C average annual precipitation is 28.82 inches. The annual recharge for the aquifer ranges from 44,000 to 2,000,000 acre-feet and averages 635,500 acre-feet per year. Thousands of springs flow from this groundwater source, including the largest springs in the state, and potable water is the primary use of the groundwater supply (Bowles and Arsuffi, 1993). Recharge of the aquifer has been monitored by the U.S. Geological Survey (USGS) since 1915, while water quality monitoring began in 1930.

Currently, more than 1.7 million people rely on the Edwards Aquifer. However, recharge of the porous karstic limestone occurs primarily during wet years when precipitation infiltrates deeply into the soils and underlying rock. As a result, new laws were introduced that changed the legal basis of ownership from "right of capture" for a demonstrated "beneficial use" of the extracted water to a new approach based on prior appropriation (i.e., senior water rights). Concern increased as several springs (Comal, San Antonio, San Pedro) in the area began to dry up following a seven-year drought in the 1950s. Groundwater storage is critical in most aquatic ecosystems to provide persistent springs and streams during drought. Diverse microbial communities and a wide range of invertebrate

Continues

BOX 3-5 Continued

and vertebrate species live in groundwaters (Gibert et al., 1994; Jones and Mulholland, 2000). Their main ecosystem functions are breaking down organic matter and turning dead materials (detritus) into live biomass that is consumed in food webs. Thus, these species recycle nutrients and are important in secondary productivity. The trade-offs in extracting groundwater include possible loss of habitat for endemic species that are protected by state and federal regulations. For example, the Edwards Aquifer-Comal Springs ecosystem provides critical habitat for the Texas blind salamander (Crowe and Sharp, 1997; Edwards et al., 1989). Moreover, 91 species and subspecies of fish are endemic in this underground ecosystem (Bowles and Arsuffi, 1993; Culver et al., 2000; Longley, 1986). Several economic values of groundwater are associated with ecosystem services such as processing of organic matter by diverse microbes and invertebrates, providing possible dilution of some types of surface-originating contaminants, and sustaining populations of rare and endangered species that are often restricted to very local habitats (Culver et al., 2000).

By 1970, new regulations were issued to protect water quality in the Edwards Aquifer. These new rules limited economic development within the recharge zone to balance the long-term average recharge rate with the extraction rate. This steady-state equilibrium, however, is often characterized by time lags in recharge and drought frequencies that complicate predictable levels of water supply. Other physical considerations include how much and what types of development occur without disrupting rapid infiltration of the recharge zone. Degradation of subsurface water quality as well as declines in rates of recharge occur when economic development increases the extent of impervious surfaces that, in turn, cause more rapid runoff and loss of infiltration during and after precipitation events. The increased surface area of roof tops, roads, parking lots, and so on changes stormwater and groundwater hydrology and water chemistry. As groundwater is depleted the cost for deeper drilling and pumping increases costs and can terminate or slow the rate of extraction. Thus, it is difficult to consistently define "overextraction." The rate of extraction depends on future values relative to current values under specific alternative uses and climatic conditions (Custodio, 2002).

The Texas legislature created the Edwards Aquifer Authority to control pumping and to reallocate water through market mechanisms (Kaiser and Phillips, 1998; McCarl et al., 1999; Schaible et al., 1999). This approach has reallocated water from lower economic uses (e.g., agricultural irrigation) to higher-valued uses (e.g., for domestic and industrial water supplies and environmental and recreational uses). Especially during dry years, it appears feasible for transfers from irrigation to offset demands for municipal water supplies. In 1997, farmers accepted an offer of $90 per acre prior to the cropping season in a pilot study of the Irrigation Suspension Program (Keplinger and McCarl, 2000; Keplin-

Continues

BOX 3-5 Continued

ger et al., 1998). Drought increases the demand for water while the supply declines. Chen et al. (2001) used a climate change model to estimate the regional loss of welfare at $2.2 million to $6.8 million per year from prolonged drought. To protect endangered species in springs and groundwater, an additional reduction of 9 to 20 percent in pumping would add $0.5 million to $2 million in costs.

Traditionally, the only costs for the use of groundwater was the expense of installing a well and paying for pumping of this "open-access, free resource." However, when rates of extraction exceed recharge, the reduction in water levels may exceed an uncertain threshold, and cause irreversible changes. For example, removal of water in the underground area may cause collapse of the overlying substrata. These collapses decrease future storage capacity below ground and can alter land values. In some areas the depleted groundwater may cause intrusion of low-quality water from other aquifers or from marine-derived salt or brackish waters that could not readily be restored for freshwater storage and use. Contamination of groundwater from landfills, leaking petroleum storage tanks, and pesticides can also makes aquifers unusable.

In 1993 the Sierra Club sued the state for failure to guarantee a minimum flow of 100 cubic feet per second (cfs) to Comal and San Marcos Springs. The State of Texas and the U.S. Fish and Wildlife Service have entered into an agreement to resolve this conflict. To avoid jeopardizing the endangered species living in these springs, the Edwards Aquifer Authority banned the use of irrigation sprinklers whenever flow declined below a threshold that limited habitat in the Comal Springs. Approximately 1.5 million people were affected when the USGS reported that the flow declined to 145 cfs in September 2002. Limited pumping also had large economic consequences on agriculture. While water markets may ultimately resolve reallocation issues among stakeholders in the Edwards Aquifer region (Chang and Griffin, 1992; Kaiser and Phillips, 1998; McCarl et al., 1999; Schaible et al., 1999), the predictability of water markets as suppliers of water for different needs is complex and will help reallocate water only if some level of supply is available.

The construction of water-transfer pipelines and additional surface storage reservoirs is under consideration along with conjunctive storage (pumping water into sub-surface storage associated with aquifers.) The estimated cost of building a surface reservoir (Applewhite) to provide an additional 170,000 acre-feet of water for sale was $317 per acre-foot compared to $67 per acre-foot if pumped from the Edwards Aquifer (John Merrifield, University of Texas-San Antonio, personal communication, 2003). The combination of climatic change (more extremes in drought and in distribution of rainfall) and increased human population growth will stress the current rules on allocation of water to maintain natural ecosystem functions and survival of endangered species.

BOX 3-6
The South Florida Ecosystem

South Florida is dominated by the waters of the Kissimmee-Ockeechobee-Everglades (KOE) ecosystem. In the late summer and fall, rainfall enters the Kissimmee River near Orlando and gradually flows south to Lake Ockeechobee. The waters gather more rainfall and continue south, flowing into agricultural fields, an extensive system of flood control canals and reservoirs, and the "river of grass" called the Everglades. Eventually, the waters flow through the Everglades to enrich the mangrove forests and estuaries on the Atlantic and Gulf Coasts (Purdum, 2002).

The KOE ecosystem covers almost 17,000 square miles in South Florida. The ecosystem is home to more than 6 million people and the dynamic regional economies of Orlando and South Florida, including the cities of Miami, Fort Lauderdale, and West Palm Beach. The ecosystem's preserves and natural areas are known throughout the world for their uniqueness and beauty: including the Everglades National Park, Big Cypress Preserve, the Florida Keys, Biscayne Bay, and the estuary of Florida Bay (NRC, 2002a, 2003).

The ecosystem is a mix of natural and human forces. Ten thousand years ago, the KOE area was dry prairie, inhabited by horses, camels, bison, and mammoths and the humans who hunted them. About 9,000 years ago, the oceans began to rise with the ending of the last ice age. The habitat shifted as the climate changed to humid subtropics in the north and tropical savannah in the south (Purdum, 2002). Swamps, marshes, pinelands, the everglades, and hardwood hammocks developed in inland areas, sustained by the gradual flow of waters. Mangroves and estuaries gained a footing in coastal areas. Tropical and subtropical wildlife grew in abundance, ranging from crocodiles to bear to birds in wide variety.

In the last 100 years, the annual tropical cycle of sun in the winter drought and dependable rain in the summer and fall attracted residents from around the world, but torrential rains caused flooding. As settlements grew, there was a steady human effort to control and redirect the annual flooding. Some redirected water went to serve urban and agricultural uses, but much was simply channeled into the ocean.

By the end of the twentieth century, the KOE ecosystem was criscrossed by more than 1,800 miles of canals and levees, controlling the floods but also cutting off the established flows of KOE water. Water became scarce in humid area such as the Everglades and Florida Bay estuaries. Some species were particularly hard hit. Nesting wading birds declined by 90 percent (Lord, 1993). Saltwater began to intrude into freshwater aquifers supplying 90 percent of potable water for the human population (Purdum, 2002).

Major investments are now being made to restore the quantity of water available and its flow through the remaining natural systems. One significant project is the $7.8 billion Everglades Restoration Plan (see NRC, 2002a; 2003). The plan proposes to remove major barriers to water flows into Everglades National Park, treat surface water runoff from urban areas, reuse wastewater, and store water from heavy rainfall rather than shunting it out to sea (Purdum, 2002). The project is expensive, but is it enough given the value of ecosystem resources and services? Methods for valuing ecosystem services would help provide an answer.

Table 3-1 Selected Findings and Results from the 2002 National Water Quality Resource Inventory

Waterbody Type	Total Size[a]	Amount[b] Assessed (% of total)	Good[c] (% of assessed)	Impaired[d] (% of assessed)	Leading Pollutants and Causes of Impairment[e]	Leading Sources of Impairment[e]	Notes
Rivers and streams	3,692,830 miles	699,946 miles (19%)	426,633 miles (61%)	269,258 miles (39%)	Pathogens (bacteria) Siltation Habitat alteration Oxygen-depleting substances Nutrients Thermal modification Metals Flow alteration	Agriculture Hydrologic modifications Urban runoff and storm sewers Forestry Municipal point sources Resource extraction	See Chapter 2 and Appendix A of EPA (2002) for further information
Lakes, reservoirs, and ponds	40,603,893 acres	17,339,080 acres (43%)	9,375,891 acres (55%)	7,702,370 acres (45%)	Nutrients Metals Siltation Total dissolved solids Oxygen-depleting substances Excess algal growth Pesticides	Agriculture Hydrologic modifications Urban runoff and storm sewers Atmospheric deposition Municipal point sources Land disposal	See Chapter 3 and Appendix B of EPA (2002) for further information
Coastal resources: Estuaries	87,369 sq. miles	31,072 sq. miles (36%)	14,873 sq. miles (49%)	15,676 sq. miles (51%)	Metals Pesticides Oxygen-depleting substances Pathogens (bacteria) Priority toxic organic chemicals Polychlorinated biphenyls (PCBs) Total dissolved solids	Municipal point sources Urban runoff/storm sewers Industrial discharges Atmospheric deposition Agriculture Hydrologic modifications Resource extraction	See Chapter 4 and Appendix C of EPA, 2002 for further information
Coastal resources: Great Lakes shoreline	5,521 miles	5,066 miles (92%)	1,095 miles (22%)	3,955 miles (78%)	Priority toxic organic chemicals Nutrients Pathogens (bacteria) Sedimentation and Siltation Oxygen-depleting substances Taste and odor PCBs	Contaminated sediments Urban runoff and storm sewers Agriculture Atmospheric deposition Habitat modification Land disposal Septic tanks	See Chapter 4 and Appendix F of EPA (2002) for further information

continues

Table 3-1 Continued

Waterbody Type	Total Size[a]	Amount[b] Assessed (% of total)	Good[c] (% of assessed)	Impaired[d] (% of assessed)	Leading Pollutants and Causes of Impairment[e]	Leading Sources of Impairment[e]	Notes
Coastal resources: Ocean shoreline waters	58,618 miles	3,221 miles (6%)	2,755 miles (86%)	434 miles (14%)	Pathogens (bacteria) Oxygen-depleting substances Turbidity Suspended solids Oil and grease Metals Nutrients	Urban runoff and storm sewers Nonpoint sources Land disposal Septic tanks Municipal point sources Industrial discharges Construction	See Chapter 4 and Appendix C of EPA (2002) for further information
Wetlands	105,500,000 acres[f]	8,282,133 acres (8%)	4,839,148 acres (58%)	3,442,985 acres (42%)	Sedimentation and siltation Flow alterations Nutrients Filling and draining Habitat alterations Metals	Agriculture Construction Hydrologic modifications Urban runoff Silviculture Habitat modifications	See Chapter 5 and Appendix D of EPA (2002) for further information

[a] Units are miles for rivers and streams; acres for lakes, reservoirs, ponds, and wetlands; square (sq.) miles for coastal resources (estuaries, Great Lake shoreline, and ocean shoreline waters).
[b] Includes waterbodies assessed as not attainable for one or more designated uses (i.e., total number of waterbody units assessed as good and impaired do not necessarily add up to total assessed).
[c] Fully supporting all designated uses or fully supporting all uses, but threatened for one or more uses.
[d] Partially or not supporting one or more designated uses.
[e] For those states and jurisdictions that reported this type of information (i.e., often a subset of the total number of states and jurisdictions that assessed and reported on various waterbodies; see EPA 2002 for further information).
[f] From *Status and Trends of Wetlands in the Conterminous United States 1986 to 1997* (Dahl, 2000).
SOURCE: Adapted from EPA (2002).

Section 401 of the Emergency Wetlands Resources Act of 1986 requires the USFWS to conduct studies of the status and trends of the nation's wetlands and report the results to Congress each decade. The third report of the USFWS's National Wetlands Inventory (NWI), *Status and Trends of the Wetlands in the Conterminous United States 1986 to 1997*, was released in 2000 (Dahl, 2000). This NWI report provides the most recent and comprehensive estimates of the areal extent (status) and trends of wetlands in the conterminous 48 United States on all public and private lands between 1986 and 1997. In that report, wetlands, deepwater, and upland (land-use) categories are divided into a wide variety of habitats and groupings; however, wetlands are classified principally as estuarine and marine wetlands and freshwater wetlands.[3] The study design included 4,375 randomly selected sample plots 4 square miles in area that were examined using remotely sensed data in conjunction with fieldwork and verification to determine wetland change. However, the report does not address water quality conditions or provide an assessment of wetland functions.

As of 1997, the lower 48 states contained about 105.5 million acres of wetlands of all types (Dahl, 2000), an area about the size of California. Of these, about 95 percent are inland freshwater wetlands, while the remaining 5 percent are saltwater (marine and estuarine) wetlands. Between 1986 and 1997, the net loss of wetlands was 644,000 acres with an annual loss rate of 58,545 acres (see also Table 1-1); 98 percent of these losses occurred in freshwater wetlands.[4]

A fourth major federal program report related to the extent and status of aquatic and related terrestrial ecosystems is the *Summary Report of the 1997 National Resources Inventory* (Revised December 2000) (USDA, 2000). The NRI is conducted every five years by the U.S. Department of Agriculture's Natural Resources Conservation Service in cooperation with the Iowa State University Statistical Laboratory. The 1997 NRI report is the fourth summary report in a series that began in 1982 and is a scientifically based, longitudinal panel survey designed to consistently assess conditions and trends of the nation's soil, water, and related resources for all nonfederal lands for all 50 states and other jurisdictions (e.g., Puerto Rico) using photo interpretation and other remote sensing methods and techniques. Thus, all values provided in the 1997 NRI report are estimates based on data collected at sample sites, not data taken from a census.[5]

---

[3]See Table 1 and Appendixes A through B in Dahl, 2000 for further information.
[4]This and other USFWS's NWI reports, their data, resources, and other information are available on-line at *http://wetlands.fws.gov.* Accessed June 11, 2004.
[5]The 1997 NRI report has detailed information on study design, data collection methods, compilation, synthesis, and analysis, in addition to the resource inventory results.

CATALOGING ECOSYSTEM STRUCTURE AND FUNCTION AND MAPPING ECOSYSTEM GOODS AND SERVICES

Ecosystem Structure and Function

As a general rule, the literature on ecosystem valuation attempts to use the terms "structure" and "function" as descriptors of natural systems (i.e., free of "value" content; see Chapter 2 for further discussion). These are features of natural systems that result in a capacity to provide goods and services, which can in turn be valued by humans (see also Box 3-7). The "value-free" distinction is ultimately blurred when considering intrinsic values of natural systems, but identification of ecosystem structure and function is a reasonable starting point for the subsequent mapping of ecosystem goods and services.

There are at least three key elements in the effective description of aquatic ecosystems: (1) geomorphology, (2) hydrology, and (3) biology. Collectively, these factors constrain the stocks of organic and inorganic materials in the system and the internal and external fluxes of those materials and energy. For this reason, many classification efforts focus on these three elements in developing taxonomies of aquatic ecosystems.

BOX 3-7
Energy Analysis and Valuation

Some ecologists use energetics (Odum, 1988, 1996) as a common currency for valuation. More specifically, energetic valuation (Odum and Odum, 2000) attempts to put the contributions of the economy on the same basis as the work of the environment by using *one kind* of energy (e.g., solar energy) as the common denominator. Accordingly, the term "emergy" was proposed to express all values in *one kind* of energy required to produce designated goods and services, for the purpose of eliminating confusion with other energetic valuation concepts (Odum, 1996). As an example, to evaluate the total worth of an estuary, the total energy flow in terms of embodied energy (which represents all of the work of the ecosystem) is determined and then this energy value is converted to monetary units on the basis of the ratio between energy and money in the production of market goods (Odum, 1993).

Energetic evaluation is presented as a strategy by which ecological data can be used to influence environmental policies (Odum and Odum, 2000) and it has served as a useful tool to examine the interface between ecosystems and economics (e.g., Odum and Turner 1990; Turner et al. 1988). However, it rejects the premise that values arise from the preferences of individuals and that the fundamental purpose of economic valuation is to estimate the change in willingness to pay (or accept compensation) for the various losses and gains experienced by individuals when confronted by changes in ecosystem services.

An example of extant classification systems is the one adopted by the NWI of the USFWS (Cowardin et al., 1979). This hierarchical system distinguishes general kinds of aquatic ecosystems (e.g., rivers, lakes, estuaries) and then places special emphasis on a site's vegetative community and hydroperiod. The method does not purport to address function. Indeed, much of the relevant literature in wetlands ecology documents the great variability of functions within and among NWI wetland types.

A newer classification scheme developed by Brinson (1993), called the HydroGeomorphic Method (HGM), is now being developed into an assessment methodology by the U.S. Army Corps of Engineers and the EPA (Smith et al., 1995). The HGM classification places emphasis on the hydrology and topographic setting of a wetland. The classification system has become the basis for development of a growing number of wetland condition assessment models. The models support evaluation of the degree of departure from ideal or "reference" conditions for specific classes of wetlands. The assumption is that stressors in the wetland or surrounding landscape (e.g., soil disturbance, grazing, pollution discharges) will affect the natural functions of the ecosystem and that this effect can be related to observable changes in the wetland. This approach begins to establish a relationship between wetland condition and capacity to perform certain functions. Nevertheless, the natural variability of wetland ecosystems confounds simple inference about functions based simply on HGM classification.

There are similar efforts to develop classifications for lakes (e.g., Busch and Sly, 1992; Maxwell et al., 1995) and streams (e.g., Rosgen, 1994; TNC, 1997; Vannote et al., 1980). Again, each of these approaches starts with structural attributes of the system being evaluated and directly or indirectly addresses some aspect(s) of function. However, none of these efforts purport to support direct inferences about a comprehensive suite of ecological functions.

The fact that there is no explicit and invariant link between structure and function of aquatic ecosystems is part of the problem in efforts to assess all goods and services provided by these natural systems. If the behavior of a particular ecosystem is dependent not only on its composition, but also on linkages to surrounding systems and the impact of stressors, then comprehensive recognition of goods and services provided is not straightforward. The constantly evolving body of work on wetlands assessment exemplifies this challenge. Describing the structure of wetland ecosystems in terms of plant community composition, soil characteristics, and water movement is a well-developed practice with generally accepted protocols. Assessing the level of function in a wetland is, however, an exceptionally complex undertaking. As noted previously, a wetland's "capacity" to perform a function interacts with its "opportunity" to perform the function.

In a simple example case of habitat function, the structural characteristics of a wetland determine its capacity to meet the requirements of amphibians. The amounts of open water, the seasonal patterns of soil saturation, the types of sheltering plant material, and the size of the wetland all combine to determine if the wetland could support amphibians (e.g., Sousa, 1985). Landscape setting, or the larger system within which the wetland system exists, determines other factors that affect a wetland's opportunity to reach its potential as amphibian habitat. Adjacent land use affects access, water quality, and the density of potential predator populations. These and other external factors have significant impacts on the level at which habitat functions are performed (e.g., Knutson et al., 1999). The point is that wetland ecosystem structure alone is not an adequate predictor of the amphibian habitat services provided. Thus, as a generality, mapping ecosystem goods and services does not proceed linearly from system structure.

The default response to the lack of a simple logic linking structure to function has been development of generalized lists of potential functions appropriate to broad categories of aquatic ecosystems. Researchers interested in describing the importance of natural systems to humans frequently begin by generating lists of things normally functioning ecosystems can do. The scope of these lists is not universally constant.

Review of extant attempts to identify the suite of potential functions performed by aquatic ecosystems indicates that the list continues to evolve. The wetlands literature provides one example of this progression. In the 1970s, important wetlands functions included production of plant biomass, provision of habitat, modification of water quality, flood storage, and sediment accumulation (e.g., Wass and Wright, 1969). At present, the list has been expanded considerably and now includes functions in global carbon cycles, maintenance of biodiversity, and global climate control, among others (e.g., Ewel, 2002). There is no reason to believe the list will not continue to evolve as understanding of wetlands and aquatic ecosystems increases.

There have been a number of efforts to develop and suggest a taxonomy for ecosystem functions, and they tend to converge on a generalized categorization suggested by de Groot et al. (2000). These authors argue that the cumulative list of ecosystem functions can be grouped into four primary categories: (1) regulation, (2) habitat, (3) production, and (4) information (see also Table 3-3 below for further information). As described by de Groot and colleagues, regulation functions include those processes affecting gas concentrations, water supply, nutrient cycling, waste assimilation, and population levels. Habitat functions are directly related to provision of suitable living space for an ecosystem's flora and fauna. Production functions include primary (autotrophic) and secondary (heterotrophic) production, as well as generation of genetic material and biochemical substances. Information functions are those that provide an opportunity for cognitive development and, as such, are functions that can be realized only through human interaction.

The committee's review of the literature and attempts to catalog ecosystem functions leads to the conclusion that the absence of a consensus taxonomy is a product of both the complexity of natural systems and the challenge of communicating across multiple disciplines. The committee could find underlying logic in many of the alternative approaches, but no single approach was without complications, and none was intuitively explanatory across disciplines or to all reviewers. For the present, this appears to be the state of the science.

Although a perfect taxonomy for ecosystem functions remains elusive, this may be less important than developing a consensus on an appropriate cumulative list of potential aquatic ecosystem functions. In this regard, de Groot et al. (2000) represent an important iteration in the process of generating a useful checklist to inform aquatic ecosystem valuation exercises. While the committee found reasons to debate aspects of the proposed listing, the value as stimulus to discussion was clear. Continued work on such compilations will enhance our ability to develop more comprehensive ecosystem valuation scenarios. In the interim, it seems that using a relatively detailed list of ecosystem functions (and goods and services; see more below) like that provided by de Groot et al. (2002) can offer guidance to help ensure some breadth to the assessment of specific ecosystems.

Unfortunately, identification of the particular functions performed by an aquatic ecosystem is only part of the assessment problem. The level at which specific ecosystem functions are performed can also vary significantly, in part because these systems can vary so widely in terms of their physical and biological composition. Thus, production functions can reach extreme levels in eutrophic ponds and estuaries or drop to very low levels in oligotrophic lakes. Climate regulation functions can occur and take on great importance at very high levels in the Great Lakes or be effectively nonexistent in small prairie potholes (wetlands). Thus, while almost all ecosystem functions can be argued to occur at some level in every aquatic ecosystem, the significance of the processes can vary from great to trivial depending on the type of system, its size, and location.

Time can be another important dimension in appropriate assessment of ecosystem function, particularly when economic valuation is the end objective. The rates at which various ecological processes occur will affect their ease of recognition and measurement. For example, habitat functions are arguably easier to identify and measure than carbon sequestration, whereas primary production is easier to assess than generation of genetic material. The frequency with which certain functions are performed can similarly influence recognition and measurement. Production may be a relatively constant or at least seasonal process, while hydroperiod modification may only occur at irregular intervals of years' duration. Finally, the developmental state of the ecosystem will affect its capacity to sustain performance of certain functions. Most aquatic ecosystems change overtime; ponds fill in or dry up, rivers meander and get dammed, and tidal marshes erode. All of these changes alter the capacity of an ecosystem to perform functions over very short to very long time periods.

As a result of the inherent variability in both structure and functions of natural systems, there is no straight forward methodology (let alone a consensus paradigm) for comprehensive assessment of each and every type of aquatic ecosystem. The practical default approach is to work from an evolving list of potential ecosystem functions (e.g., de Groot et al., 2002; MEA, 2003) and evaluate the capacity of the system under consideration to perform each function. Essential to the process is incorporation of both spatial and temporal considerations in developing the ecosystem assessment.

Ecosystem Goods and Services

Daily (1997) states that "ecosystem services are the conditions and processes through which natural ecosystems, and the species that make them up, sustain and fulfill human life. They maintain biodiversity and the production of ecosystem goods. . ." Many of the goods and services provided by aquatic ecosystems are intuitive, such as potable water sources, food production, transportation, waste removal, and contributing to landscape aesthetics. To a great extent ecologists are able to catalogue and estimate these kinds of goods and services at both small and large spatial scales. Extending those assessments of goods and services through time is more challenging as ecosystems are constantly changing.

Other, less intuitive, goods and services have been recognized only as knowledge of the global ecosystem has evolved. Some of these include maintenance of biodiversity, and contributing to biogeochemical cycles and global climate. As noted previously, it is likely that the list of potential ecosystem goods and services will continue to evolve.

Reviewers of the subject area have tried to catalog ecosystem goods and services in a variety of ways. Services are sometimes grouped from the perspective of human users into categories such as extractive and nonextractive or consumptive and nonconsumptive. A compilation of some sample lists is included in Table 3-2. Reviewers have also attempted to articulate the link between ecosystem functions and the derived goods and services. One previously noted example of this approach is the de Groot et al. (2002) taxonomy for ecosystem functions, goods, and services shown in Table 3-3.

The state of the science is such that there is no broad consensus on a comprehensive list of potential goods and services derived from aquatic ecosystems. However, there is enough similarity among proposed lists to suggest that full valuation of any particular ecosystem's goods and services must look well beyond the amounts of water, fish, waste assimilation, and recreational use provided to individuals in direct contact with the system. At present, ecologists can quantify many of the more readily accepted goods and services, although methods may vary. It is noteworthy that the international Millennium Ecosystem Assessment (MEA) being coordinated by the United Nations Environment Programme has adopted a taxonomy of ecosystem services drawn from the de Groot et al. (2002) construct (Available on-line at *http://www.millenniumassessment.org/en/index.asp*). After considering a number of alternative schemes for grouping ecosystem services, the approach based on function was selected for use in the MEA. In this particular iteration, services are classified as provisioning, regulating, cultural, or supporting.

TABLE 3-2 Lists of Ecosystem Services

Ecosystem Services (Daily, 1997)
Purification of air and water
Mitigation of floods and droughts
Detoxification and decomposition of wastes
Generation and renewal of soil and soil fertility
Pollination of crops and natural vegetation
Control of the vast majority of potential agricultural pests
Dispersal of seeds and translocation of nutrients
Maintenance of biodiversity, from which humanity has derived key elements of its agricultural, medicinal, and industrial enterprises
Protection from the sun's harmful ultraviolet rays
Partial stabilization of climate
Moderation of temperature extremes and the force of winds and waves
Support of diverse human cultures
Providing aesthetic beauty and intellectual stimulation that lift the human spirit
Services Provided by Rivers, Lakes, Aquifers, and Wetlands (Postel and Carpenter, 1997)
Water Supply
Drinking, cooking, washing, and other household uses
Manufacturing, thermoelectric power generation, and other industrial uses
Irrigation of crops, parks, golf courses, etc.
Aquaculture
Supply of Goods Other Than Water
Fish
Waterfowl
Clams and mussels
Pelts
Nonextractive or Instream Benefits
Flood control
Transportation
Recreational swimming, boating, etc.
Pollution dilution and water quality protection
Hydroelectric generation
Bird and wildlife habitat
Soil fertilization
Enhanced property values
Nonuser values
Wetland Ecosystem Services (Ewel, 2002)
Biodiversity: Sustenance of Plant and Animal Life
Evolution of unique species
Production of harvested wildlife:
Water birds, especially waterfowl
Fur-bearing mammals (e.g., muskrats)
Reptiles (e.g., alligators)
Fish and shellfish
Production of wildlife for nonexploitative recreation
Production of wood and other fibers
Water Resources: Provision of Production Inputs
Water quality improvements
Flood mitigation and abatement
Water conservation
Global Biogeochemical Cycles: Provision of Existence Values
Carbon accumulation
Methane production
Denitrification
Sulfur reduction

continues

TABLE 3-2 Continued

Ocean Ecosystem Services (Peterson and Lubchenco, 2002)
- Global materials cycling
- Transformation, detoxification, and sequestration of pollutants and societal wastes
- Support of the coastal ocean-based recreation, tourism, and retirement industries
- Coastal land development and valuation
- Provision of cultural and future scientific values

SOURCE: Adapted from Daily (1997); Ewel (2002); Peterson and Lubchenco (2002); Postel and Carpenter (1997).

TABLE 3-3 Functions, Goods, and Services of Natural and Seminatural Ecosystems

Functions	Ecosystem Processes and Components	Goods and Services
Regulation	Maintenance of essential ecological processes and life support systems	
Gas regulation	Role of ecosystems in biogeochemical cycles	Ultraviolet-B protection Maintenance of air quality Influence on climate
Climate regulation	Influence of land cover and biologically mediated processes	Maintenance of temperature, precipitation
Disturbance prevention	Influence of system structure on dampening environmental disturbance	Storm protection Flood dampening
Water regulation	Role of land cover in regulating runoff and river discharge	Drainage and natural irrigation Medium for transport
Water supply	Filtering, retention, and storage of freshwater (e.g., in aquifers)	Provision of water for consumptive use
Soil retention	Role of vegetation root matrix and soil biota in soil retention	Maintenance of arable land Prevention of damage from erosion and siltation
Soil formation	Weathering of rock, accumulation of organic matter	Maintenance of productivity on arable land
Nutrient regulation	Role of biota in storage and recycling of nutrients	Maintenance of productive ecosystems
Waste treatment	Role of vegetation and biota in removal or breakdown of xenic nutrients and compounds	Pollution control and detoxification

continues

TABLE 3-3 Continued

Functions	Ecosystem Processes and Components	Goods and Services
Pollination	Role of biota in movement of floral gametes	Pollination of wild plants species
Biological control	Population control through trophic-dynamic relations	Control of pests and diseases
Habitat	Providing habitat (suitable living space) for wild plant and animal species	
Refugium	Suitable living space for wild plants and animals	Maintenance of biological and genetic diversity Maintenance of commercially Harvested species
Nursery	Suitable reproductive habitat	Hunting; gathering of fish, game, fruit, etc. Aquaculture
Production	Provision of natural resources	
Food	Conversion of solar energy into edible plants and animals	Building and manufacturing Fuel and energy Fodder and fertilizer
Raw materials	Conversion of solar energy into biomass for human construction and other uses	Improve crop resistance to pathogens and pests
Genetic resources	Genetic material and evolution in wild plants and animals	Drugs and pharmaceuticals Chemical models and tools Test and assay organisms
Medicinal resources	Variety of (bio)chemical substances in, and other medicinal uses of, natural biota	
Ornamental resources	Variety of biota in natural ecosystems with (potential) ornamental use	Resources for fashion, handicraft, worship, decoration, etc.
Information	Providing opportunities for cognitive development	
Aesthetic	Attractive landscape features	Enjoyment of scenery
Recreation	Variety in landscapes with (potential) recreational uses	Ecotourism

continues

TABLE 3-3 Continued

Functions	Ecosystem Processes and Components	Goods and Services
Cultural and artistic	Variety in natural features with cultural and artistic value	Inspiration for creative activities
Spiritual and historic	Variety in natural features with spiritual and historic value	Use of nature for religious or historic purposes
Science and education	Variety in nature with scientific and educational value	Use of nature for education and research

SOURCE: Adapted from de Groot et al. (2002).

ISSUES AFFECTING IDENTIFICATION OF GOODS AND SERVICES

Ecosystems vary in time and space. As ecologists extend their analyses of ecosystem structure and function to include potential goods and services, the uncertainty affecting assessments increases across both time and space. The interaction of ecological and social systems makes extrapolation of observations and prediction of future conditions exceptionally complex (Berkes et al., 2003; Gunderson and Holling, 2002; Gunderson and Pritchard, 2002). The challenges arise from the heterogeneity of ecosystems and values across space which complicates aggregation for assessment at larger scales, and from nonlinear system behavior that confounds forecasting. Recognition of the thresholds of change in both space and time is one of the principal challenges in ecological research.

Scale

It may be argued that almost all ecosystem functions can be performed by aquatic ecosystems at any scale. Indeed, Limburg et al. (2002) found that scaling rules describing production and delivery of ecosystem services are yet to be formulated and quantified (as noted in the preceding sections). However, there are clearly thresholds in the level of their relative importance. For example, individual wetlands in a watershed may each have the capacity to slow the flow of waters moving through them, but this function becomes important only when there are a sufficient number of wetlands in a watershed to significantly alter the flow of floodwaters downstream.

The complication in assessment of ecosystem goods and services arises because the scale at which functions become important is not always the same.

Continuing with the watershed example above, each wetland may have the capacity to accrete organic matter, sequestering carbon. However, the significance of this function for carbon cycles may not be realized at any scale less than all of the nation's wetlands. Alternatively, the provision of suitable habitat for a rare plant may be regionally significant at the scale of a single wetland.

Some generalizations regarding recognition of ecosystem services across scales may be possible (see Table 3-4 for one example). The problem is recognition of the thresholds at resolution sufficient to inform management and policy decisions. Knowing precisely the scale at which services can be realized is a practical challenge. Success in identification of these scale thresholds would increase opportunities for accurate recognition and appropriate economic valuation of ecosystem services.

Another challenge in valuing ecosystem services across scales arises in attempts to aggregate such information. The complex nature of ecosystems means that many interrelationships and feedback loops may operate at scales above the level of individual service assessment. Protection of wetlands important as habitat for migrating waterfowl may be undermined by loss of wetlands at other critical points on the flyway. Restoration of wetlands as nursery grounds for fish along the Louisiana coast may be less successful if nutrient pollution in the Mississippi River degrades open water habitat for the adult populations. The implication is that aggregation of service values to larger scales or composite system evaluations will almost axiomatically misrepresent the processes at the

TABLE 3-4 Examples of the Generation of Ecosystem Services at Different Scales for Aquatic Ecosystems

Time or Space Scale (day) (meters)	Aquatic Ecosystem	Example of Ecosystem Service	Scale at Which Service is Valued
10^{-6} to 10^{-5}	Bacteria	Nutrient uptake and production of organic matter	Local/regional
10^{-3} to 10^{-1}	Plankton	Trophic transfer of energy and nutrients	Local/regional
10^{0} to 10^{1}	Water column and/or sediments, small streams	Provision of habitat	Local
10^{2} to 10^{4}	Lakes, rivers, bays	Fish and plant production	Local/regional
$\geq 10^{5}$	Ocean basins, major rivers, and lakes	Nutrient regulation, CO_2 regulation	Global

SOURCE: Adapted from Limburg et al. (2002).

target scale. This is a particularly difficult problem since it is assumed to exist and yet can be managed only by comprehensive knowledge of the system under study.

The uncertainties associated with consideration of scale in assessment of ecosystem goods and services will only be resolved by continuing investigation of natural systems. At present the practical solution is upfront recognition of the potential for aggregation errors and careful framing of the assessment question. Explicit identification of the ecosystem goods and services being evaluated, careful definition of the scale at which those services are generally realized, and comparison to the scale of the assessment being undertaken can at least bound the valuation process and inform subsequent decisions.

System Dynamics

Natural systems are increasingly understood as dynamic constructs that may exist in a number of alternate states (also referred to as "regimes" or "domains of ecological attraction" depending on the terminology being used). A system may move, or "flip," from one state to another if it passes a threshold of some controlling variable. The transition to an alternate state may be rapid or gradual, and may or may not reflect a change in the trajectory of the system. The concept of alternative states with boundary thresholds is used to explain the nonlinear behavior of natural systems. Indeed, examples of thresholds and regime shifts in aquatic ecosystems have been a significant part of the evolving understanding of nonlinear ecosystem behavior (Muradian, 2001; Scheffer and Carpenter, 2003; Scheffer et al., 2001; Walker and Meyers, 2004).

Many ecosystems can persist in a particular state or regime for some time because they exhibit resistance or resilience. Resistance is measured by the capacity to withstand disturbance without significant change, while resilience is indicated by the capacity to return to the original state after perturbation toward an alternate state. Resilience was originally described by Holling (1978) and persists as an important concept in the analysis of social-ecological system dynamics today (Walker and Myers, 2004; Walker et al. 2004).

The nonlinear system behavior that emerges in response to thresholds and regime shifts can be problematic for assessment of ecosystem services. Recognition of the points at which alternative behavior will emerge is difficult in many systems. (See Figure 3-1 for a conceptual representation of the nonlinear ecosystem response to stress.) As noted by Chavas (2000) ". . . ecosystem dynamics can be highly nonlinear, meaning that knowing the path of a system in some particular situation may not tell us much about its behavior under alternative scenarios."

An example of this type of behavior can be found in the waste assimilation and transport services of lakes, rivers, and estuaries. Increased nutrient loads in an aquatic ecosystem may simply increase productivity of the resident biota up to the point of harmful eutrophication. At that point, the high levels of primary

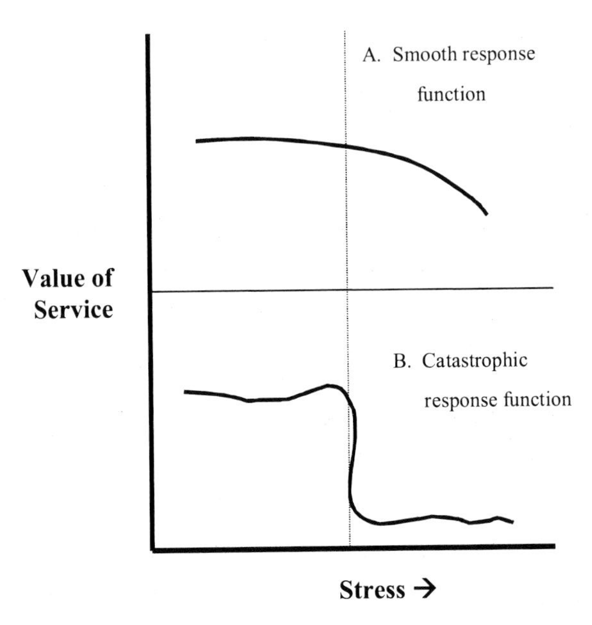

FIGURE 3-1 Value responses to stress under marginal (well-behaved dynamics) and nonmarginal (nonlinear, threshold dynamics) system behaviors. SOURCE: Reprinted, with permission, from Limburg et al. (2002) © 2000 by Elsevier.

production overwhelm secondary production and decomposition processes, resulting in excessive accumulation of organic matter, depletion of oxygen in the water column, and a change in the trophic structure. The change can represent a new and undesirable condition that may persist even if nutrient loads are reduced (see Carpenter, 2003; Carpenter et al., 1998). From the perspective of ecosystem service assessment, waste assimilation may still be occurring, but habitat services, recreational services, and maintenance of biodiversity may all be significantly changed. The point at which this abrupt shift in services occurs may be controversial and unpredictable.

In some circumstances the abrupt shift, or flip to an alternate regime in state may be part of a hysteretic system behavior. In this case the stress threshold that generated the response may be significantly higher than the stress threshold that will allow a recovery. This type of response can be found in many dense and highly productive aquatic communities, such as seagrass beds (Batuik et al., 2000). Often these communities can tolerate significant levels of physical stress simply because there are a sufficient number of individuals to moderate physical conditions inside the community and enough reproductive potential to offset the continual losses. When the physical stresses surpass a community's capacity to withstand them, reestablishment can often succeed only in conditions significantly less stressful than the robust community could tolerate (Molles, 2002). In essence, the recovery threshold differs from the impact threshold such that the state of the system will lag in response to changes in controlling forces.

Cascading effects are another example of ecosystem dynamics that can be difficult to predict (Molles, 2002). Harvest of top-level predators can result in increases in lower-level predators, decreases in herbivore prey, and resultant changes in vegetation. Alterations in river flows can change the timing of nutrient introductions to downstream waterbodies, resulting in modified phytoplankton and zooplankton communities, and culminating in shifts in habitat quality for higher-trophic-level fish communities.

There is considerable ongoing research to define thresholds and develop indicators of system condition that will assess proximity of thresholds. While understanding of these system dynamics continues to expand, this knowledge can inform assessment of ecosystem functions only if the assessment occurs at appropriate spatial and temporal scales, and appropriate spatial and temporal scales can be identified only if the dynamics are already understood. In the face of this apparent conundrum the practical solution to the need to complete an assessment of ecosystem function and/or provision of services is to proceed with caution. Observations of a system's behavior through time are an obvious first step, but such monitoring data can only confirm the existence of nonlinear behavior, not prove its absence. Simply considering the possibilities for threshold responses may be adequate to inform some assessments, and is certainly preferable to ignoring the issue.

Intrinsic Values

Many people believe that ecosystems have value quite apart from any human interest in explicit goods or services (see Chapter 2 for further information). The fact that ecosystems exhibit emergent behaviors and operate to sustain themselves is sufficient to argue that they have value to their components. Although comprehending this intrinsic value does not trouble most individuals, assessing it is problematic. Farber et al. (2002) state, "As humans are only one of many species in an ecosystem, the values they place on ecosystem functions, structures and processes may differ significantly from the values of those ecosystem characteristics to species or the maintenance (health) of the ecosystem itself."

Incomplete Knowledge

Comprehensive valuation of aquatic ecosystems should be viewed as a practical improbability. The assumption that our knowledge is imperfect is at the root of the concern for aggregation of assessments to larger scales and composite valuation of whole ecosystems. As a consequence, unforeseen behaviors and services are anticipated, and valuations are automatically caveated with concern for the state of the science. This does not imply no ecosystem valuation can be accomplished, simply that comprehensive valuation should not be presumed. Many decisions using economic or other valuation techniques can be made without a comprehensive assessment of ecosystem goods and services

An example of how the state of our understanding can impact the capacity to value an ecosystem service involves the relationship between biodiversity and aquatic ecosystem functions. In efforts to identify ecosystem services, researchers typically acknowledge the importance of habitat functions for maintenance of biodiversity. For some time, high biodiversity was assumed to confer some inherent resistance and/or resilience to a system, allowing it to sustain performance of other valued services in the face of disturbance. However, researchers are not of a single mind about the nature of the relationship between biodiversity and ecosystem functioning (e.g., Duarte, 2000; Ghilarov, 2000; Hulot et al., 2000; Schwartz et al., 2000; Ulanowicz, 1996). It can be difficult, if not impossible, therefore to accurately assess the importance of any particular ecosystem's contribution to maintenance of biodiversity, or conversely the role of biodiversity in the functioning of the ecosystem.

Another area in which a lack of comprehensive knowledge limits full recognition of services provided by aquatic ecosystems is the continual growth in the number of ways humans can use aquatic resources. The continually expanding lists of medicinal and industrial products found in aquatic ecosystems provide obvious examples, while the evolving number of aquatic recreational activities is another. The point is that the list of services is not determined entirely by the suite of natural functions in aquatic ecosystems, but also by human ingenuity in deriving benefits.

SUMMARY: CONCLUSIONS AND RECOMMENDATIONS

In review and discussion of the state of the science in the identification of aquatic ecosystem functions and their linkage to goods and services, the committee arrived at several specific conclusions:

- Ecologists understand the uncertainties in ecosystem analysis and accept them as inherent caveats in all discussions of system performance.
- As the committee pursued its charge, the problems of developing an interdisciplinary terminology and/or a universally applicable protocol for valuing aquatic ecosystems were illuminated, but ultimately identified as unnecessary objectives.
- From an ecological perspective, the value of specific ecosystem functions/services is entirely relative. The spatial and temporal scales of analysis are critical determinants of potential value.

- Potentially useful classification and inventories of aquatic ecosystems as well as their functional condition exist at both regional and national levels, though the relevance of these classification and inventory systems to assessing and valuing aquatic ecosystems is not always clear.
- Ecologists have qualitatively described the structure and function of most types of aquatic ecosystems. However, the complexity of ecosystems remains a barrier to quantification of these features, particularly their interrelationships.
- General concepts regarding the linkages between ecosystem function and services have been developed. Although precise quantification of these relationships remains elusive, the general concepts seem to offer sufficient guidance for valuation to proceed with careful attention to the limitations of any ecosystem assessment.
- Many, but not all, of the goods and services provided by aquatic ecosystems are recognized by both ecologists and economists. These goods and services can be classified according to their spatial and temporal importance.
- Complex ecosystem dynamics and incomplete knowledge of ecosystems will have to be resolved before comprehensive valuation of ecosystems is tractable, but comprehensive ecosystem valuation is not generally essential to inform many management decisions.
- Further integration of the sciences of economics and ecology at both intellectual and practical scales will improve ecologists' ability to provide useful information for assessing and valuing aquatic ecosystems.

There remains a significant amount of research and work to be done in the ongoing effort to codify the linkage between ecosystem structure and function and the provision of goods and services for subsequent valuation. The complexity, variability, and dynamic nature of aquatic ecosystems make it likely that a comprehensive identification of all functions and derived services may never be achieved. Nevertheless, comprehensive information is not generally necessary to inform management decisions. Despite this unresolved state, future ecosystem valuation efforts can be improved through use of several general guidelines and research conducted in the following areas:

- Aquatic ecosystems generally have some capacity to provide consumable resources (e.g., water, food); habitat for plants and animals; regulation of the environment (e.g., hydrologic cycles, nutrient cycles, climate, waste accumulation); and support for nonconsumptive uses (e.g., recreation, aesthetics, research). Considerable work remains to be done in documentation of the potential that various aquatic ecosystems have for contribution in each of these broad areas.
- Delivery of ecosystem goods and services occurs in both space and time. Local and short-term services may be most easily observed and documented, but the less intuitive accumulation of services over larger areas and time

intervals may also be significant. Alternatively, services that are significant only when performed over large areas or long time intervals may be beyond the capacity of some ecosystems. Investigation of the spatial and temporal thresholds of significance for various ecosystem services is necessary to inform valuation efforts.

- Natural systems are dynamic and frequently exhibit nonlinear behavior. For this reason, caution should be used in extrapolation of measurements in both space and time. Although it is not possible to avoid all mistakes in extrapolation, the uncertainty warrants explicit acknowledgment. Methods are needed to assess and articulate this uncertainty as part of system valuations.

REFERENCES

Batuik, R.A., P. Bergstrom, M. Kemp, E. Koch, L. Murray, J.C. Stevenson, R. Bartleson, V. Carter, N.B. Rybicki, J.M. Landwehr, C. Gallegos, L. Karrh, M. Naylor, D. Wilcox, K.A. Moore, and S. Ailstock. 2000. Chesapeake Bay Submerged Aquatic Vegetation Water Quality and Habitat-Based Requirements and Restoration Targets: A second technical synthesis. Annapolis, Md.: U.S. Environmental Protection Agency, Chesapeake Bay Program.

Berkes, F., J. Colding, and C. Folke (eds.). 2003. Navigating Social-Ecological Systems: Building Resiliency for Complexity and Change. Cambridge, U.K.: Cambridge University Press. Bowles, D.E., and T.L. Arsuffi. 1993. Karst aquatic ecosystems of the Edwards Plateau region of central Texas, USA—A consideration of their importance, threats to their existence, and efforts for their conservation. Aquatic Conservation-Marine and Freshwater Ecosystems 3:317-329.

Brinson, M.M. 1993. A Hydrogeomorphic Classification for Wetlands. U.S. Army Corps of Engineers, Wetlands Research Program Technical Report WRP-DE-4. Vicksburg, Miss.: U.S. Corps of Engineers.

Busch, W.D.N., and P.G. Sly. 1997. The Development of an Aquatic Habitat Classification System for Lakes. Boca Raton, Fla.: CRC Press.

Carpenter, S.R. 2003. Regime Shifts in Lake Ecosystems: Pattern and Variation. Book 15 in the Excellence in Ecology Series. Ecology Institute: Olendorf/Luhe, Germany.

Carpenter, S., N. Caraco, D.L. Correll, R.W. Howarth, A.N. Sharpley, and V.H. Smith. 1998. Nonpoint pollution of surface waters with phosphorus and nitrogen. Ecological Applications 8:559-568.

Chang, C., and R.C. Griffin. 1992. Water marketing as a reallocative institution in Texas. Water Resources Research 28:879-890.

Chavas, J. 2000. Ecosystem valuation under uncertainty and irreversibility. Ecosystems 3:11-15.

Chen, C.C., D. Gillig, and B.A. McCarl. 2001. Effects of climatic change over a water dependent regional economy: A study of the Texas Edwards Aquifer. Climatic Change 49:397-409.

Cowardin, L.M., V. Carter, F.C. Golet, and E.T. LaRoe. 1979. Classification of Wetlands and Deepwater Habitats of the United States. FWS/OBS-79/31. Corvallis, Ore.: U.S. Fish and Wildlife Service.

Crowe J.C., and J. M. Sharp, Jr. 1997. Hydrogeologic delineation of habitats for endangered species—The Comal Springs/River system: Berlin. Environmental

Geology 30(1-2): 17-33.

Culver, D.C., L.L. Master, M.C. Christman, and H.H. Hobbs. 2000. Obligate cave fauna of the 48 continuous United States. Conservation Biology 14:386-401.

Custodio, E. 2002. Aquifer over-exploitation: What does it mean? Hydrogeology Journal 10: 254-277.

Dahl, T.E. 2000. Status and Trends of Wetlands in the Conterminous United States 1986 to 1997. Washington, D.C.: U.S. Department of the Interior, Fish and Wildlife Service.

Daily, G.C. 1997. Introduction: What are ecosystem services? Pp. 1-10 in Nature's Services: Societal Dependence on Natural Ecosystems, G.C. Daily (ed.). Washington, D.C.: Island Press.

De Groot, R.S., M.A. Wilson, and R.M.J. Boumans. 2002. A typology for the classification, description and valuation of ecosystem functions, goods and services. Ecological Economics 41:393-408.

Duarte, C.M. 2000. Marine biodiversity and ecosystem services: An elusive link. Journal of Experimental Marine Biology and Ecology 250(1-2):117-131.

Edwards, R.J., G. Longley, R. Moss, J. Ward, R. Matthews, and B. Stewart. 1989. A classification of Texas aquatic communities with special consideration toward the conservation of endangered and threatened taxa. Texas Journal of Science 41:231-240.

EPA (U.S. Environmental Protection Agency). 2001. National Coastal Condition Report. EPA-620/R-01/005. Washington, D.C.: U.S. EPA, Office of Research and Development/Office of Water. Also available on-line at *http://www.epa/gov/owow/oceans/NCCR/index* Accessed October 2002.

EPA. 2002. 2000 National Water Quality Inventory. EPA-841-R-2-001. Washington, D.C.: Office of Water.

Ewel, K.C. 2002. Water quality improvement by wetlands. Pp. 329-344 in Nature's Services: Societal Dependence on Natural Ecosystems, G.C. Daily (ed.). Washington, D.C.: Island Press.

Farber, S.C., R. Costanza, and M.A. Wilson. 2002. Economic and ecological concepts for valuing ecosystem services. Ecological Economics 41:375-392.

Ghilarov, A.M. 2000. Ecosystem functioning and intrinsic value of biodiversity. Oikos 90(2): 408-412.

Gibert, J., D.L. Danielopol, and J.A. Standford (eds.). 1994. Groundwater Ecology. San Diego, Calif.: Academic Press.

Great Lakes National Program Office. 2001. Great Lakes Ecosystem Report. Chicago, Ill.: U.S. Environmental Protection Agency.

Great Lakes National Program Office. 2002. The Great Lakes: An Environmental Atlas and Resource Book. Washington, D.C.: U.S. Environmental Protection Agency.

Gunderson, L.H. and C.S. Holling (eds). 2002. Panarchy: Understanding Transformations in Human and Natural Systems. Washington, D.C.: Island Press.

Gunderson, L.H. and L. Pritchard, Jr. (eds). 2002. Resilience and the Behavior of Large Scale Systems. Washington, D.C.: Island Press.

Holling, C.S. (ed.). 1978. Adaptive Environmental Assessment and Management. New York: John Wiley and Sons.

Hulot, F.D., G. Lacroix, F. Lescher-Moutoue, and M. Loreau. 2000. Functional diversity governs ecosystem response to nutrient enrichment. Nature (London) 405(6784): 340-344.

Jones, J.B., and P.J. Mulholland (eds.). 2000. Streams and Ground Waters. San Diego, Calif.: Academic Press.

Junk, W.J., P.B. Bayley, and R.E. Sparks. 1989. The flood pulse concept in river-floodplain systems. In Proceedings International Large River Symposium (LARS), C.P. Dodge (ed.). Canadian Special Publication in Fisheries and Aquatic Sciences 106:110-127.

Kaiser, R.A., and L.M. Phillips. 1998. Dividing the waters: Water marketing as a conflict resolution strategy in the Edwards Aquifer region. Natural Resources Journal 38:411-444.

Keplinger, K.O., and B.A. McCarl. 2000. An evaluation of the 1997 Edwards Aquifer irrigation suspension. Journal of the American Water Resources Association 36:889-901.

Keplinger, K.O., B.A. McCarl, M.E. Chowdhury, and R.D. Lacewell. 1998. Economic and hydrologic implications of suspending irrigation in dry years. Journal of Agricultural and Resource Economics 23:191-205.

Knutson, M.G., J.R. Sauer, D.A. Olsen, M.J. Mossman, L.M. Hemensath, and M.J. Lannoo. 1999. Effects of landscape composition and wetland fragmentation on frog and toad abundance and species richness in Iowa and Wisconsin, U.S.A. Conservation Biology 13(6):1437-1446.

Limburg, K.E., R.V. O'Neill, R. Costanza, and S. Farber. 2002. Complex systems and valuation. Ecological Economics 41:409-420.

Longley, G. 1986. The biota of the Edwards aquifer and the implications for paleozoogeography. Pp. 51-54 in The Balcones Escarpment, Central Texas, P.L. Abbott and C.M. Woodruff, Jr. (eds.). Boulder, Colo.: Geological Society of America.

Lord, L.A. 1993. Guide to Florida Environmental Issues and Information. Winter Park, Fla.: Florida Conservation Foundation.

Maxwell, J.R., C.J. Edwards, M.E. Jensen, S.J. Paustian, J. Parrot, and D.M. Hill. 1995. A Hierarchical Framework of Aquatic Ecological Units in North America. GTR-NC-176. St. Paul, Minn.: U.S. Department of Agriculture, Forest Service, North Central Forestry Experimental Station.

McCarl, B.A., C.R. Dillon, K.O. Keplinger, and R.L. Williams. 1999. Limiting pumping from the Edwards Aquifer: An economic investigation of proposals, water markets, and spring flow guarantees. Water Resources Research 35:1257-1268.

MEA (Millenium Ecosystem Assessment). 2003. Ecosystems and Human Well-being: A Framework for Assessment. Washington, D.C.: Island Press.

Molles, M.C. Jr. 2002. Ecology: Concepts and Applications. New York: McGraw Hill.

Muradian, R. 2001. Ecological thresholds: A survey. Ecological Economics 38:7-24.

NRC (National Research Council). 2002a. Florida Bay Research Program and Their Relation to the Comprehensive Everglades Restoration Plan. Washington, D.C.: National Academy Press.

NRC. 2002b. The Missouri River Ecosystem: Exploring Prospects for Recovery. Washington, D.C.: National Academy Press.

NRC. 2003. Adaptive Monitoring and Assessment of the Comprehensive Everglades Restoration Plan. Washington, D.C.: The National Academies Press.

Odum, E.P. 1993. Ecology and Our Endangered Life-Support Systems. Second Edition. Sunderland, Mass.: Sinauer Associates Incorporated.

Odum, E.P., and M.G. Turner. 1990. The Georgia landscape: A changing resource. Pp. 131-163 in Changing Landscapes: An Ecological Perspective, I.S. Zonnevald and R.T.T. Forman (eds.), New York: Springer-Verlag.

Odum, H.T. 1988. Self-organization, transformity, and information. Science 242:1132-1139.

Odum, H.T. 1996. Environmental Accounting: Energy and Environmental Decision-Making. New York: John Wiley.

Odum, H.T., and E.P. Odum. 2000. The energetic basis for valuation of ecosystem services. Ecosystems 3:21-23.

Peterson, C.H., and J. Lubchenco. 2002. Marine ecosystem services. Pp. 177-194 in Nature's Services: Societal Dependence on Natural Ecosystems, G.C. Daily (ed.). Washington, D.C.: Island Press.

Postel, S.L., and S. Carpenter. 1997. Freshwater ecosystem service. Pp. 195-214 in Nature's Services: Societal Dependence on Natural Ecosystems, G.C. Daily (ed.). Washington, D.C.: Island Press.

Purdum, E.D. 2002. Florida Waters: A Water Resources Manual from Florida's Water Management Districts. Orlando, Fla.: Institute of Science and Public Affairs, Florida State University for Florida's Water Management Districts.

Rosgen, D.L. 1994. A classification of natural rivers. Catena 22:169-199.

Schaible, G.D., B.A. McCarl, and R.D. Lacewell. 1999. The Edwards Aquifer water resource conflict: USDA farm program resource-use incentives? Water Resources Research 35: 3171-3183.

Scheffer, M., and S.R. Carpenter. 2003. Catastrophic regime shifts in ecosystems: Linking theory to observation. Trends in Ecology and Evolution 18(12):648-656.

Scheffer, M., S. Carpenter, J.S. Foley, C. Folke, and B. Walker. 2001. Catastrophic shifts in ecosystems. Nature 413:591-596.

Schwartz, M.W., C.A. Brigham, J.D. Hoeksema, K.G. Lyons, M.H. Mills, and P.J. van-Mantgem. 2000. Linking biodiversity to ecosystem function: Implications for conservation ecology. Oecologia 122(3):297-305.

Scientific and Technical Advisory Committee. 2003. Chesapeake Futures: Choices for the 21st Century. STAC Publication Number 03-001. Edgewater, Md.: CRC, Inc.

Smith, R.D., A. Ammann, C. Bartoldus, and M.M. Brinson. 1995. An Approach for Assessing Wetland Functions Using Hydrogeomorphic Classification, Reference Wetlands, and Functional Indices. Technical Report WRP-DE-9. Vicksburg, Miss.: U.S. Army Engineer Waterways Experiment Station.

Sousa, P.J. 1985. Habitat Suitability Index Models: Red-Spotted Newt. Biological Report 82 (10.111). Washington, D.C.: U.S. Fish and Wildlife Service

Stanford, J. S., J.V. Ward, W.J. Liss, C.A. Frissell, R.N. Williams, J.A. Lichatowich, and C.C. Coutant. 1996. A general protocol for restoration of regulated rivers. Regulated Rivers: Research and Management 12:391-413.

TNC (The Nature Conservancy). 1997. A classification framework for freshwater communities. In Proceedings of the Nature Conservancy's Aquatic Community Classification Workshop. Chicago, Ill.: TNC, Great Lakes Program Office.

Turner, M.G., E.P. Odum, R. Constanza, and T.M. Springer. 1988. Market and non-market values of the Georgia landscape. Environmental Management 12(2):209-217.

Ulanowicz, R.E. 1996. The propensities of evolving systems. In Social and Natural Complexity, Khalil, E.L., K.E. Boulding (eds.). London: Routledge.

USDA (U.S. Department of Agriculture-Natural Resources Conservation Services). 2000. Summary Report of the 1997 National Resources Inventory. Washington, D.C.: Iowa State University Statistical Laboratory and USDA Natural Resources Conservation Services.

Vannote, R.L., G.W. Minshall, J.R. Cummins, J.R. Sedell, and C.E. Cushing. 1980. The river continuum concept. Canadian Journal of Fisheries and Aquatic Sciences 37:130-137.

Walker, B., C. S. Holling, S.R. Carpenter, and A. Kinzig. 2004. Resilience, adaptability and transformability in social-ecological systems. Ecology and Society 9(2):5. Available on-line at *http://www.ecologyandsociety.org/vol9/iss2/art5*.

Walker, B. and J.A. Meyers. 2004. Thresholds in ecological and social-ecological systems: A developing database. Ecology and Society 9(2):3. Available on-line at: *http://www.ecologyandsociety.org/vol9/iss2/art3*.

Wass, M.L., and T.D. Wright. 1969. Coastal Wetlands of Virginia. Special Report in Applied Marine Science. Gloucester Point, Va.: Virginia Institute of Marine Science.

Wetzel, R.G. 2001. Freshwater and wetland ecology: challenges and future frontiers. In Sustainability of Wetlands and Water Resources: Achieving Sustainable Systems in the 21dst Century, M.M. Holland, E. Blodd, and L. Shaffer (eds.). Covelo, Calif.: Island Press.

4
Methods of Nonmarket Valuation

INTRODUCTION

This chapter outlines the major methods that are currently available for estimating economic (monetary) values for aquatic and related terrestrial ecosystem services. Within the chapter is a review of the economic approach to valuation, which is based on a total economic value framework. In addition to presenting the valuation approaches, the chapter discusses the applicability of each method to valuing ecosystem services. It is important to note that the chapter does not instruct the reader on how to apply each of the methods, but rather provides a rich listing of references that can be used to develop a greater understanding of any of the methods. Based on this review, the chapter includes a summary of its conclusions and recommendations.

The substance of this chapter differs from the various books and chapters that provide overviews of nonmarket valuation methods (e.g., Braden and Kolstad, 1991; Champ et al., 2003; Herriges and Kling, 1999; Mäler and Vincent, 2003; Mitchell and Carson, 1989; Ward and Beal, 2000) because these prior contributions were designed to summarize the state of the art in the literature or to teach novices how to apply the various methods. This chapter also differs from government reports that provide guidance for implementing nonmarket valuation methods (EPA, 2000a; NOAA, 1993). The purpose of this chapter is to carefully lay out the basic valuation approaches and explain their linkages to valuing aquatic ecosystems. This is done within the context of the committees' implicit objective (see Box ES-1) of assessing the literature in order to facilitate original studies that will develop a closer link between aquatic ecosystem functions, services, and value estimates.

ECONOMIC APPROACH TO VALUATION

Economic Valuation Concepts

As discussed in Chapter 2, the concept of economic valuation adopted in this report is very broad. That is, the committee was concerned with how to estimate the impacts of changes in ecosystem services on the welfare, or utility (satisfaction or enjoyment), of individuals. If ecosystem changes result in individuals feeling "worse off," then one would like to have some measure of the loss of economic value to these individuals. Alternatively, if the changes make people "better off," one would like to estimate the resulting value gain.

The basic concepts that economists use to measure such gains and losses are economic values measured as a *monetary payment* or a *monetary compensation*. The essence of this approach is to estimate values as subtractions from or additions to income that leave people equally economically satisfied with or without a change in the services provided by an aquatic ecosystem. For example, suppose a lake was contaminated with polychlorinated biphenyls (PCBs) discharged by a nearby factory. In such a case, the logical valuation concept is an estimate of the monetary compensation that is required to bring the affected people back to the same level of satisfaction they enjoyed prior to the contamination event. Such a measure of value, when aggregated over all affected people, could be used to assign a damage payment to the factory responsible for the pollution. Funds collected from the polluter would not typically be paid directly to the affected people, but would be used for restoration projects that would return services to the lake.

Another type of application would be a project to enhance a freshwater wetland to improve sportfishing opportunities. In this example, one group of people consists of the direct beneficiaries, people who fish recreationally. Valuation would be used to estimate the "maximum" that anglers would pay for this improvement in fishing. Although no money would actually be collected from the anglers, each angler's expression of his or her maximum willingness to pay represents how much the angler is prepared to compensate the rest of society for the increased individual enjoyment gained from the improved recreational fishing. Maximum willingness to pay is aggregated for all anglers who benefit to determine whether the benefits of the wetland project exceed the costs, which facilitates an assessment of whether public funds should be spent on the project.

These two examples provided several insights:

1. Values arise from the preferences of individual people; thus, values are estimated for individuals or households and then aggregated to obtain the values that society places on changes in aquatic ecosystems.
2. Valuation methods are used to estimate the gains or losses that people may experience as a result of changes in aquatic ecosystems in order to inform policy discussions and decisions.
3. Different types of changes in aquatic ecosystems affect different groups of people, which, as discussed in more detail below, may influence the choice of valuation methods used.
4. There are two basic concepts of value (noted elsewhere in this report), willingness to accept (WTA) (compensation) and willingness to pay (WTP).[1]

Whether WTA or WTP is conceptually the appropriate measure of value for changes in aquatic ecosystems depends on the presumed endowment of property rights. In the case of PCB contamination, the presumed property right of society was to a lake that is free of PCBs. This implies that the conceptually appropriate

---

[1] For further discussion of measurements of WTP and WTA, see Chapter 2.

value measure that would restore people to their original level of satisfaction is WTA compensation. In contrast, in the freshwater wetland restoration example, the presumed property right is in the existing fishing conditions and the appropriate value measure is WTP to obtain the improvement in fishing conditions. Unfortunately, economists have had difficulty in measuring WTA (Boyce et al., 1992; Brown and Gregory, 1999; Coursey et al., 1987; Hanemann, 1991) and most empirical work for policy applications involve measures of WTP. This issue arises for a variety of reasons, such as survey respondents not being familiar with WTA questions and because most respondents have incomplete knowledge of relative prices. Thus, most of the following discussion focuses on the use of valuation methods to estimate WTP.

Why Valuation Is Required

Chapter 2 discusses the importance of economic valuation as input into decision-making and, in particular, for aiding the assessment of policy choices or trade-offs concerning various management options for aquatic ecosystems. As Chapter 3 has illustrated, given the complex structure and functioning of aquatic and related terrestrial ecosystems, these systems often yield a vast array of continually changing goods and services. The quality and quantity of these services are in turn affected by changes to ecosystem structure and functioning. Thus, alternative policy and management options can have profoundly different implications for the supply of aquatic ecosystem services, and it is the task of economic valuation to provide estimates to decision-makers of the aggregate value of gains or losses arising from each policy alternative.

Valuation is especially important because many services provided by aquatic ecosystems have attributes of public goods. Public goods are are nonrival and nonexcludable in consumption, which prevents markets from efficiently operating to allocate the services. An example would be wetland filtration of groundwater. As long as the quantity of groundwater is not limiting, everyone who has a well in the area can enjoy the benefits of unlimited potable groundwater. However, in the absence of any market for the provision of water through wetland filtration, there is no observed price to reveal how much each household or individual is willing to pay for the benefits of this service. Although everyone is free to use the aquifer, no one is responsible for protecting it from contamination. This is not an action that could be undertaken by a company and provided for a fee (price) because no individual has ownership of the wetland filtration process or the aquifer. However, nonmarket values can be estimated to assess whether the benefits of collective action—perhaps through a state environmental agency or the U.S. Environmental Protection Agency (EPA)—exceed the cost of the proposed actions to protect the wetland, and consequently the wetland filtration process and the quality of the water in the aquifer for drinking purposes.

It is also the case that some aquatic ecosystem services indirectly contribute to other services that are provided through a market, but the value of this ecological service itself is not traded or exchanged in a market. For example, an estuarine marshland may provide an important "input" into a commercial coastal fishery by serving as the breeding ground and nursery habitat for fry (juvenile fish). Although disruption or conversion of marshland may affect the biological productivity of the marsh, and thus its commercial fishery, a market does not exist for the commercial fishery to pay to maintain the habitat service of the marshland. The problem is also one of transaction costs. It is costly for participants in the commercial fishery to get together to negotiate with marshland owners and there may be many owners of for which protection agreements must be sought. Estimation of the implicit (nonmarket) value to the fishery of marsh habitat can be used to understand whether laws and rules to protect the breeding and nursery functions of the marsh.

Aquatic ecosystem services that do not have market prices are excluded from explicit consideration in cost-benefit analyses and other economic assessments, and are therefore likely to not get full consideration in policy decisions. As noted in Chapter 2, Executive Order 13258, which supersedes Executive Orders 12866[2] and EO 12291,[3] requires government agencies to demonstrate that the benefits of regulations outweigh the costs. (All of the benefit-cost discussion occurs in Executive Order 12866 and federal agencies still reference this order.) This mandate is followed by the EPA (2000a) *Guidelines for Preparing Economic Analyses*, which emphasizes the importance of valuation to decision-making on the environment. Thus, if monetary values of ecosystem services are not estimated, many of the major benefits of aquatic ecosystems will be excluded in benefit-cost computations. The likely outcome of such an omission would be too little protection for aquatic ecosystems, and as a consequence the services that people directly and indirectly enjoy would be undersupplied. Valuation, therefore, can help to ensure that ecosystem services that are not traded in markets and do not have market prices receive explicit treatment in economic assessments. The goal is not to create values for aquatic ecosystems. Rather, the purpose of valuation is to formally estimate the "nonmarket" values that people already hold with respect to aquatic ecosystems. Such information on nonmarket values will in turn assist in assessments of whether to protect certain types of aquatic ecosystems, to enhance the provision of selected ecosystem services, and to restore damaged ecosystems.

Finally, economic values are often used in litigation involving damage to aquatic ecosystems from pollution or other human actions. For evidence to be credible, including ecosystem modeling and economic values, it must pass a Daubert test,[4] the essential points of which are whether the following apply:

- the theories and techniques employed by the scientific expert have been

[2] Executive Order 12866. October 4, 1993. Federal Register 58 (190).
[3] Executive Order 12291. February 19, 1981. Federal Register 46(33).
[4] For further information about the Daubert test, see *http://www.daubertontheweb.com/Chapter_2.htm.*

tested;

- they have been subjected to *peer review* and *publication*;
- the techniques employed by the expert have a *known error rate*;
- they are subject to *standards* governing their application; and
- the theories and techniques employed by the expert enjoy widespread acceptance.

All of the nonmarket valuation methods discussed in this chapter meet these conditions in general. A key issue, and thus theme of this chapter is which of the methods are applicable to valuing the services of aquatic and related terrestrial ecosystems and under what conditions and circumstances? Issues raised throughout this chapter suggest areas in need of original research between ecologists and economists that will ultimately provide better aquatic ecosystem value estimates to support policy evaluations and decision-making that are defensible.

The Total Economic Value Framework

As discussed in Chapter 2, the total economic value (TEV) framework is based on the presumption that individuals can hold multiple values for ecosystems and is developed for categorizing these various multiple benefits. Although any taxonomy of values is somewhat arbitrary and may differ from one use to another, the TEV framework is necessary to ensure that some components of value are not omitted in empirical analyses and that double counting of values does not occur when multiple valuation methods are employed. For example, Table 3-2 presents several categorizations of ecosystem services. In any empirical application it is necessary to map these services to how they affect humans and then select an appropriate valuation method. This chapter presents information that helps with the selection of a valuation approach, while Chapter 5 discusses the mapping of changes in ecosystem to effects upon humans through a series of case studies. The TEV approach presents a road map that facilitates this mapping of ecosystem services to effects and the selection of valuation methods.

Valuation Under Uncertainty

Estimation of use and nonuse values (see Chapter 2 for a detailed discussion of use and nonuse values; see also Table 2-1) is often associated with uncertainty. For example, current efforts to restore portions of the Florida Everglades (see also Chapter 5 and Box 3-6) do not imply that the original services of this wetland area can be restored with certainty. It is also impossible to predict with certainty the changes in service provided by aquatic ecosystems due to global warming. These situations are not unique when aquatic ecosystem services are valued. In addition, individuals may be uncertain about their future demand for the services provided by restoration of the Everglades or the services affected by global warming. For example, someone living in New York may be unsure if they will ever visit the Everglades, which affects how they might value the improvements in opportunities to watch birds in the Everglades. Someone who lives in the Rocky Mountain states may be unsure about whether they will ever visit the Outer Banks in North Carolina, which affects the value they place on losing this coastal area to erosion.

These uncertainties can affect the estimation of use and nonuse values from an ex ante ("beforehand") perspective. The economist's concept of TEV for ex ante valuation under uncertainty, from either the supply or the demand side, is *option price* (Bishop, 1983; Freeman, 1985; Larson and Flacco, 1992; Smith, 1983; Weisbrod, 1964).[5] The notion of option price follows that of TEV, whereas option value is simply the concept of TEV when uncertainty is present and includes all use and nonuse values an individual holds for a change in an aquatic ecosystem. Option price is the amount of money that an individual will pay or must be compensated to be indifferent between the status quo condition of the ecosystem and the new, proposed condition. Option prices can be estimated for removing the uncertainty or for simply changing probabilities; reducing the probability of an uncertain event (beach erosion); or increasing the probability of a desirable event (e.g., increased quality of bird watching). Option prices are also estimated for conditions where probabilities do not change, but the quantity or quality associated with a probability changes.

The following section of the chapter focuses on the micro-sense of uncertainty in the estimation of individual, or perhaps household, values, whereas Chapter 6 takes a broader perspective of uncertainty that includes how values estimated in the presence of uncertainty are used to inform policy decisions. The discussion in Chapter 6 includes concepts such as "quasi-option value" and its relationship to option values.

CLASSIFICATION OF VALUATION APPROACHES

Since economists often employ a variety of methods to estimate the various use and nonuse values depicted in Table 2-1, another common classification is by *measurement approaches*. As shown in Table 4-1, this type of categorization is usually organized according to two criteria:

[5] Another component of value, *option value*, is commonly referred to as a nonuse value in the literature (see Chapter 6 for further information). Option value arises from the difference between valuation under conditions of certainty and uncertainty and is a numerical calculation, not a value held by people. The literature cited above makes this distinction and does not mistakenly include option value as a component of TEV.

TABLE 4-1 Classification of Valuation Approaches

	Revealed Preferences	Stated Preferences
Direct	Competitive market prices Simulated market prices	Contingent valuation, open-ended response format
Indirect	Household production function models Time allocation Random utility and travel cost Averting behavior Hedonics Production function models Referendum votes	Contingent valuation, discrete-choice and interval response formats Contingent behavior Conjoint analysis (attribute based)

SOURCE: Adapted from Freeman (1993a).

1. whether the valuation method is to be based on *observed* economic behavior, from which individual preferences can be inferred, or whether the valuation method is to be based on responses to survey questions that reveal *stated preferences* by individuals, and
2. whether monetary estimates of values are observed directly or inferred through some indirect method of data analysis.

Because of the public good nature of many of the services described previously, market prices do not exist. Simulated markets are typically used as a benchmark to judge the validity of value estimates derived from indirect methods, but simulated markets are rarely used to develop policy-relevant estimates of value. The open-ended format is not commonly used in contingent valuation studies due to problems with zero bids and protest responses (Bateman et al., 2002; Boyle, 2003). Indirect methods are the most commonly used approaches to valuing aquatic ecosystem services, and the discussion below focuses on these approaches.

Household Production Function Methods

Household production function (HPF) approaches involve modeling consumer behavior, based on the assumption of a substitutional or complementary relationship between an ecosystem service and one or more marketed commodities. The combination of the environmental service and the marketed commodities, through a household production process, results in the "production" of a utility-yielding good or service (Bockstael and McConnell, 1983; Freeman, 1993a; Mäler, 1974; Smith, 1991, 1997). Examples of these approaches include time allocation models for collecting water, travel-cost methods for estimating the demand for visits to a recreation site, averting behavior models that are frequently used to measure the health impacts of pollution, and hedonic property value or wage models.

The inspiration for HPF approaches is the "full income" framework for determining household resource allocation and consumption decisions as developed by Becker (1965), although the HPF model can be applied to a valuation problem without assuming a single, "full income" constraint. The HPF provides a framework for examining interactions between purchases of marketed goods and the availability of nonmarket environmental services, which are combined by the household through a set of technical relationships to "produce" a utility-yielding final good or service. For example, in the documented presence of contaminated drinking water a household would be expected to invest time and purchased inputs (e.g., an averting technology, bottled water) to provide a desired service, namely potable water. This is the essence of the averting behavior approach, and in the above example the household is attempting to avoid exposure to a degraded drinking water system.

Appendix B, using travel-cost models, averting behavior approaches, and hedonic price methods, illustrates that the assumptions underlying the "household production function" will vary depending on the environmental problem and the valuation approach. Nevertheless, the common theme in all applications of the HPF approach is the derivation of derived demand for the environmental asset in question. Thus, information on the value of environmental quality can be extracted from information on the household's purchases of marketed goods. The following section illustrates the HPF framework with three examples applied to aquatic ecosystems: (1) random utility or travel-cost models, (2) averting behavior models, and (3) hedonic models.

Random Utility and Travel-Cost Models

The modern variants of travel-cost models are known as random utility models (RUMs). Random utility models arise from the empirical assumption that people know their preferences (utility) with certainty, but there are elements of these preferences that are not accessible to the empirical observer (Herriges and Kling, 1999; Parsons, 2003a). Thus, parameters of peoples' preferences can be recovered statistically up to a random error component. This econometric approach is used to estimate modern travel-cost models. The most common application of this modeling framework has been valuing recreational fishing in freshwater lakes and rivers and marine waters.

Travel-cost studies attempt to infer nonmarket values of ecological services by using the travel and time costs that an individual incurs to visit a recreation site (Bockstael, 1995). Out-of-pocket travel costs and the opportunity cost travel time are used as the implicit price of visiting a site, perhaps a lake to fish or swim. Traditional travel-cost studies utilized the implicit price of travel and the number of times each individual in a sample visited a site to estimate the demand for visits to the site. If the site is a lake and the recreation activity is fishing, this approach yields an in situ value for fishing at the site, only part of which is attributable to the aquatic ecosystem services. The values of ecosystem services are fixed for any given lake at a specific point in time and cannot be identified statistically.

In the case of qualitative differences in the ecological attributes and thus the recreational potential of different sites, random utility models have been employed to value changes in the desirable ecological characteristics that make each site attractive for recreation. The advantage of the RUM approach over traditional travel-cost studies is that, by assuming each recreational site option is mutually exclusive, it is possible to determine how ecological characteristics or attributes of each site affect the decision of an individual to select one particular site for recreation. Thus, the RUM approach is uniquely designed to estimate values for attributes of recreation sites, which for fishing include the quantity and quality of the aquatic ecosystem services. The RUM approach looks at peoples' choices of recreation sites among the menu of available sites and determines the implied values people hold for site attributes by making choices between sites that vary in terms of the cost of visiting the sites and their component attributes, which include aquatic ecosystem characteristics. All other factors being equal, the basic premise of the travel-cost approach is that people will choose the site with the lowest travel cost. When two sites have equal travel costs, people will choose the site with higher quality. If one site has more desirable species of fish, say native trout, then that site will be chosen. Alternatively, if one site has degraded water quality that results in a fish consumption advisory, this site would not be chosen. RUMs use information on these revealed choices to estimate the values people place on aquatic ecosystem services that support recreational opportunities. That is, people will travel further to improve the quality of their visit to an aquatic ecosystem. This behavior allows the empirical investigator to infer the value that individuals place on an improvement or degradation in an aquatic ecosystem.

Another aspect of RUMs is that they can be designed to allow the number of participants to increase (or decrease) as an ecosystem is enhanced (or diminished). The individual actually faces three choices: (1) whether to participate in an activity (e.g., sportfishing), (2) where to go fishing on any particular occasion, and (3) how often to participate in fishing. This is important because both the average value per visit per person, the number of visits an individual makes, and the number of affected people determine aggregate, societal values. While travel-cost models and their modern RUM variants are based on the conceptual framework of household production technology, the production is generally assumed to be undertaken on an individual basis and values are estimated for individuals, not households.

A common concern of human interactions with ecosystems is the potential for the extinction of species through pollution, destruction of habitat, and overuse by humans. All of these factors come into play for the Atlantic salmon in Maine rivers. The rivers in Maine have been heavily dammed to provide hydroelectric power, which diminishes and destroys salmon habitat. There is a long history of pollution by the timber industry and communities, which diminishes water quality for salmon. There has also been substantial fishing pressure, both commercial and recreational, on Atlantic salmon. Morey et al. (1993) employed a RUM to estimate the values that recreational anglers place on salmon fishing. They used a model in which anglers choose among eight salmon fishing rivers in Maine and the Canadian provinces of New Brunswick, Nova Scotia, and Quebec. This area includes all of the major salmon fishing rivers in the northeastern United States and eastern Canada readily accessible to U.S. citizens by car. The authors estimated values for a scenario that asked what the loss per angler would be if salmon numbers fell to the point that anglers are no longer able to fish the Penobscot River in Maine. The Penobscot River is the major salmon fishing river in Maine and this scenario would estimate losses if the river was closed to fishing, for example, because Atlantic salmon in the Penobscot River were listed as endangered so that fishing would be prohibited. The annual loss per angler of not being able to fish the Penobscot River, but still being able to fish one of the other seven sites in the model was about $800. They also estimated a model that asked what would happen if restoration of salmon to the Penobscot River increased the salmon population so that catch rates doubled. The annual benefit per angler was about $650 per year. The first scenario estimates the value for loss of an ecosystem service, and no specific information from ecologists was needed to estimate this value. The second scenario estimates a value from an improvement in ecosystem services. To develop the estimate for the latter scenario, Morey et al. (1993) included angler catch rates in their model and sportfishing as an indicator of the quality of the ecosystem services enjoyed by people.

Two important considerations arise here. First, in order to simulate a doubling of catch rates on the Penobscot River it is necessary for other fishing sites to have catch rates that approximate a doubling of the catch rate for the Penobscot. This means that value predictions are within the range of quality over which anglers have exhibited revealed behavior. This provides observations of revealed choice for this change in quality. Second, absent from the model was a link between salmon populations in the Penobscot River and catch rates. To make the latter scenario realistic for policy analyses it would be necessary to model the relationship between catch rates and population to know what population of salmon is necessary in the Penobscot River to support this doubling of service. Although there is nothing technically wrong with the value estimates reported, there is no direct ecosystem link to indicate how a biological intervention would affect catch rate and the subsequent catch rate could be used to estimate a policy-relevant value. At present, the values reported are simply illustrative. This also leads to the question of what has to be undertaken from an ecological perspective to enhance the population of Atlantic salmon in the river.

Another interesting RUM application is also a sportfishing study. In this study, researchers looked at the effect of fish consumption advisories on choices of sportfishing site (Jakus et al., 1997; see also Jakus et al., 1998). Here the ecosystem service is the effect on human health from consumption of fish. However, this service has been diminished by pollution at some sites, which has been signaled to anglers through consumption advisories (i.e., official warnings not to fish). This study considered fishing on 22 reservoirs in Tennessee, 6 of which had consumption advisories against fishing. Only reservoirs that were within 200 miles of an angler's residence were considered possible fishing sites in the model. Jakus and colleagues found that removing fish consumption advisories from the two reservoirs within 200 miles of residents of central Tennessee had a value of $22 per angler per year. Likewise, removing the advisories from six reservoirs within 200 miles of residents of east Tennessee would have a value of $47 per angler per year. These are estimates of the damages from pollution as signaled by fish consumption advisories. From a policy perspective, to compute aggregate losses it is necessary to know whether ecological restoration will allow removal of the advisories and when this might occur. Thus, the losses of $22 and $47 per angler per year will continue to accumulate each year that the advisories remain in place.

Other studies that have used RUMs to estimate values for aquatic ecosystem services include the following:

- effects of river and reservoir water levels on recreation in the Columbia River basin (Cameron et al., 1996);
- fishing in the Great Lakes (Phaneuf et al., 1998);
- fishing in freshwater lakes (Montgomery and Needleman, 1997);
- river fishing (Morey and Waldman, 1998);
- fishing and viewing wildlife in wetlands (Creel and Loomis, 1992);
- fishing in coastal estuaries (Greene et al., 1997);
- swimming in lakes (Needleman, and Kealy, 1995);
- beach use (Haab and Hicks, 1997);
- boating on lakes (Siderelis et al., 1995); and
- effects of climate change on fishing (Pendleton and Mendlesohn, 1998).

The largest majority of RUMs have valued recreational fishing in lakes (Parsons, 2003b), but as the above examples indicate, there have been applications to other types of aquatic ecosystems and services. Even some terrestrial applications may have relevance to aquatic ecosystem services valuation. For example, one of the early RUM applications was to downhill skiing (Morey, 1981). As ski areas continue to draw more surface water to make snow, there are likely to be increasing impacts on nearby aquatic ecosystems. Thus, policies that affect how much surface water can be used to make snow will have an effect on the value people place on downhill skiing.

The most common use of RUMs is to estimate the in situ value of visiting a recreational site that is related to an aquatic ecosystem. The typical effects of ecosystem services valued in RUMs are changes in fish catch rates, the presence of fish consumption advisories, and degradation of surface waters due to eutrophication from nonpoint pollution. Rarely are other dimensions of ecological services of aquatic ecosystems valued. The key element of applications of

RUMs to aquatic ecosystems is that there must be a service that affects the sites people choose to visit. This could include fish catch rates, fish consumption advisories, or waters levels, as demonstrated in the studies cited above. This is by no means an exhaustive list of services, just the obvious services that have been commonly used in developing RUMs.

RUMs have typically been applied to single-day recreation trips and have not examined multiple-day trips. The reason for ignoring multiple-day trips is that these may be multiple-site, multiple-length, and multiple-purpose trips, which makes it extremely difficult to estimate values for ecosystem services at specific sites. Ignoring multiple-day trips serves to underestimate the aggregate value that people who engage in recreation place on aquatic ecosystem services. Estimates for day trips can be affected by several key elements of any application. The first is the researcher's choice of the measurement of travel cost including the opportunity cost of travel time. A subjective decision by an analyst to include or exclude elements from the measurement of travel cost will increase or decrease the measurement of travel cost and affect value estimates.

The second factor of particular concern for applications to aquatic ecosystems is the degree to which aquatic ecosystem services are correlated with each other and with other physical attributes of a site. This multicollinearity makes it difficult to identify aquatic ecosystem attributes that people value and omitting relevant ecosystem attributes may lead to biased estimates. For example, if the environmental variable of concern is binary and represents the presence of native trout and native trout occur in beautiful mountain streams, then the value estimate for native trout may also capture a value for scenic beauty. On the other hand, if a fish consumption advisory is place on an industrial river and is modeled as a binary variable in the RUM, then the value of removing the fish consumption advisory may also capture the value of fishing at a nonindustrial location.

A third key element affecting the quality of an application is the lack of consistent data on attributes that measure the same given attribute across all the sites in the choice set. Most of the RUMs employ the small set of attributes that are available for all sites. A related issue is the distinction between objective and subjective measures of site attributes—what matters is not how the attributes are measured by the experts but how they are perceived by the individual making the choice of recreation sites. It is much harder to obtain data on perceptions of site attributes.

Averting Behavior Models

Averting behavior models have been increasingly used as an indirect method to evaluate the willingness of individuals to pay for improved health or to avoid undesirable health consequences (Dickie, 2003). In terms of aquatic ecosystems there are only two notable averting behavior applications: (1) a study of averting behavior in the presence of a waterborne disease giardiasis

(Harrington et al., 1989) and (2) groundwater contamination by the solvent tricholoroethylene (TCE) (Abdalla et al., 1992).

Averting behavior models are based on the presumption that people will change their behavior and invest money to avoid an undesirable health outcome. Thus, averting behavior analyzes the rate of substitution between changes in behavior and expenditures on and changes in environmental quality in order to infer the value of certain nonmarketed environmental attributes (see Appendix B). For example, in the presence of water pollution, a household may install a filter on the primary tap in the house to remove or reduce the pollutant. This involves a capital expenditure by the household and changes in behavior because potable water can now be safely obtained only from the primary tap, not from other taps in the house. Rather than producing a fishing trip or other type of recreational experience, as is the household production that underlies the estimation of a RUM, the household production here is protection from an undesirable outcome that is commonly health-related (Bartik, 1988; Courant and Porter, 1981; Cropper, 1981).

The giardiasis study by Harrington et al. (1989) is one of the best known averting behavior applications and one of the few applied to water. This study differs conceptually from the replacement cost studies for public water supplies discussed in Chapter 5, which are not based on individual preferences. The approach here is to measure people's actual averting expenditures to estimate a household value for avoiding an undesirable situation (i.e., contaminated drinking water exposure). The model was applied to estimate the losses due to an outbreak of waterborne giardiasis in Luzerne County, Pennsylvania, that took place from 1983 to 1984. The outbreak occurred as a result of microbial contamination of the reservoir supplying drinking water to households in that county. Such contamination is typically caused by the ingestion of cysts of the enteric protozoan parasite *Giardia lamblia*, which is often found in animal (and sometimes human) feces deposited in upland watersheds that are subsequently transported to reservoirs used a source of drinking water. During the nine-month period of the Luzerne County outbreak, households were advised to boil their drinking water, but many also bought bottled water at supermarkets or collected free water supplied by some public facilities. The authors' "best estimate" of the average costs of these actions taken to avoid contaminated water ranged from $485 to $1,540 per household, or $1.13 to $3.59 per person per day for the duration of the outbreak.

In another averting behavior study conducted in Pennsylvania, Abdalla et al. (1992) investigated behavior by the Borough of Perkaise in response to TCE in well water. Of the households in the borough, 43 percent indicated that they were aware of TCE in their water and 44 percent undertook actions to avoid exposure. The averting actions included purchasing bottled water, installing a home water treatment system, obtaining water from an uncontaminated source, and boiling water. Each of these actions required households to change their behavior and make out-of-pocket expenditures. The investigators found that households were more likely to undertake averting behavior if their perceived risk of consuming water with TCE was higher, if they knew more about TCE, or if they had children the household had between the ages of 3 and 17. Of the households that averted, those with children less than three years of age spent more on averting activities than did other households. The average daily expenditure per household undertaking averting behaviors was about $0.06 during the 88 weeks that the TCE contamination persisted.

For an averting behavior study on water quality to be successful, four conditions are necessary:

1. households must be aware of compromised water quality;
2. households must believe that the compromised water quality will adversely affect the health of at least one household member;
3. there must be activities that a household can undertake to avoid or reduce exposure to the compromised water; and
4. households must be able to make expenditures that result in optimal protection.

The fourth element is rarely met however, so that total expenditures generally underestimate value and marginal expenditures should cautiously be interpreted as a measure of marginal willingness to pay.

Thus, an averting behavior study provides an estimate of the value households place on improving water quality. However, averting behavior studies rarely provide estimates of economic values of ecosystem services as defined in Chapter 2 and discussed at the beginning of this chapter. Averting expenditures generally are not the same as subtractions to income that leave people equally satisfied from an economic perspective as they would be if water quality were not improved. Averting behavior can underestimate or overestimate this value. An averting-behavior study would underestimate the economic value of clean water because averting behavior studies do not include the inconvenience of having to undertake the averting behavior. Economic value can also be underestimated if households cannot fully remove the diminished water quality. For example, onsite reverse-osmosis treatment systems do not fully remove arsenic in drinking water (EPA, 2000b; Sargent-Michaud and Boyle, 2002). Averting behavior overestimates economic values when joint production is present, which could arise when contamination is present and the natural taste of the water is undesirable. Averting behavior would be undertaken to avoid the contamination and to obtain potable (more palatable) water. In this case, averting expenditures overstate what would be spent just to avoid the contamination.

Although averting behavior studies will generally provide a lower or upper bound on the damages to compromised drinking water, they are not likely to be useful in measuring other economic values of aquatic ecosystem services. Certainly, potable water is an important service of aquatic ecosystems to humans. Protected water for human consumption will have additional benefits of the clean water for other living organisms. As with RUMs, modeling is needed to understand how actions taken to protect or improve aquatic ecosystems will affect potable water.

Hedonic Methods

Hedonic methods analyze how the different characteristics of a marketed good, including environmental quality, might affect the price people pay for the good or factor. This type of analysis provides estimates of the implicit prices paid for each characteristic. The most common application of hedonic methods in environmental economics is to real estate sales (Palmquist, 1991, 2003; Taylor, 2003). For example, the hedonic price function for residential property sales might decompose sale prices into implicit prices for the characteristics of the lot (e.g., acreage), characteristics of the house (e.g., structural attributes such as square footage of living area), and neighborhood and environmental quality characteristics. In terms of aquatic ecosystems, properties with lake frontage sell for more than similar properties that do not have lake frontage. Among properties with lake frontage, those located on lakes with good water quality would be expected to sell for more than those located on lakes with poor water quality. In this regard, a hedonic analysis is simply a statistical procedure for disentangling estimates of the premium people pay for lake frontage or for higher water quality, which is the revealed value for these ecological services.

There are two stages in the estimation of a hedonic model (Bartik, 1987; Epple, 1987). The first stage, which is commonly undertaken, simply decomposes sale prices of properties to estimate the implicit prices of property characteristics as described above. The implicit price estimates provide the marginal prices that people would pay for a small change in each characteristic. For example, if the attribute of interest was feet of frontage that the property had on a lake, the first-stage analysis provides the implicit price of a 1-foot increase in frontage. What if the policy question was how much value 100 feet of frontage would add to a property? However, the marginal price cannot provide this value estimate. The second-stage analysis uses either restrictions on the underlying utility function to derive value estimates (Chattopadhayay, 1999) or implicit price estimates from a number of different lakefront markets (Palmquist, 1984).

The application of a hedonic analysis requires a large number of property sales where characteristics of the properties vary. For example, data from a single lake might be used to estimate a first-stage equation for lake frontage if the amount of frontage varies for different properties on the lake. However, data from one lake probably cannot be used to estimate the value of water quality because all properties on a lake likely experience the same level of water quality. To estimate an implicit price for water quality it is necessary to have sales from a number of different lakes that differ in ambient water quality.

In order to operationalize a hedonic model to estimate values for aquatic ecosystem services, it must be assumed that buyers and sellers of properties have knowledge of the services and have access to the same information. For example, one problem in examining the effects of water pollution on property prices is that the use of water quality indices developed by natural scientists to measure pollutants, such as dissolved oxygen, nitrogen, and phosphorus, may not provide relevant information. As such, the physical measures of quality are not observable to homeowners, test results may not be generally available or easily obtained, and diminished water quality may not directly impair the enjoyment that households derive from waterfront homes (Leggett and Bockstael, 2000).

Consider groundwater contamination as an example. The water that comes through a household tap may appear clean and taste fine but, if contaminated (e.g., by arsenic), may not be safe to drink. A hedonic model can be operational only if buyers and sellers are aware of arsenic levels in tap water and what levels are considered safe. Such information would be available if the public were generally aware of arsenic contamination, if sellers were required to reveal test results, or if buyers were advised to have the water tested if test results were not provided by the seller. In this example, since there is no obvious clue to the public that water quality is compromised, public information is necessary to prompt buyers and sellers to react to potential contamination. Another example is eutrophication of lakes. Although buyers and sellers cannot directly observe elements of the water chemistry that is compromised, they can certainly observe the physical manifestations of eutrophication. Thus, a summary measure of eutrophication (e.g., Secchi disk measurement of water clarity; see more below) may more be more closely aligned with buyer and seller perceptions than actual measures of water chemistry. This means that Secchi disk measurements may do a better job of explaining changes in sale prices of properties than measurements of dissolved oxygen, which implies a more accurate estimate of the implicit price placed on eutrophication by homeowners.

As noted above, most hedonic studies just estimate the first-stage, hedonic price function. Several of these studies have estimated implicit prices for water and coastal quality in the Chesapeake Bay area (Feitelson, 1992; Leggett and Bockstael, 2000; Parsons, 1992). Leggett and Bockstael (2000) showed that the concentration of fecal coliforms (a commonly used bacterial indicator of the potential presence of waterborne pathogens; see also NRC, 2004) in water has a significant effect on property values along the bay. They found that a change in fecal coliform counts of 100 colony forming units (CFUs) of water per 100 mL would affect sale prices of properties by about 1.5 percent, with the dollar amount ranging from about $5,000 to nearly $10,000. The average sale prices of properties in the study were $378,000 dollars, and the fecal contamination index ranged from 10 to 1,762, with a mean of 108 CFUs.

Parsons (1992) used a repeated-sale analysis to observe price changes on houses sold before and after the State of Maryland imposed building restrictions in critical coastal areas of the Chesapeake Bay. Prices for waterfront properties increased by 46-62 percent due to the restrictions, between 13 and 27 percent for houses nearby but not on the waterfront, and between 4 and 11 percent for houses as far as 3 miles away. Parsons noted however, that the price increases may be due to the increasing scarcity of near-coastal land as a result of the state restrictions. The Parsons study is interesting for two reasons. First, although a water quality attribute does not directly enter the hedonic price function, the benefits of the building restrictions include protection of aquatic and related coastal ecosystems along the coast. However, the second interesting feature is a complication of many hedonic studies—that environmental attributes may be highly correlated. Thus, it may be impossible to statistically disentangle the implicit price for the protection of aquatic ecosystems along the coast and other benefits of building restrictions.

Other applications of hedonic models to estimate implicit prices for aquatic ecosystems include the following:

- effects of water clarity on sale prices of lakefront properties (Michael et al., 2000; Steinnes, 1992; Wilson and Carpenter, 1999);
- effect of the potential for surface water contamination on farmer purchases of herbicides (Beach and Carlson, 1993);
- proximity of properties to hazardous waste sites that pollute groundwater (Kiel, 1995);
- extent of aquatic area proximate to properties (Paterson and Boyle, 2002);
- proximity of properties to wetlands (Doss and Taff, 1996; Mahan et al. 2000);
- effects of various measures of lake water quality (e.g., summer turbidity, chlorophyll concentrations, suspended solids, dissolved oxygen) on sale prices (Brasheres, 1985);
- effect of minimum lake frontage on sale prices of property to preserve lake amenities (Spalatro and Provencher, 2001);
- effect of coastal beach pollution on property prices (Wilman, 1984); and
- effect of pH levels in streams on property sale prices (Epp and Al-Ani, 1979).

A notable consideration of these studies is that the services of aquatic ecosystems have been included in the first-stage hedonic price equations in three ways. The first is a measure of ecosystem quality as it affects the desirability of human use. The second is simply proximity to the aquatic ecosystem, and the third, which has been made possible with enhanced geographic information system (GIS) databases, measures the physical size of an aquatic ecosystem. All of the listed studies assessed surface water, with a primary focus on water quality in lakes. Furthermore, the Beach and Carlson (1993) study was the only hedonic analysis that considered an aquatic ecosystem that was not based on sales of residential properties.

Only one study has estimated the second-stage demand for an aquatic ecosystem service. Boyle et al. (1999) estimated the demand for water clarity in lakes using the multiple-market method. Clarity is measured by the depth at which a Secchi disk[6] disappears from sight as it is lowered into the water. Given an initial clarity reading of 3.78 meters, an increase in clarity to 5.15 meters results in a one-time value estimate of about $4,000 per household. Conversely, a decline of clarity from 3.78 meters to 2.41 meters results in a loss of value of at least $25,000 per household.

While hedonic models provide a useful method of estimating values for aquatic ecosystem services, the collinearity of attributes in hedonic price equation is a serious issue. In the Michael et al. (2000) study, Secchi disk measurements were used as a summary measure of lake eutrophication that is observable to property owners. Other lake attributes are highly correlated with reduced Secchi disk measurements, such as lake area and lake depths, and small shallow lakes are more likely than larger lakes to be eutrophic. Eutrophic lakes are also typically warmer than oligotrophic lakes for swimming and support warm-water species of sportfish, including bass and perch, that are typically less desirable than trout and salmon. Thus, although the Secchi disk measurements are a summary measure of water quality, it is likely that estimated implicit prices include the effects of other lake attributes on sale prices.

For a hedonic study to be operational there are two important conditions: (1) the effects of aquatic ecosystems must be observable to property owners, and (2) there should be minimal correlation between aquatic ecosystem services that affect sale price of properties and other attributes that affect sale prices. A key feature in the modeling of aquatic ecosystem services is that the variable included in the hedonic price equation to reflect the ecosystem service being valued must be observable to property owners. As noted above, measured elements of water chemistry such as dissolved oxygen and chlorophyll levels may be less important than a summary measure such as Secchi disk readings. However, the question remains of whether homeowners' subjective perceptions of clarity are a better measure of service quality than physical Secchi disk measures. Poor et al. (2001) demonstrated that Secchi disk measurements of water clarity do a better job of explaining differences in sale prices than did property owners subjective ratings of water clarity. Thus, while aquatic ecosystem characteristics must be observable to homeowners, some type of objective measure of the characteristics is likely to be better than self-reports of the quantity or quality of services by homeowners. Finally, as long as aquatic ecosystem services are correlated with other attributes of property, hedonic analyses are likely to overestimate implicit prices and values.

---

[6] A Secchi disk is most commonly an 8-inch metal disk painted with alternating black and white quadrants and is used to see how far a person can see into the water (see *http://www.mlswa.org/secchi.htm* for further information).

Production Function Methods

Production function (PF) approaches, also called "valuing the environment as input," assume that an environmental good or service essentially serves as a factor input into the production of a marketed good that yields utility. Thus, changes in the availability of the environmental good or service can affect the costs and supply of the marketed good, the returns to other factor inputs, or both. Applying PF approaches therefore requires modeling the behavior of producers and their response to changes in environmental quality that influence production (see Appendix C for further information about the general PF approach). Dose-response and change-in-productivity models, which have been used for some time, can be considered special cases of the PF approach in which the production responses to environmental quality changes are greatly simplified.

However, more sophisticated PF approaches are being increasingly employed for a diverse range of environmental quality impacts and ecosystem services, including the effects of flood control, habitat-fishery linkages, storm protection functions, pollution mitigation, and water purification. A two-step procedure is generally invoked (Barbier, 1994). First, the physical effects of changes in a biological resource or ecological service on an economic activity are determined. Second, the impact of these environmental changes is valued in terms of the corresponding change in the marketed output of the relevant activity. In other words, the biological resource or ecological service is treated as an "input" into the economic activity, and like any other input, its value can be equated with its impact on the productivity of any marketed output.

For some ecological services that are difficult to measure, an estimate of ecosystem area may be included in the production function of marketed output as a proxy for the ecological service input. For example, in models of coastal habitat-fishery linkages, allowing wetland area to be a determinant of fish catch is thought to "capture" some element of the economic contribution of this important ecological support function (Barbier and Strand, 1998; Barbier et al., 2002; Ellis and Fisher, 1987; Freeman, 1991; Lynne et al., 1981). That is, if the impacts of the change in the wetland area input can be estimated, it may be possible to indicate how these impacts influence the marginal costs of production. As shown in Figure 4-1, for example, an increase in wetland area increases the abundance of crabs and thus lowers the cost of catch. The value of the wetlands support for the fishery—which in this case is equivalent to the value of increments to wetland area—can then be imputed from the resulting changes in consumer and producer value.

For the PF approach to be applied effectively, it is important that the underlying ecological and economic relationships are well understood. When production is measurable and either there is a market price for this output or one can be imputed, determining the marginal value of the ecological service is relatively straightforward. If the output of the affected economic activity cannot be measured directly, then either a marketed substitute has to be found or possible complementarity or substitutability between the ecological service and one or more

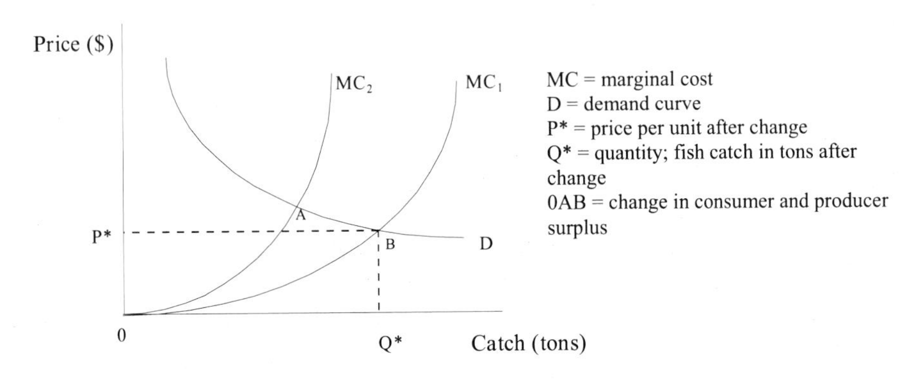

FIGURE 4-1 The economic value effects of increased wetland area on an optimally managed fishery. For optimally managed fishery a change in wetland area that serves as a breeding ground and nursery results in a shift in the marginal cost curve (MC) of the fishery. The welfare impact is the change in consumer and producer surplus (represented by area 0AB). SOURCE: Adapted from Freeman (1991).

of the other (marketed) inputs has to be explicitly specified. All of these applications require detailed knowledge of the physical effects on production of changes in the ecological service. However, applications that assume complementarity or substitutability between the service and other inputs are particularly stringent in terms of the information required on physical relationships in production. Clearly, cooperation is required between economists, ecologists, and other researchers to determine the precise nature of these relationships.

In addition, as pointed out by Freeman (1991), market conditions and regulatory policies for the marketed output will influence the values imputed to the environmental input. For instance, in the previous example of coastal wetlands supporting an offshore crab fishery, the fishery may be subject to open-access conditions. Under these conditions, profits in the fishery would be dissipated, and price would be equated to average and not marginal costs. As a consequence, producer values are zero and only consumer values determine the value of increased wetland area (see Figure 4-2).

A further issue is whether a static or dynamic model of the relationship between the ecological service and the economic activity is required. As discussed in Appendix B, this usually depends on whether or not it is more appropriate to characterize this relationship as affecting production of the economic activity over time. Figures 4-1 and 4-2 represent PF models that are essentially static. The value of changes in the environmental input is determined through producer and consumer value measures of any corresponding changes in the one-period market equilibrium for the output of crabs. In dynamic approaches, the ecological service is considered to affect an intertemporal, or "bioeconomic," production relationship. For example, a coastal wetland that serves as a breeding and nursery habitat for fisheries could be modeled as part of the growth function of the fish stock, and any value impacts of a change in this habitat support function can be determined in terms of changes in the long-run equilibrium conditions of the fishery or in the harvesting path to this equilibrium (see Appendix B). Figure 4-3 shows that the long-run supply curve for an open-access fishery is typically backward-bending (Clark, 1976). Since coastal wetland habitat affects the biological growth of the fishery, a decline in wetland area will shift back the long-run supply curve of the fishery and thus reduce long-run harvest levels. The corresponding losses can be measured by the fall in economic value, which will be greater if the demand curve is more inelastic (i.e., steeper).

A number of recent studies have used PF models to estimate the economic benefits of coastal wetland-fishery linkages. Much of this literature owes its development to the approach of Lynne et al. (1981) who suggested that the support provided by the marshlands of southern Florida for the Gulf Coast fisheries could be modeled by assuming that marshland area supports biological growth of the fishery. For the blue crab fishery in western Florida salt marshes, the authors estimated that each acre of marshland increased productivity of the fishery by 2.3 pounds per year. Others have applied the Lynne et al. approach to additional Gulf Coast fisheries in western Florida (Bell, 1997) and in southern Louisiana (Farber and Costanza, 1987). Using data from the Lynne et al. (1981) case study, Ellis and Fisher (1987) determined the impacts of changes in the

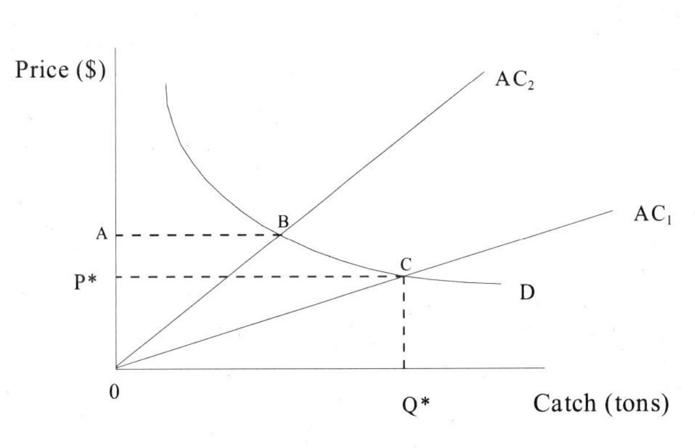

AC = average cost
D = demand curve
P* = price per unit after change
Q* = quantity; fish catch in tons after change
P*ABC = change in consumer and producer surplus after increase in wetlands area

FIGURE 4-2 The economic value effects of increased wetland area on an open-access fishery. For an open-access fishery, a change in wetland area that serves as a breeding ground and nursery results in a shift in the average cost curve, AC, of the fishery. The welfare impact is the change in consumer surplus (area P*ABC). SOURCE: Adapted from Freeman (1991).

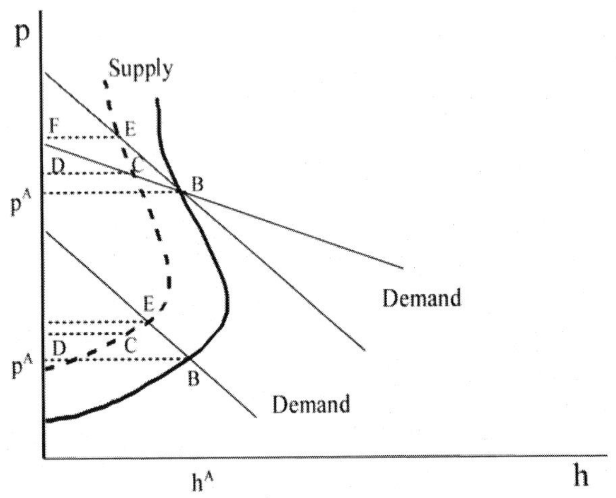

FIGURE 4-3 Wetland loss and the long-run market equilibrium of an open-access fishery. The effect of a fall in wetland area is to shift the long-run equilibrium supply curve of an open access fishery to the left. The result is a decline in fish harvest h^A. The loss in consumer value will be greater if the demand curve is more inelastic (area P^ABEF*)* than elastic (area P^ABCD). SOURCE: Adapted from Barbier et al. (2002).

Florida Gulf Coast marshlands on the supply-and-demand relationships of the commercial blue crab fishery. They demonstrated that an increase in wetland area increases the abundance of crabs and thus lowers the cost of catch. The value of the wetlands' support for the fishery—which in this case is equivalent to the value of increments to wetland area—can then be imputed. Freeman (1991) has extended Ellis and Fisher's approach to show how the values imputed to wetlands are influenced by market conditions and regulatory policies that affect harvesting decisions in the fishery. In assuming an open-access crab fishery supported by Louisiana coastal wetland habitat, the value of an increase in wetland acreage from 25,000 to 100,000 acres could range from $47,898 to $269,436. If the fishery is optimally managed, the increase in coastal wetland is valued from $116,464 to $248,009.

More "dynamic," or long-term, approaches to analyzing habitat-fishery linkages have also been developed (e.g., see Barbier and Strand, 1998; Barbier et al., 2002; Kahn and Kemp, 1985; McConnell and Strand, 1989). For example, in their case study of valuing mangrove-shrimp fishery linkages in the coastal regions of Campeche, Mexico, Barbier and Strand (1998) analyzed the effects of a change in mangrove area in terms of influencing the long-term equilibrium of an open-access fishery (i.e., one in which there are no restrictions on additional fishermen entering to harvest the resource). Their results indicate that the economic losses associated with mangrove deforestation appear to vary with long-term management of the open-access fishery. During the first two years of the simulation (1980-1981), which were characterized by much lower levels of fishing effort and higher harvests, a 1 km^2 decline in mangrove area was estimated to reduce annual shrimp harvests by around 18.6 tons, or a loss of about $153,300 per year. In contrast, during the last two years of the analysis (e.g., 1989-1990), which saw much higher levels of effort and lower harvests in the fishery, a marginal decline in mangrove area resulted in annual harvest losses of 8.4 tons, or $86,345 each year.

Kahn and Kemp (1985) and McConnell and Strand (1989) considered the impacts of water quality on fisheries in the Chesapeake Bay. Kahn and Kemp related the environmental carrying capacity of fish populations to the level of subaquatic vegetation, which is in turn affected by the runoff of agricultural chemicals, discharges from waste treatment plants, and soil erosion. Based on this analysis, the authors were able to determine marginal and total damage functions for various finfish and shellfish species in the bay.

Swallow (1994) modeled the impacts of developing "high-quality" and "normal-quality" freshwater pocosin (peat-bog) estuarine wetlands on the Pamlico Sound, North Carolina, shrimp fishery. Drainage of the pocosin wetlands for forestry and agricultural uses irreversibly alters the local hydrological system by eliminating the vegetative and peat-bog structure that inhibits water flow, causing a decline in the salinity of the estuarine shrimp nursery areas. The result is a decline in the juvenile shrimp stock necessary to replenish the Pamlico Sound fishery each year. Through his production function model linking development to salinity changes in the pocosin and fishery declines, Swallow estimated that the greatest losses to the shrimp fishery are estimated as $3.37 per acre per year for developing agriculture that affects high-quality wetlands near the southwestern shore of the sound. However, losses in other areas of the estuary with normal-quality wetlands are much lower. Based on these estimates, Swallow was able to determine the net opportunity cost of development of different-quality wetlands in the sound. The efficient policy would be to halt agricultural development when the marginal value of development net of the offshore fishery impacts fell to an annualized $1.12 per acre ($14 in present value). For the pocosin wetlands of the sound, this implies that 9,800 of the 11,009 acres of normal-quality southeastern wetlands could be safely developed, but all 1,209 high-quality southwestern wetlands should be preserved.

As these preceding examples illustrate, most uses of the production function approach have been concerned with valuing single ecosystem services. However, there have been a number of recent attempts to extend this approach to the ecosystem level through integrated economic-ecological modeling. The PF approach has the advantage of capturing more fully the ecosystem functioning and dynamics underlying the provision of key services and can be used to value multiple services arising from aquatic ecosystems.

For example, Wu et al. (2003) examined the effectiveness of alternative salmon habitat restoration strategies in the John Day River Basin, Oregon, through employment of integrated biological, hydrologic, and economic models. The purpose of the modeling was to shed light on two sets of unknown factors affecting salmon restoration investments: (1) the effects of uncertain environmental factors, such as weather and ocean conditions; and (2) the limited information on the potential ecological and hydrological threshold effects that can affect the potential payoffs on restoration investments. In an ideal salmon habitat, stream temperature must be below a certain threshold level. When water temperature exceeds this level, reducing temperature by one or two degrees will have no impact on fish survival. Other ecological factors, such as streamside vegetation, soil sedimentation, and species interaction, should also be modeled to examine trade-offs between different conservation benefits through investments targeted at one benefit (e.g., salmon habitat restoration). For example, Wu and colleagues demonstrated that for cold water-adapted fish species (e.g., rainbow trout, Chinook salmon), provided water temperature is maintained below its critical threshold, the number of fish increases as the vegetative use index improves. However, for speckled dace, the number of fish per kilometer of stream decreases as vegetative use improves and temperature decreases. In their fully integrated model, the authors were able to show the trade-offs of different salmon restoration investments in terms of the decline of speckled dace and the estimated marginal social value of increased numbers of cold-water fish species. This is a trade off between quantity in one aspect of the ecosystem and quality in another aspect. A three-degree drop in stream temperature, from 26˚C to 23˚C, will result in an estimated social benefit of $22,129 from increases in cold-water sportfish species, but a reduction of 506 speckled dace per kilometer of stream.

Carpenter et al. (1999) demonstrated how an integrated ecological-economic model of eutrophication of small shallow lakes can demonstrate the value impacts of irreversible ecological change (see also Chapter 5). Tschirhart and Finhoff (2001) developed a general equilibrium ecosystem with a regulated open-access fishery to analyze simulations of an eight-species Alaskan marine ecosystem that is affected by fish harvesting. Fishing impacts the commercial fish population as well as the populations of other species, including Steller sea lions, an endangered species. Settle and Shogren (2002) developed an integrated ecological-economic model to analyze the impacts of the introduction of exotic lake trout into Yellowstone Lake, which pose a risk to the native cutthroat trout. The authors demonstrated that an integrated model leads to different policy results than treating the ecological and economic systems separately. Under the best case scenario, the U.S. Park Service eliminates lake trout immediately and without cost, while under the worst-case scenario lake trout are left alone. An integrated model has little effect on the worst-case scenario, because the likely outcome is elimination of cutthroat trout. However, under the best-case scenario without feedbacks, the steady-state population of cutthroat trout is about 2.7 million. With feedbacks, the steady-state population is about 3.4 million. The integrated model predicts that the maximum optimal fixed budget for lake trout control is $169,000.

Other applications of production function models to estimate the value of services of aquatic ecosystems include the following:

- habitat-fishery linkages (Barbier, 2000 and 2003; Batie and Wilson, 1978; Bell, 1989; Costanza et al., 1989; Danielson and Leitch, 1986; Hammack and Brown, 1974; Sathirathai and Barbier, 2001);

- coastal erosion control and storm protection (Costanza and Farber, 1987; Costanza et al. 1989; Sathirathai and Barbier, 2001);
- groundwater recharge of wetlands (Acharya 2000; Acharya and Barbier 2000, 2002);
- water quality-fishery linkages (Kahn, 1987; Loomis, 1988; Wu et al., 2000); and
- general equilibrium modeling of integrated ecological-economic systems (Tschirhart, 2000).

Stated-Preference Methods

Stated-preference methods have been commonly used to value aquatic ecosystem services. There are two variants of stated-preference methods, contingent valuation (e.g., Bateman et al., 2002; Boyle, 2003; Mitchell and Carson, 1989) and conjoint analysis (e.g., Holmes and Adamowicz, 2003; Louviere, 1988; Louviere et al., 2000). Contingent valuation was developed by economists and is the more commonly used approach, whereas conjoint analysis was developed in the marketing literature (Green and Srinivasan, 1978). Contingent valuation attempts to measure the value people place on a particular environmental item taken as a specific bundle of attributes; conjoint analysis aims to develop valuation functions for the component attributes viewed both separately and in alternative potential combinations.

Contingent valuation is used to estimate values for applications, such as aquatic ecosystem services, where neither explicit nor implicit market prices exist. The first known application of contingent valuation was by Davis (1964) for hunters and other visitors to the woods of Maine. About 10 years later, the third application of contingent valuation (Hammack and Brown, 1974) estimated the value of waterfowl and wetlands. Through the 1980s and 1990s, the quality and extent of contingent valuation studies appear to have increased steadily.

While conjoint analysis was developed in the marketing literature to estimate prices for new products or modifications of existing products, it is conceptually similar to contingent valuation, and economists have come to recognize that it is another stated-preference approach to estimating economic value when market prices are unavailable. The first known environmental application was by Rae (1983) to value air quality in national parks. The number of environmental applications of conjoint analysis increased throughout the 1990s.

Both contingent valuation and conjoint analysis use survey questions to elicit statements of value from people with two key distinctions. First, contingent valuation studies generally pose written or verbal descriptions of the environmental change to be valued, while conjoint analysis poses the change in terms of changes in the attributes of the item to be valued. Consider a wetland restoration project as an example—the Macquarie Marshes in New South Wales, Australia (Morrison et al., 1999; also discussed below). A contingent valuation survey would contain a description of the wetland in its current condition and the wetland after restoration, whereas a conjoint survey would describe the wetland in terms of key attributes. These might be acres of wetland, number of species of breeding birds, and frequency with which birds breed. A contingent valuation study may contain this same information, but it would not be presented to estimate component values for each of these attributes. In terms of valuation, the contingent valuation study provides an estimate of the value of change in the marsh due to restoration, while the conjoint study provides a similar estimate and also estimates the amount of value contributed by each attribute. Thus, like a hedonic model, the attribute-based approach of conjoint analysis provides implicit prices for key attributes of the aquatic ecosystem.

The second key difference between these stated-preference methods involves the response formats. Contingent valuation studies typically ask respondents to state their value directly or to indicate a range in which the value resides (Welsh and Poe, 1998). In the latter case, econometric procedures are used to estimate the latent value based on the monetary intervals that respondents indicate. In conjoint analysis, survey respondents would be given alternatives to consider (e.g., three marsh restoration programs) and asked to choose the preferred alternative or to rank the alternatives (Boyle et al., 2001). Again, econometric procedures are used to estimate values from the choices or ranks.

Of the many contingent valuation studies that have been conducted, perhaps the two most well known involve aquatic ecosystems. In one of the earliest large-scale, contingent valuation studies, Mitchell and Carson (1981) estimated total national values for inland waters that are swimmable, fishable, and drinkable. They found that people who use freshwater for recreation were willing to pay \$237 annually to obtain swimmable, fishable, and drinkable freshwater, while the comparable estimate for nonusers was \$111.

The second study examined the value that a national sample would place on protecting Prince William Sound from an oil spill of the magnitude of the *Exxon Valdez* spill (Carson et al., 1992). In this study, a national survey was also conducted and total values were estimated, although the estimates were assumed to be primarily nonuse values because most people in the nationwide sample would never actually visit Price William Sound. The median value estimated was about \$33 per household for a one-time payment to protect Prince William Sound from a large-scale oil spill.

Many contingent valuation studies have investigated values for aquatic ecosystem services. So many, in fact, that several meta-analyses of these studies have been conducted, including protection of groundwater from contamination (Boyle et al., 1994); wetland values (Woodward and Wui, 2001); and sportfishing (Boyle et al.,1998a,b).

The primary application of the contingent valuation groundwater studies is protection from nitrate contamination resulting from agricultural practices. A particularly interesting attribute of the wetland meta-analysis is that the authors attempted to determine how values for wetlands vary with the services they provide. Lastly, the vast majority of sportfishing contingent valuation studies have investigated values of a single-day fishing trip—some focusing on individual species and others addressing some type of contamination.

The use of conjoint analysis is relatively new for nonmarket valuation and very few conjoint studies of aquatic ecosystems services have been undertaken. The best example is the aforementioned study of the Macquarie Marshes by Morrison et al. (1999). This study found that households in the area of New South Wales, Australia (near the marshes), would pay about $150 (Australian dollars) per year to restore the marshes to part of their original area. This change included increasing the number of species of marsh birds and the frequency at which they breed (Morrison and Boyle, 2001). Other examples include waterfowl hunting (Gan and Luzar, 1993) and salmon fishing (Roe et al., 1996).

The use of conjoint analysis in other types of applications in the literature is growing, and conjoint analysis is likely to become more prominent in the valuation of aquatic ecosystems in the future because of its ability to estimate values for multiple services. Most aquatic ecosystems provide multiple services (see also Chapter 3), and the ability to estimate marginal values for specific services is important for policy analyses.

To implement a stated-preference study two key conditions are necessary: (1) the information must be available to describe the change in an aquatic ecosystem in terms of services that people care about, in order to place a value on those services; and (2) the change in the aquatic ecosystem must be explained in the survey instrument in such a way that people will understand and not reject the valuation scenario. However, achieving these two conditions is easier said than done. Identifying the services that people care about with respect to a resource is not always a simple task because aquatic ecosystems such as wetlands provide a wide variety of services. People may care about wetland birds and animals and have no difficulty linking these to wetlands; however, potential respondents may have greater difficulty linking a wetland policy to changes in flood risk or the cost of potable water. Even if respondents identify and consider all relevant services, they may misinterpret policy descriptions or misperceive the impact of policy described in a questionnaire (Johnston et al., 1995; Lupi et al., 2002).

It is now common for valuation research to use qualitative methods to identify valued services and develop stated-choice questionnaires. Valuation questionnaires pose a cognitive problem to respondents, and the design of the questionnaire may facilitate or detract from respondents' solutions to the problem (Sudman et al., 1996; Tourangeau et al., 2000). Focus groups and individual interviews are both effective in understanding ecosystem services and the valuation problem from respondents' points-of-view (Johnston et al., 1995; Kaplowitz and Hoehn, 2001). Draft questionnaires may be tested and refined through individual pretest interviews, followed by careful debriefing by interviewers especially trained to identify questionnaire miscues (Kaplowitz et al., 2003).

The development of a questionnaire can be problematic with regard to obtaining the information necessary to explain the change in an aquatic ecosystem in lay terms. In the case of potential groundwater contamination, it may be difficult to develop the probability that an aquifer will become contaminated and even more difficult to inform individual survey respondents of the likelihood that their wells will become contaminated. Poe and Bishop (1999) demonstrated that this type of respondent-specific information is crucial to the development of valid value estimates. There are also cases in which respondents might reject a valuation scenario outright. Using Lake Onondaga in Syracuse, New York, as an example, the long-term contamination of this site and the severity of the contamination might lead survey respondents to reject any scenario that elicited values for cleaning up pollution damages.

Having noted and provided some examples of the limitations of stated-preference methods however, the vast number of stated-preference methods in the literature is testimony to the wide array of aquatic ecosystem applications in which contingent valuation and conjoint analysis can be employed. Nevertheless, it is also important to note that much of the criticism of stated-preference methods has arisen because they are not based on actual behavior (e.g., Diamond and Hausman, 1994; Hanemann, 1994; Portney, 1994). The debate has centered mainly on the validity of employing contingent valuation techniques to estimate nonuse values (NOAA, 1993). In contrast, the validity of conjoint estimates of value is a relatively unexplored area of research. However, there is a basic concern regarding the accuracy of stated-preference estimates of value. Do stated-preference methods result in overestimates of value? Studies conducted in controlled experimental settings suggest that both contingent valuation and conjoint methods may overestimate values (Boyle, 2003; Cummings and Taylor, 1998, 1999). Although this concern exists, the absolute magnitude of overestimation has not been established, nor has if been established that this error is any greater that the errors identified for stated-preference methods elsewhere in this chapter.

Another issue that has not received enough attention in the stated-preference literature concerns the accuracy of this approach and what level of accuracy is acceptable. Whereas stated-preference methods have been criticized because experimental design features affect value estimates, context effects have been largely ignored in revealed-preference studies. Some of the features that are problematic in stated-preference studies (e.g., information, sequencing, starting prices) also perturb markets (Randall and Hoehn, 1996). In fact, this is essentially the substance of the marketing literature. Thus, although stated-preference methods have been much maligned, revealed-preference methods have not received the comparable scrutiny that they should receive. This dichotomy of evaluation perspectives occurs simply because stated-preference methods are based on behavioral intentions, while revealed-preference methods are based on actual behavior.

The bottom line is that some real biases have been identified in contingent valuation studies, and many of these same biases carry over to conjoint studies. These biases imply that careful study design and interpretation of value estimates are required, but these biases do not appear to be specific to aquatic ecosystem applications.

Pooling Revealed-Preference and Stated-Preference Data

A number of recent valuation studies have used both revealed-preference and stated-preference data to estimate values. These analyses have pooled travel-cost data with stated-preference data that asks respondents to reveal intended visitation under specific environmental conditions (Adamowicz et al., 1994; Cameron, 1992). Pooling involves taking data from different valuation methods and using the combined data, typically from two valuation methods, to estimate a single model of preferences. Travel-cost data provide information on people's actual choice to inform the model estimation, but respondents may not have experienced the new environmental condition to be valued. These studies have used a hypothetical scenario to elicit statements of behavior, not willingness to pay, if the new condition occurred. These stated behaviors are added to the travel-cost data to estimate the preference model. This type of stated-preference data is sometimes referred to as "behavioral intentions." Some studies have framed the behavioral intention questions similar to contingent valuation questions, and visitation—not a dollar value—is the requested response (Cameron, 1992). Other studies have framed the behavioral intention question in a conjoint framework, asking people to indicate what type of trip they would take given the levels of different trip attributes (Adamowicz et al., 1994). The advantage of data pooling is the consistency imposed by actual choices, and the stated-preference data allow for environmental conditions where revealed behavior does not exist.

Cameron et al. (1996) used data pooling to investigate the values people place on recreation in the rivers and reservoirs in the Columbia River Basin. Data pooling was necessary because the policy question required values for water levels that were not represented in the current management regime. They found that the average consumer value for a flow management that enhanced recreation was about $72 per person for the months of July and August. If, however, the management strategy changed to facilitating fish passage for migration and spawning, the consumer value estimate fell to $40.

Almost all of the data-pooling studies to date have been conducted in the context of valuing sportfishing on freshwater lakes and rivers. The primary motivation has been to develop values where long-term contamination precludes the use of revealed-preference data to estimate values for ecosystem losses or improvements. The committee feels that these types of valuation studies will become more prevalent in the future. The issues discussed for the travel-cost method and stated-preference methods still persist in these analyses. In addition, another important issue arises that can substantially affect value estimates. That is, the empirical investigator must decide what weight to place on the stated-preference data and the revealed-preference data in the model estimation.

The existing literature has largely ignored this important issue.

Benefit Transfers

It is impossible to discuss economic valuation methods without also discussing benefit transfers. A benefit transfer is the process of taking an existing value estimate and transferring it to a new application that is different from the original one (Boyle and Bergstrom, 1992). There are two types of benefit transfers, value transfers and function transfers. A value transfer takes a single point estimate, or an average of point estimates from multiple studies, to transfer to a new policy application. A function transfers uses an estimated equation to predict a customized value for a new policy application. Benefit transfers are commonly used in policy analyses because off-the-shelf value estimates are rarely a perfect fit for specific policy questions. The EPA, recognizing the practical need to conduct benefit transfer, has developed the only peer-reviewed guidelines for conduct of these analyses (EPA, 2000a).

However, the committee does not advocate the use of benefit transfers for many types of aquatic ecosystem service valuation applications. First, with the exception of a few types of applications (e.g., travel-cost and contingent valuation estimates of sportfishing values), there are not a lot of studies that have investigated values of aquatic ecosystem services. Second, most nonmarket valuation studies have been undertaken by economists in the abstract from specific information that links the resulting estimates of values to specific changes in aquatic ecosystem services and functions. Finally, studies that have investigated the validity of benefit transfers in valuing ecosystem services have demonstrated that this approach is not highly accurate (Desvouges et al., 1998; Kirchhoff et al., 1997; Vandenberg et al., 2001). Because benefit transfers involve reusing existing data, a benefit transfer does not provide an error bound for the value in the new application after the transfer. For these reasons, benefit transfer is generally considered a "second best" valuation method by economists. The three studies cited above not only investigate the accuracy of benefit transfer, but also provide an idea of how large the error might be in using a benefit transfer to value aquatic ecosystem services.

As stated previously, the purpose of this chapter is to lay out carefully the currently available basic nonmarket valuation approaches, whereas the purpose of the report as a whole is to facilitate original research and studies that will develop a closer link between aquatic ecosystem functions, services, and value estimates that ultimately lead to improved environmental decision-making. The committee recommends that although benefit transfer is in common use, it should be employed with discretion and caution. Future research should focus on enhancing the reliability of off-the-shelf value estimates that are available for use in benefit transfer applied to valuing the services of aquatic ecosystems.

Replacement Cost and Cost of Treatment

In circumstances where an ecological service is unique to a specific ecosystem and is difficult to value by any of the above methods, and there are no reliable existing value estimates elsewhere to apply the benefit transfer approach, analysts have sometimes resorted to using the cost of replacing the service or treating the damages arising from loss of the service as a valuation approach.

Such an approach to approximating the benefits of a service by the cost of providing it is not used exclusively in environmental valuation. For example, in the health economics literature this approach is referred to as "cost of illness" (Dickie, 2003). This involves adding up the costs of treating a patient for an illness as the measure of benefit. Such an approach is not preference-based and is not a measure of economic value. If the treatment is not fully successful, then the patient might be willing to pay even more to avoid or treat an illness. On the other hand, market disturbances, often caused by government policies, might create conditions where more service is provided than an individual is actually willing to pay for. This information should be on the cost side of the benefit-cost ledger, not counted as a benefit.

Because of the lack of data for many ecological services arising from aquatic ecosystems, valuation studies may consider resorting to a similar *replacement cost* or *cost of treatment* approach. For example, the presence of a wetland may reduce the cost of municipal water treatment for drinking water because the wetland system filters and removes pollutants. It is therefore tempting to use the cost of an alternative treatment method, such as the building and operation of an industrial water treatment plant, to represent the value of the wetland's natural water treatment service. As with the health example, this is not a preference-based approach, and does not measure value; it is the cost of providing the aquatic ecosystem service that people value.

In general, economists consider that the replacement cost approach to estimating the value of a service should be used with great caution if at all. However, Shabman and Batie (1978) suggest that this method can serve as a last resort "proxy" valuation estimation for an ecological service if the following conditions are met: (1) the alternative considered provides the same services; (2) the alternative used for cost comparison should be the least-cost alternative; and (3) there should be substantial evidence that the service would be demanded by society if it were provided by that least-cost alternative. In the absence of any information on benefits, when a decision has to be made to take some action, then treatment costs become a way of looking for a cost-effective policy action.

Chapter 5 (see also Chapter 6) provides a case study discussion of the provision of clean drinking water to New York City by the Catskills watershed, in which the decision to restore the watershed was based on a comparison of the cost of replacing the water purification services of the watershed with a new drinking water filtration system. Thus, this application of the replacement cost method appears to fulfill the criteria of appropriate use of this method for valuation as suggested by Shabman and Batie (1978).

Summary of Valuation Approaches and Methods: Pros and Cons

Thus far, this chapter has discussed a variety of environmental valuation methods and provided some examples of their application to aquatic ecosystem services. Table 4-2 summarizes this discussion of nonmarket valuation method and approaches and their applicability to key aquatic ecosystem services. The last column in Table 4-2 is perhaps the most important link in moving from this chapter to Chapter 5 because it identifies ways that aquatic ecosystem services have been included in empirical valuation studies to date.

For revealed-preference methods, the key issue is whether ecosystem services affect peoples' behavior. If a service of an aquatic ecosystem does not affect peoples' choices, there are three alternative means of addressing this in a valuation analysis.

1. The service that does not affect site choice may affect a service that does affect site choice. In this case, ecological modeling is needed to establish the link between services, which is the essence of the production function approach.
2. Another valuation approach may be needed. For example, if a wetland provides filtration to yield potable groundwater, then a RUM is not the approach to capture this value. The value of potable groundwater might be better estimated using a hedonic model or a stated-preference study.
3. If currently available methods of economic valuation or ecological knowledge are not capable of modeling the ecosystem service relationship of interest, then consideration of the service has to be acknowledged outside the empirical benefit analysis.

Although the above conditions apply to all revealed-preference methods discussed in this chapter, they are best illustrated in conjunction with the production function approach. As discussed earlier, the production function approach is reliant on actual market behavior or value estimates from revealed-preference or stated-preference studies. This approach is important because many changes in important functions and service of aquatic ecosystems do not directly affect humans (e.g., water quality and habitat changes that influence coastal and riparian fisheries; eutrophication; biological invasions). The production function approach is therefore a means of identifying values for these indirect relationships. However, to date, the applicability of production function approaches has been limited to a few types of aquatic ecosystem services, such as habitat effects on fisheries, coastal erosion, lake habitat quality, and the resilience of aquatic systems to invasive species. There are two reasons for this. First, for this approach to be applied effectively, it is important that the underlying ecological and economic relationships are well understood. Unfortunately, our knowledge of the ecological functions underlying many key aquatic ecosystem services is not fully developed (see Chapter 3). Second, effective applica-

TABLE 4-2 Integrating Nonmarket Valuation Methods of Aquatic Ecosystem Applications

Valuation Methods	Types of Values Estimated	Common Types of Applications	Ecosystem Services
Travel cost	Use	Recreational fishing	Site visitation Fish catch rates Fish consumption advisories
Averting behavior	Use	Human health	Waterborne disease Toxic contamination
Hedonics	Use	Residential property	Proximity (distance) to aquatic ecosystems Water clarity Various measures of water chemistry (e.g., pH, dissolved oxygen) Area of aquatic ecosystems proximate to a property
Production function	Use	Commercial and recreational fishing; Hydrological functions; Residential property; Ecological-economic modeling of the effects of invasions	Habitat-fishery linkages Water quality-fishery linkages Habitat restoration Groundwater recharge by wetlands Biological invasions Eutrophication Storm protection
Stated preferences	Use and nonuse	Recreation, Human health and any other activity, including passive use, that affects peoples' economic values	Groundwater protection Wetland values Sportfishing Waterfowl hunting
Benefit transfer	Use and nonuse	Recreation and passive use	Sportfishing

tion of production function approaches also requires detailed knowledge of the physical effects on production of changes in the ecological service. Threshold effects and other nonlinearities in the underlying hydrology and ecology of aquatic systems, and the need to consider trade-offs between two or more environmental benefits generated by ecological services, complicate this task. Recent progress in developing dynamic production function approaches to modeling ecosystem services, such as habitat-fishery linkages and integrated ecological-economic analysis to incorporate multiple services and environmental benefit trade-offs, have illustrated that the production function approach may have a wider application to valuing the services of aquatic ecosystems as our knowledge of the ecological, hydrological, and economic features of these systems improves.

In comparison to revealed-preference methods, stated-preference methods exhibit the following advantages, they are: (1) the only methods available for estimating nonuse values; (2) employed when environmental conditions have not or cannot be experienced so that revealed-preference data are not available; and (3) used to estimate values for ecosystem services that do not affect peoples' behavior.

The first advantage is quite obvious, nonuse values by definition do not have a behavioral link that would allow a revealed-preference method to be employed. People do not have to exhibit any type of use behavior or monetary transaction to hold nonuse values. More importantly, a second advantage of stated-preference approaches is that they can be employed in situations where people may not have experienced the new environmental condition. For example, Lake Onondoga in New York has experienced sufficient long-term contamination to preclude uses such as fishing and swimming. Thus, it would be impossible to estimate travel-cost models for these activities. However, it might be possible to develop a stated-preference survey to elicit values if it were possible to improve water quality in the lake. Finally, there may be ecosystem services that serve important ecological functions (see Tables 3-2 and 3-3), but do not affect peoples' use of aquatic ecosystems in a directly observable manner. If the ecological link were explained to people it might be possible to use a stated-preference study to elicit values for such services. For example, people might not understand the role that wetlands play in the purification of groundwater recharge from surface waters. It would be possible, however, to design a stated-preference study to elicit values for the protection of wetlands to protect water purification services.

Despite these advantages of stated-preference methods, the above discussion highlights a number of concerns and problems identified in the literature, including issues of identifying the relevant ecological services, questionnaire development, overestimation of values, and issues of accuracy. However, in some instances, criticisms of stated-preference methods have arisen simply because they are based on behavioral intentions, and they have been scrutinized more carefully than revealed-preference methods, which are based on actual behavior. As the committee has sought to indicate in this chapter and summarized in Table 4-2, both revealed- and stated-preference methods have their advantages and disadvantages, and the choice of method will depend largely on what aquatic ecosystem service is being valued, as well as the policy or management issue that requires valuation.

Lastly, it is important to recognize that each of the economic valuation methods reviewed in this chapter can result in an overestimate or underestimate of individual values for a specific application. Before any empirical study is used in a policy application it is important for the analyst to consider whether the point estimate(s) used underestimate or overestimate the "true" value (see Chapters 6 and 7 for further information).

APPLICABILITY OF METHODS TO VALUING ECOSYSTEM SERVICES

Given the wide variety of economic methods that are currently available to value aquatic ecosystem services, it may be useful to examine how various methods could be used to value a range of services provided by a single but vitally important aquatic ecosystem. One such ecosystem that has generated several valuation studies of key ecological services is the Great Lakes. The following section reviews these Great Lake studies as an illustration of many of the nonmarket valuation methods and approaches described in this chapter.

Valuation Case Study: The Great Lakes

The Great Lakes ecosystem covers 94,000 square miles (see also Box 3-2). Collectively, the tributaries to the five Great Lakes drain a territory of 201,000 square miles. Key native species include black bear, bald eagle, wolves, moose, lake trout, and sturgeon, and the lakes surround major migratory flyways for waterfowl, songbirds, and raptors. Thirty-three million people live within the ecosystem and tourism is a major industry year-round. Recreational fishing is annually a multibillion-dollar activity in the regional economy.

In the last 50 years, regional economic changes and pollution control have restored much of the natural beauty of the Great Lakes. However, restoring the ecosystem functions of the Great Lakes remains a priority. Invasive species, such as zebra mussels and lamprey, and exotic fish, such as ruffe and goby, continue to displace and threaten native species. Significant efforts are under way to strengthen populations of Lake Superior walleye, native clams, brook trout, and sturgeon populations.

The ecosystem is also challenged by its industrial history. There are more than 30 areas of concern (AOCs) within the Great Lakes that are burdened with tons of toxic materials (International Joint Commission, 2003). These areas tend to be old industrial areas, harbors, and shipping points. While the mean concentrations tend to be low, these toxic contaminants are typically ingested by small organisms that are in turn successively eaten by other larger organisms. At each stage of the food web, these concentrations become more elevated. The results are excessive (toxic) concentrations of metals and PCBs in fish, waterfowl, and birds of prey. For example, fish consumption advisories for recreational anglers remain in effect in many popular fishing areas across the region.

Like its biological features, the physical character of the Great Lakes ecosystem changes over time. Water levels and volumes have steadily increased over thousands of years (Lewis, 1999), but water levels over the course of decades fluctuate by several feet (Boutin, 2000). The rocky, high shorelines on Lake Superior are fairly stable from a human perspective, but the softer, aggregate and sandy shorelines are susceptible to short-term flooding and long-term erosion. Living in a dynamic ecosystem poses economic risks for managing longer-term investments such as housing, harbor structures, bridges, and roads.

The following three sets of studies address these management issues. The first examines the economic benefits of controlling an exotic species that preys on native fish. The second examines the damages from PCB concentrations in Wisconsin's Fox River, one of the ecosystem's 31 areas of concern. The third explores the economic consequences of ecosystem changes over time.

Controlling an Exotic Species: Sea Lamprey Invasion

Sea lampreys are nonnative, eel-like fish that prey on lake trout, sturgeon, salmon, and other large fish in the Great Lakes. Lampreys attach themselves to prey and feed on the bodily liquids of the host fish. The host fish usually dies from infection after the lamprey feeds and detaches. Lamprey were first observed in Lake Ontario in the 1800s and arrived in Lake Michigan by the 1930s (Peeters, 1998).

Lake trout are particularly susceptible to lamprey predation. By the 1950s, lampreys had almost eliminated the self-sustaining lake trout populations in Lake Michigan and Lake Huron (Peeters, 1998). Since the 1950s, vigorous control programs have reduced lamprey populations by 90 percent and led to the restoration of lake trout in Lake Michigan (Great Lakes Fishery Commission, 2002).

However, the lamprey population remains high in Lake Huron. The St. Mary's River is the major uncontrolled spawning area on Lake Huron. The size and volume of the St. Mary's made past control efforts ineffective. Recent improvements in control technology promise much better results at lower costs (Gaden, 1997). An analysis was completed to determine whether the control costs were in line with the recreational fishing benefits of lake trout restoration. The Michigan angling demand model is a statewide travel-cost model of anglers' choices (Hoehn et al., 1996). The model divides the 30-week, non-winter fishing season into 60 fishing choice occasions. Within each occasion, anglers choose whether to go fishing and, if they do, whether they take a day trip or a multiple-day trip. Anglers also choose one of 12 different fishing types, such as cold-water Great Lakes fishing, and fishing location by destination county. Destinations vary in quality by catch rate and other features relevant to fishing choices. In all, the model incorporates 850 distinct choices on each choice occasion.

The model was estimated using a repeated logit statistical framework and data on anglers' choices (Hoehn et al., 1996). The data were obtained from a sample of more than 2,000 Michigan anglers. Sampled anglers were selected randomly from the general population to ensure that the data represented the broad spectrum of Michigan anglers. The sampled anglers were contacted initially at the beginning of the fishing season and then interviewed again (at least) several times over its course. The serial interview approach was used to minimize errors that arise when anglers try to remember a long series of trips. Anglers were also provided with fishing logs to keep track of their trips. Anglers who took frequent trips were interviewed more frequently.

The model estimated the probability of choosing a particular fishing location and type of fishing trip. Trip choices were a function of the distance and travel cost to the location and the quality of fishing. The model was used to estimate benefits for policies that might change fishing quality at a particular site and aggregation of sites, such as inland regions and lakes. For example, an initial analysis indicated that a 10 percent improvement in Lake Michigan and Lake Huron salmon and trout catch rates would result in angler benefits of $3.3 million per year (Lupi and Hoehn, 1998). The analysis considered three alternative ways of controlling lamprey in the St. Mary's River: (1) annual lampricide treatment, (2) annual lampricide and a one-time release of sterile males, and (3) annual lampricide and sterile male release every five years. Treatment costs were several times higher with the third treatment relative to the first, while the trout population and catch rates were only 30 percent higher. Trout populations and catch rates were forecast to increase by 30 to 45 percent in northern Lake Huron and 3 to 7 percent in the central and southern portions of the lake.

The Michigan travel-cost model was used to calculate the benefits of permanent programs of lamprey control using the three different treatments. As the trout population recovers, the third program of continuing lampricide and sterile male releases results in the greatest annual benefits, while the lampricide-only program has the lowest level of annual benefits. However, costs increased with each sterile male release. Although costs increased with treatment, benefits also varied with the geography of catch rate impacts.

Catch rate increases were greatest in the northern region where fewer anglers live and the least in southern Lake Huron nearer the urban areas of Macomb and Wayne Counties in Michigan. As a result, the improvements in catch rates were forecast to occur in areas relatively distant from users. Annual benefits were calculated to be almost twice as large as in the forecast case if the catch rate increase was equal to the same mean but evenly distributed across the entire lake. The result showed that use values decline as the improvement in services was more distant from the users.

The economic outcome of each control alternative was evaluated by examining net benefits. Net benefits were calculated as the present value of benefits minus the present value of costs. Net benefits were positive for each alternative. Using discount rates (see Chapters 2 and 6 for further information) between 3 and 4 percent, net benefits were greatest for annual lampricide and a one-time release of sterile males to quickly reduce the breeding population of lamprey. Net benefits for the first and third alternatives were about the same, meaning that the benefits of continuing sterile male release after the first treatment were just about offset by the costs.

Fox River Damage from PCBs

The Fox River enters Green Bay, Wisconsin, on the northwestern shoreline of Lake Michigan. It is the lake's largest tributary. Water, waterpower, and nearby forests supported the early development of the paper industry. By the 1950s, the local paper industry focused on the production of carbonless copy paper. A by-product of its production was the discharge of thousands of pounds of PCBs annually. An estimated 700,000 pounds of PCBs entered the Fox River before PCB use was stopped nationally in 1971. About 20 percent of the PCBs have been deposited in Green Bay and Lake Michigan (Wisconsin DNR, 2001).

Although the human health effects of PCBs are difficult to quantify and measure, the EPA has determined that PCBs cause a range of adverse health effects in animals and that there is "supportive evidence potential carcinogenic and non-carcinogenic effects" in humans (EPA, 2003). To avoid potential adverse health effects in humans, the State of Wisconsin advises anglers to limit their consumption of fish and to prepare fish for consumption so as to avoid fatty tissue that biomagnifies PCBs (Wisconsin DNR, 2001). The primary human use damages are the limitations on eating fish and the increased health risks for anglers and others who choose to eat the fish. Nonuse damages include the impacts on ecosystem functions and other native organisms.

The Wisconsin Department of Natural Resource is conducting a series of studies to estimate economic damages resulting from PCB contamination (Bishop et al., 2000; Breffle et al., 1999; Stratus, 1997). Initial studies focused on injuries to ecosystem functions and services through systematic data collection and analysis (Stratus, 1997). In many cases, it was possible to detect a type of injury but not to quantify its impact on a particular ecosystem service. For instance, PCBs were suspected of injury to fish populations, but it was not possible to quantitatively translate population injuries into estimates of changes in catch rates for sport and subsistence anglers.

The uncertainties regarding service flow injuries led several investigators to two types of damage estimation studies. The first study (Breffle et al., 1999) combined the travel-cost method with stated-preference analysis to estimate use values for anglers. Fishing services to anglers were impaired as a result of both fish consumption advisories (FCAs) and the elevated health risk of eating local fish that FCAs imply. Previous research demonstrated that fishing behaviors change and fishing benefits are reduced by FCAs. The second study (Bishop et al., 2000) used stated-preference analysis to estimate the total values of damages for households in the region. Total value was the sum of both use value damages for anglers and nonuse damages for all households in the study area.

Pollution Damages to Recreational Fishing

Breffle et al. (1999) designed a study to estimate the damages to anglers due to FCAs that applied to the Fox River and Green Bay as a result of past PCB releases. Damage estimates were derived from the loss of enjoyment of fishing in an area covered by an FCA and the loss of well-being as a result of fishing at another site, perhaps not covered by an FCA. The study held the number of days of fishing constant at the current, estimated level and did not attempt to estimate damages due to the reduction in the amount of overall fishing.

The analysis estimated the economic demand for fishing as a function of travel cost, whether an FCA was in force at a given site, and other fishing site quality variables. The FCA effect on demand allowed researchers to estimate the shift in fishing demand and the change in consumer value due to presence of the FCA. The reduction in value served as the measure of damages to angling use services.

Data for estimating the demand model were obtained through telephone and mail surveys. The telephone survey used random sample methods to contact a total of 3,190 anglers in northeastern Wisconsin. Respondents were asked to think back over the 1998 angling season and recall their fishing activities. Based on respondents' recollections, the interviewers obtained data on total days spent fishing during 1998, number of days spent fishing in the study area, and attitudes about actions to improve fishing. The mail survey asked respondents to make stated-preference choices across fishing sites that varied in quality. The combined data set allowed researchers to estimate a random utility model of fishing demand conditional on the presence or absence of FCAs in the study area.

The analysis estimated that the 48,600 anglers in the study area fished a total of 641,000 days in 1998. The mean value of damages was $4.17 per trip (1998 dollars). The present value of fishing use damages was estimated to be $148 million for a baseline scenario in which natural processes required 100 years to reduce PCBs to levels where FCAs are unnecessary. Restoration efforts that reduced recovery time to 40 years reduced damages to $123 million, resulting in benefits of $25 million. Restoration efforts that reduced recovery time to 20 years reduced damages to $106 million, resulting in cleanup benefits of $42 million.

Total Value of Lost Ecosystem Services

Bishop et al. (2000) investigated the total value of ecosystem services lost due to PCB contamination of the Fox River and Green Bay. That study examined the monetary value of damages as well as the in-kind restoration programs that residents might view as alternatives to removing and containing PCBs. Alternative restoration choices included projects to remove PCB-laden sediments, restore wetlands, enhance recreation, and reduce nonpoint source pollution.

Stated preferences for the restoration alternatives were elicited in a random sample, mail-based survey of 470 households in the study area. The survey questionnaire presented PCB removal as one of several projects to improve natural resources in northeast Wisconsin. The questionnaire also presented six alternative pairs of natural resource programs. Each program within a pair offered different levels of PCB removal, wetland restoration, recreation enhancement, pollution control, and annual tax cost per household.

Respondents were asked to consider each pair and identify their preferred program for each pair. Factorial design methods were used to vary the plans and costs across respondents in the sample. A probit-type discrete choice statistical model was used to estimate the influence of restoration and tax cost on the probability of acceptance. The probit model parameters were then used to calculate willingness to pay a tax cost as a function of the quality of restoration.

The estimates showed that wetlands restoration, improvements in recreational facilities and nonpoint pollution control were poor substitutes for removing and safely containing the PCB-laden sediments. Setting the wetland, recreation, and pollution projects at their maximum levels made up for only 40 years of PCB damages. Natural processes alone were expected to take more than 100 years to reduce PCBs to safe levels.

The present value of PCB damages was estimated to be $610 million (1999 dollars). A restoration that reduced PCBs to safe levels in 40 years resulted in benefits of $248 million by reducing PCB damages to $362 million over the 40-year cleanup interval. An intensive restoration that reduced PCBs to safe levels in 20 years resulted in benefits of $356 million by reducing damages to $254 million over the 20-year cleanup interval.

The final step in the analysis compared the estimated total ecosystem damages with fishing use damages for the 11 percent of households that included at least one angler. This comparison found that estimated total values were 8 to 28 percent greater than use values alone, suggesting that nonuse value was about 8 to 28 percent of use value in angler households.

Lakeshore Erosion

Shoreline erosion offers a short-term laboratory for examining the economic consequences of aquatic ecosystem change. As noted previously, shoreline is valued by property owners for its views, for its proximity to water, and as a location for residential and commercial structures and development. Erosion rates of one to three feet per year do not appreciably affect the amount of shoreline for views, and proximity views and are passed on to the adjacent parcels.

However, erosion does pose a risk of loss of residential and commercial structures, and reducing the risk of loss involves a number of trade-offs. Structures degrade from use and changes in technology in a manner analogous to automobiles and machinery. Locating newly constructed structures far enough away from the existing shoreline so that a building is dilapidated and obsolete before it is threatened by erosion can minimize the risk of erosion to the structure. Increasing the distance to the shore, however, reduces amenities such as panoramic views and increases the time required to get to the beach. Thus, there is a trade-off between the value of these amenities and the economic risk of erosion.

Erosion may be offset for existing structures by physical protection. Rock and concrete armoring protects the shoreline to some extent. However, wave action will eventually undercut such protection. Eroded beaches may sometimes be maintained by dredging offshore sand deposits and using them to replace eroded material. These types of physical protection measures, however, may have impacts on shoreline and coastal ecosystem functions. For instance, armor may reduce erosion of the shoreline, while also reducing sand and sediment flows along the shoreline. Reduced material flows may increase erosion or reduce beach accretion in nearby, unprotected shoreline areas (USACE, 2000).

Economic processes may moderate the risk of erosion to manmade structures by spreading out its consequences over time. In this regard, markets in real property tend to be forward-looking. If there are significant risks from erosion over time, these may be gradually entered into the prices of properties as the risks increase. Buyers are likely to pay more for lower-risk properties and less for higher-risk properties. Property owners may sell a property before the erosion discount becomes higher than the value they place on being near the shore. The annual incremental discount associated with erosion risk might be viewed as part of the cost of a shoreline property, similar to the ordinary costs of depreciation and obsolescence.

Two studies use hedonic methods to examine the impact of erosion risk on the values of shoreline, residential properties. The first examined shoreline property values on Lake Erie (Kriesel et al., 1993), and the second combined data for homes on both Lake Erie and Lake Michigan (Heinz, 2000).

Both studies estimated hedonic regressions where the dependent variable was the logarithm of the sales prices of an individual residential property and the independent variables were the physical characteristics of the property. Physical characteristics included features such as floor area of the structure, parcel size, number of rooms, number of bathrooms, and erosion risk. Erosion risk was measured by the estimated number of years until the shoreline reached the leading, shoreward edge of a structure. The Lake Erie study analyzed data for approximately 300 structures. The combined study used data for 139 structures from the Lake Erie study and data obtained in a mail survey for about 150 Lake Michigan residences.

The results of the two hedonic analyses show that residential property markets are, indeed, forward looking. The major share of erosion's economic cost is incurred long before the actual loss of a residential structure. One way to illustrate this impact uses the estimated hedonic coefficients to calculate the percentage change in property values as years to erosion loss decline. As time to loss declines by 1 year, the property value of a home with a loss in 100 years is discounted by about one-tenth of a percent of its value. At 60 years, a home has lost an accumulated 20 percent of its value due to erosion risk and loses further value at the rate of about 0.6 percent per year. At 20 years, the cumulative discount is 40 percent of the value at 100 years, and the annual discount rate is about 2 percent. At 10 years, the residence is discounted by 60 percent relative to a structure with a risk of 100 years to loss, and the annual rate of loss is 5 percent. At 5 years to loss, a residential structure has lost more than 70 percent of its value relative to the same structure with 100 years to loss.

The analyses show that the cost of erosion is incurred gradually over a long period of time. More than 60 percent of the value of a residence is lost before a residence is within 10 years of the date of its estimated loss. The annual cost of erosion is about $1,400 for a $500,000 residence with an erosion risk of 100 years. For the same structure, the annual cost is about $2,500 at 50 years, $10,400 at 10 years, and $18,400 at 5 years.

Valuation Case Study: Conclusions

The above studies from the Great Lakes ecosystem illustrate both the strengths and the weaknesses of different valuation methods and approaches. First, the studies show that valuation is a useful tool for assessing a wide range of ecological services and key policy issues concerning management of the Great Lakes, including control of a damaging biological invasion, water pollution by toxic waste, pollution damages to recreational fishing, and the impacts of shoreline erosion. As the extended case study demonstrates, a variety of nonmarket valuation methods are available for assessing these ecosystem management concerns, and if applied correctly, they can yield reliable estimates of the value of key aquatic services. If valuation methods can be applied successfully to a complex and geographically extensive aquatic ecosystem such as the Great Lakes, then nonmarket valuation can also be implemented for equally important aquatic ecosystems elsewhere.

Second, the studies illustrate some of the limitations of revealed- and stated-preference valuation methods discussed earlier in the chapter. For example, the applicability of revealed-preferences methods of valuation depends on whether the ecological service affects peoples' behavior, and whether both the changed environmental condition and the resulting modification in human behavior can be directly or indirectly observed. Thus, for example, the effect of the lamprey invasion could be assessed only in terms of the impact on the recreational fishing benefits of lake trout restoration, which in turn was assessed through the application of a travel-cost model to calculate the possible benefits of alternative lamprey control programs. Clearly, such a valuation estimate can capture only one of many possible complex ecological and economic impacts of the lamprey invasion, although in this instance assessing this recreational benefit was sufficient to determine that the net benefits of lamprey control were positive for all treatments and to identify the preferred treatment method. Similarly, various studies of the health impacts of PCB contamination in the Fox River indicate that the lack of ecological data meant that it was not always possible to quantify how damages to fish populations translate into estimates of changes in catch rates for sport and subsistence anglers, thus limiting reliance on the travel-cost method alone as a method of valuing such impacts. Instead, researchers had to rely either on combined travel-cost and stated-preference methods or on stated-preference methods alone to estimate the total values for households in the region. Although the latter study attempted to separate the households' estimates of use values compared to nonuse values in their overall valuation of the benefits of PCB removal, some of the concerns about the validity of employing contingent valuation techniques to estimate nonuse values may be applicable in this case (NOAA, 1993).

ISSUES

In describing and discussing currently available nonmarket valuation methods and their applicability to aquatic ecosystem services, a number of key issues have emerged, these include assessing ecological disturbance and threshold effects, limitations to ex ante and ex post valuation, partial versus general equilibrium approaches, and the problem of scope. The following section discusses each of these issues in turn.

Ecological Disturbance and Threshold Effects

Severe disturbance of an aquatic ecosystem may lead to an abrupt, and possibly very substantial disruption in the supply of one or more ecological services (see Chapter 3 for further information). This "break" in supply is often referred to as a *threshold effect*. The problem for economic valuation is that before the threshold is reached, the marginal benefits associated with a particular ecological service may either be fairly constant or change in a fairly predictable manner with the provision of that service. However, once the threshold is reached, not only may there be a large "jump" in the value of an ecological service, but how the supply of the service changes may be less predictable. Such ecosystem threshold effects pose a considerable challenge, especially for ex ante economic valuation using revealed-preference methods—that is, when one wants to estimate the value of an ecological service that takes into account any potential threshold effects. Since such severe and abrupt changes have not been experienced, peoples' choices in response to them have not been observed. This means that stated-preference methods are the only tool for measuring such values, but there are two complications that warrant discussion.

The first is that there is likely to be considerable uncertainty surrounding both the magnitude and the timing of any threshold effect associated with ecosystem disturbance. Thus, the ecological information may not be available to accurately develop a scenario to describe the ecosystem change in a stated-preference survey. In such a case, a stated-preference survey might be designed to value a variety of plausible ecosystem changes so that it is possible to describe the sensitivity of value estimates to likely outcomes.

The second complication may be that survey respondents will simply reject the valuation scenario as implausible or unbelievable. A large-scale oil spill is one example when survey respondents may reject the valuation scenario out of hand and state that the responsible company should pay for damages, not the general public. Carson et al. (1992) avoided this problem by asking survey respondents to value a public program to prevent an oil spill of the magnitude of the *Exxon Valdez*. Thus, substantial creativity and design effort may be required to develop plausible stated-preference valuation scenarios for large-scale disturbances to aquatic ecosystems that have threshold effects.

Threshold effects can also occur in peoples' preferences. Over some range of change in ecosystem services, marginal values may be quite small, but change dramatically when a drastic change occurs (e.g., listing of an aquatic species as endangered). This suggests that threshold changes in an aquatic ecosystem may stimulate threshold changes in preferences. This issue further complicates the valuation of threshold changes because stated-preference valuation methods must be designed to convey the threshold change and motivate people to think how their values would change with the different set of relative prices that would be present after the ecosystem threshold change occurs.

Limitations of Ex Ante and Ex Post Valuation

The limitations of ex ante valuation using stated-preference methods and real choices are not limited to large-scale, threshold effects. There are many common instances in which people may not have experienced an ecological improvement or degradation and revealed-preference valuation methods are not applicable. Although stated-preference methods are applicable to such changes, it may be difficult for individuals to value trade-offs implied by changes they have not personally experienced. Thus, while stated-preferences are very helpful for ex ante valuation, they are not a complete or infallible solution. There will be circumstances in which nonmarket valuation methods cannot develop accurate value estimates in an ex ante setting.

In the ex post situation, the change has been observed but does not always translate to the revealed choices. For example, the market price of fish may reflect a change in the underlying ecological service, such as the loss of coastal nursery grounds, and thus, there appears to be no value assigned to this ecosystem service. Again, stated-preference methods are the alternative, but they may not be applicable in all situations.

Partial Versus General Equilibrium Approaches

Most valuation methods and valuation studies represent a partial equilibrium approach to a particular policy question. However, as is clear from Chapter 3, the ecological functioning and dynamics that result in most aquatic ecosystem services suggest that to more fully capture the affects of ecosystem changes on the provision of these services, a more general equilibrium approach may be required. A series of independent value estimates for different ecosystem services, when added together, could substantially understate or overstate the full value of changes in all services. The key issue is whether there is substitute or complementary relationships between the services (Hoehn and Loomis, 1993).

As discussed above, there have been a number of recent attempts to use such an approach, or integrated economic-ecological modeling, to value various services of aquatic ecosystems. In essence, these approaches represent the extension of the production function approach to a full ecosystem level.

Scope

Insensitivity to scope is a major issue in contingent valuation studies of nonuse values of ecosystem services. This issue was raised by the National Oceanic and Atmospheric Administration Panel on Contingent Valuation (1993), which stated that this problem demonstrates "inconsistency with rational choice." Insensitivity to scope is exhibited by value estimates' being insensitive to the magnitude of the ecosystems change being valued. For example, if values estimated for restoring 100 and 1,000 acres of wetlands were statistically identical, this would indicate lack of sensitivity to scope. The inconsistency with rational choice arises because it is expected that people would pay more for the larger restoration project, all other factors being equal. The basis for the NOAA panel's concern was a study by Boyle et al. (1994) who found that estimates of nonuse values were not sensitive to whether 2,000, 20,000, or 200,000 bird deaths were prevented in waste oil holding ponds. While this study was criticized in a variety of public fora, Ahearn et al. (2004) reported a similar result in another study of grassland bird numbers. Notably, this latter study generally followed the NOAA panel's (1993) guidelines for the design of a credible contingent valuation study of nonuse values.

Insensitivity to scope is a major issue for valuing aquatic ecosystems services because stated-preference methods, which include contingent valuation, are likely to be important in estimating many component values in a TEV framework. There are many instances in which there is no visible behavior that supports the use of revealed-preference methods, although two important caveats should be considered.

First, the NOAA panel focused on the use of contingent valuation to estimate nonuse values. There will be many cases in which stated-preference methods are needed to estimate use values for aquatic ecosystem services. Sensitivity to scope has been demonstrated clearly in the estimation of use values in the literature, and some of these studies are applications to aquatic ecosystems (e.g., Boyle et al., 1993). In fact, Carson (1997) provides a list of contingent valuation studies that have demonstrated scope effects when use values are involved, and the vast majority of these studies have implications for valuing aquatic ecosystem services. Moreover, Carson et al. (1996) show that contingent valuation estimates are comparable to similar revealed-preference estimates—thereby, demonstrating the convergent validity of the stated-preference and revealed-preference estimates. Thus, the literature supports the use of contingent valuation for estimating use values for aquatic ecosystem services.

The second caveat applies to the use of contingent valuation to estimate nonuse values. Although the NOAA panel stated that contingent valuation can provide useful information on nonuse values, the ability of contingent valuation methods to demonstrate scope effects has not been shown clearly in the literature. This a major concern for valuing aquatic ecosystems because nonuse values would be expected to be an important and large component of any total economic value assessment. In this regard, attribute-based, conjoint analysis provides a promising option. This approach presents the description of the aquatic ecosystem to be valued in component services and clearly informs survey respondents that there are different levels of these services. Respondents are then asked to select alternatives that differ in terms of the component services. This relative context has been shown to demonstrate scope effects (Boyle et al., 2001). The key difference is that contingent valuation has used a between-subjects design where independent samples are asked to value each of the different levels of the ecosystem. Conjoint analysis uses a within-subjects design where each respondent sees multiple levels of the ecosystem. Although a between-subjects design is appealing from an experimental design perspective, this is not the way real-world decisions are made. People make revealed choices where they observe ecosystem goods and services with different levels of attributes, and whereas conjoint analysis mimics this choice framework, contingent valuation does not. A question then arises as to what standard contingent valuation should be held. A between-subjects design to test for scope holds contingent valuation to a higher standard than market decisions are based upon (Randall and Hoehn, 1996), whereas the within-subject design of conjoint analysis mimics the relative choices that occur in markets. These results imply that conjoint analysis may be the better method to employ in estimating nonuse values for aquatic ecosystem services.

SUMMARY: CONCLUSIONS AND RECOMMENDATIONS

This chapter demonstrated that there is a variety of nonmarket valuation approaches that can be applied to valuing aquatic and related terrestrial ecosystem services.

For revealed-preference methods, the types of applications are limited to a set number of specific aquatic ecosystem services. However, both the range and the number of services that can potentially be valued are increasing with the development of new methods, such as dynamic production function approaches, general equilibrium modeling of integrated ecological-economic systems, conjoint analysis, and combined revealed- and stated-preference approaches.

Stated-preference methods can be applied more widely, and certain values can be estimated only through the application of such techniques. On the other hand, the credibility of estimated values for ecosystem services derived from stated-preference methods has often been criticized in the literature. For example, contingent valuation methods have come under such scrutiny that it led to the NOAA panel guidelines of "good practice" for these methods.

Benefit transfers and replacement cost/cost of treatment methods are increasingly being used in environmental valuation, although their application to aquatic ecosystem services is still limited. Economists generally consider benefit transfers to be a "second-best" valuation method and have devised guidelines governing their use. In contrast, replacement cost and cost of treatment methods should be used with great caution if at all. Although economists have attempted to design strict guidelines for using replacement cost as a last resort "proxy" valuation estimation for an ecological service, in practice estimates employing the replacement cost or cost of treatment approach rarely conform to the conditions outlined by such guidelines.

Although the focus of this chapter has been on presenting the array of valuation methods and approaches currently available for estimating monetary values of aquatic and related terrestrial ecosystem services, it is important to remember that the purpose of such valuation is to aid decision-making and the effective management of these ecosystems. Building on this critical point, at least three basic questions arise for any method that is chosen to value aquatic ecosystem services:

1. Are the services that have been valued those that are the most important for supporting environmental decision-making and policy analyses involving benefit-cost analysis, regulatory impact analysis, legal judgments, and so on?
2. Can the services of the aquatic ecosystem that are valued be linked in some substantial way to changes in the functioning of the system?
3. Are there important services provided by aquatic ecosystems that have not yet been valued so that they are not being given full consideration in policy decisions that affect the quantity and quality of these systems?

In many ways, the answers to these questions are the most important criteria for judging the overall validity of the valuation method chosen.

It is clear that economists and ecologists should work together to develop valid estimates of the values of various aquatic ecosystem services that are useful to inform policy decision-making. The committee's assessment of the literature is that this has not been done adequately in the past and most valuation studies appear to have been designed and implemented without any such collaboration. Chapter 5 helps to begin to build this bridge.

The range of ecosystem services that have been valued to date are very limited, and effective treatment of aquatic ecosystem services in benefit-cost analyses requires that more services be subject to valuation. Chapter 3 begins to develop this broad perspective of aquatic ecosystem services.

Nonuse values require special consideration; these may be the largest component of total economic value for aquatic ecosystem services. Unfortunately, nonuse values can be estimated only with stated-preference methods, and this is the application in which these methods have been soundly criticized. This is a clear mandate for improved valuation study designs and more validity research.

There is a variety of nonmarket valuation methods that are available and have been presented in this chapter. However, no single method can be considered the best at all times and for all types of aquatic ecosystem valuation applications. In each application it is necessary to consider what method(s) is the most appropriate.

In presenting the various nonmarket valuation methods available for estimating monetary values of aquatic and related terrestrial ecosystem services, this chapter has also sought to provide some guidance on the appropriateness of the various methods available for a range of different services. Based on this review of the current literature and the preceding conclusions, the committee makes the following recommendations:

- There should be greater funding for economists and ecologists to work together to develop estimates of the monetary value of the services of aquatic and related terrestrial ecosystems that are important in policymaking.
- Specific attention should be given to funding research at the "cutting edge" of the valuation field, such as dynamic production function approaches, general equilibrium modeling of integrated ecological-economic systems, conjoint analysis, and combined stated-preference and revealed-preference methods.
- Specific attention should be given to funding research on improved valuation study designs and validity tests for stated-preference methods applied to determine the nonuse values associated with aquatic and related terrestrial ecosystem services.
- Benefit transfers should be considered a "second-best" method of ecosystem services valuation and should be used with caution, and only if appropriate guidelines are followed.
- The replacement cost method and estimates of the cost of treatment are not valid approaches to determining benefits and should not be employed to value aquatic ecosystem services. In the absence of any information on benefits,

and under strict guidelines, treatment costs could help determine cost-effective policy action.

REFERENCES

Abdalla, C.A., B.A. Roach, and D.J. Epp. 1992. Valuing environmental groundwater changes using averting expenditures: An application to groundwater contamination. Land Economics 68:163-169.

Acharya, G. 2000. Approaches to valuing the hidden hydrological services of wetland ecosystems. Ecological Economics 35:63-74.

Acharya, G., and E.B. Barbier. 2000. Valuing groundwater recharge through agricultural production in the Hadejia-Jama'are wetlands in northern Nigeria. Agricultural Economics 22:247-259.

Acharya, G., and E.B. Barbier. 2002. Using domestic water analysis to value groundwater recharge in the Hadejia-Jama'are floodplain in Northern Nigeria. American Journal of Agricultural Economics 84(2):415-426.

Adamowicz, W.L., J. Louviere, and M. Williams. 1994. Combining revealed and stated preference methods for valuing environmental amenities. Journal of Environmental Economics and Management 26:271-292.

Ahearn, M., K.J. Boyle, and D. Hellerstein. 2004. Designing a contingent-valuation study to estimate the benefits of the conservation reserve program on grassland bird populations. In The Contingent-Valuation Handbook, J. Kahn, D. Bjornstad, and A. Alberini (eds.). Cheltenhan, U.K.: Edward Elger.

Barbier, E.B. 1994. Valuing environmental functions: Tropical wetlands. Land Economics 70(2):155-173.

Barbier, E.B. 2000. Valuing the environment as input: Applications to mangrove-fishery linkages. Ecological Economics 35:47-61.

Barbier, E.B. 2003. Habitat-fishery linkages and mangrove loss in Thailand. Contemporary Economic Policy 21(1):59-77.

Barbier, E.B., and I. Strand. 1998. Valuing mangrove-fishery linkages: A case study of Campeche, Mexico. Environmental and Resource Economics 12:151-166.

Barbier, E.B., M. Acreman, and D. Knowler. 1997. Economic Valuation of Wetlands: A Guide for Policy Makers and Planners. Geneva, Switzerland: Ramsar Convention Bureau.

Barbier, E.B., I. Strand, and S. Sathirathai. 2002. Do open access conditions affect the valuation of an externality? Estimating the welfare effects of mangrove-fishery linkages. Environmental and Resource Economics 21(4):343-367.

Bartik, T.J. 1987. The estimation of demand parameters in hedonic price models. Journal of Political Economy 95: 81-88.

Bartik, T.J. 1988. Evaluating the benefits of non-marginal reductions in pollution using information on defensive expenditures. Journal of Environmental Economics and Management 15: 111-27.

Bateman, I.J., R.C. Carson, B. Day, M. Hanemann, N. Hanley, T. Hett, M.J. Lee, G. Loomes, S. Mourato, E. Özdemiroğlu, D. Pearce, R. Sugden, and J. Swanson. 2002. Economic Valuation with Stated Preference Techniques: A Manual. Cheltneham, U.K.: Edward Elger.

Batie, S.S., and J.R. Wilson. 1978. Economic values attributable to Virginia's coastal wetlands as inputs in oyster production. Southern Journal of Agricultural Econom-

ics 10(1):111-118.
Beach, E.D., and G.A. Carlson. 1993. A hedonic analysis of herbicides: Do user safety and water quality matter? American Journal of Agricultural Economics 75(3):612-623.
Becker, G. 1965. A theory of the allocation of time. Economic Journal 75:493-517.
Bell, F.W. 1989. Application of Wetland Valuation Theory to Florida Fisheries. Report No. 95, Florida Sea Grant Program. Tallahassee: Florida State University.
Bell, F.W. 1997. The economic value of saltwater marsh supporting marine recreational fishing in the Southeastern United States. Ecological Economics 21(3):243-254.
Bishop, R.C. 1983. Option value: An exposition and extension. Land Economics 59(1):1-15.
Bishop, R.C., and M.P. Welsh. 1992. Existence values in benefit-cost analysis and damage assessment. Land Economics 68(4):405-417.
Bishop, R.C., K.J. Boyle, and M.P. Welsh. 1987. Toward total economic value of Great Lakes fishery resources. Transactions of the American Fisheries Society 116 (3):339-345.
Bishop, R.C., W.S. Breffle, J.K. Lazo, R.D. Rowe, and S.M. Wytinck. 2000. Restoration Scaling Based on Total Value Equivalency: Green Bay Natural Resource Damage Assessment. Boulder, Colo.: Stratus Consulting Inc.
Bockstael, N.E. 1995. Travel cost models. In The Handbook of Environmental Economics, D.W. Bromley (ed.). Cambridge, Mass.: Blackwell Publishers.
Bockstael, N.E., and K.E. McConnell. 1983. Welfare measurement in the household production function framework. American Economic Review 73(4):806-814.
Boutin, C. 2000. Great Lakes Water levels. Ann Arbor, Mich.: Great Lakes Environmental Research Laboratory, National Oceanic and Atmospheric Administration.
Boyce, R.R., T.C. Brown, G.D. McClelland, G.L. Peterson, and W.D. Schulze. 1992. An experimental examination of intrinsic environmental values as a source of the WTA-WTP disparity. American Economic Review 82:1366-1373.
Boyle, K.J. 2003. Contingent valuation in practice. Chapter 5 in A Primer on Nonmarket Valuation, P. Champ, K. J. Boyle and T.C. Brown (eds.). Boston, Mass.: Kluwer.
Boyle, K.J., and J.C. Bergstrom. 1992. Benefit transfer studies: Myths, pragmatism and idealism. Water Resources Research 28(3):657-663.
Boyle, K.J., and R.C. Bishop. 1987. Valuing wildlife in benefit-cost analyses: A case study involving endangered species. Water Resources Research 23(May):943-950.
Boyle, K.J., R. Bishop, J. Caudill, J. Charbonneau, D. Larson, M.A. Markowski, R.E. Unsworth, and R.W. Paterson. 1998a. A Database of Sportfishing Values. Washington, D.C.: U.S. Department of the Interior, Fish and Wildlife Service Economics Division.
Boyle, K.J., R. Bishop, J. Caudill, J. Charbonneau, D. Larson, M.A. Markowski, R.E. Unsworth, and R.W. Paterson. 1998b. A Meta Analysis of Sportfishing Values. Washington, D.C.: U.S. Department of the Interior, Fish and Wildlife Service.
Boyle, K.J., M.P. Welsh, and R.C. Bishop. 1993. The role of question order and respondent experience in contingent-valuation studies. Journal of Environmental Economics and Management 25(1):S80-S99.
Boyle, K.J., W.H. Desvousges, F.R. Johnson, R.W. Dunford, and S.P. Hudson. 1994. An investigation of part-whole biases in contingent-valuation studies. Journal of Environmental Economics and Management 27:64-83.
Boyle, K.J., G.L. Poe, and J.C. Bergstrom. 1994. What do we know about groundwater values? Preliminary implications from a meta analysis of contingent-valuation studies. American Journal of Agricultural Economics 76:1055-1061.

Boyle, K.J., T.P. Holmes, M.F. Teisl, and Brian Roe. 2001. A comparison of conjoint analysis response formats. American Journal of Agricultural Economics 83(2):441-454.

Boyle, K.J., J. Poor, and L.O. Taylor. 1999. Estimating the demand for protecting freshwater lakes from eutrophication. American Journal of Agricultural Economics 81(5):1118-1122.

Boyle, K.J., T. P. Holmes, M. F. Teisl, and B. Roe. 2001. A comparison of conjoint analysis response formats. American Journal of Agricultural Economics 83:441-454.

Boyle, K.J., M.D. Morrison, and L.O. Taylor. 2002. Provision rules and the incentive compatibility of choice surveys. Unpublished paper, Department of Economics, Georgia State University.

Braden, J.B., and C.D. Kolstad (eds.). 1991. Measuring the Demand for Environmental Quality. North Holland: Amsterdam

Brasheres, E.N. 1985. Estimating the instream value of lake water quality in southeast Michigan. Ph.D. Dissertation, University of Michigan.

Breffle, W.S., E.R. Morey, R.D. Rowe, D.M. Waldman, and S.M. Wytinck. 1999. Recreational fishing damages from fish consumption advisories in the waters of Green Bay. Boulder, Colo.: Stratus Consulting Inc.

Brown, T.C., and R. Gregory. 1999. Why the WTA-WTP disparity matters. Ecological Economics 28:323-335.

Cameron, T.A. 1992. Combining contingent valuation and travel cost data for the valuation of nonmarket goods. Land Economics 68:302-317.

Cameron, T.A., W.D. Shaw, S.E. Ragland, J.M. Callaway, and S. Keefe. 1996. Using actual and contingent behavior data with different levels of time aggregation to model recreational demand. Journal of Agricultural and Resource Economics 21(1):130-149.

Carpenter, S.R., D. Ludwig, and W.A. Brock. 1999. Management of eutrophication for lakes subject to potentially irreversible change. Ecological Applications 9(3):751-771.

Carson, R.T. 1997. Contingent valuation survey and test for insensitivity to scope. In Determining the Value of Non-marketed Goods: Economics, Psychological, and Policy Relevant Aspects of Contingent Valuation Methods, R. Kopp, J., W.W. Pommerehne, and N. Schwarz (eds.). Boston, Mass.: Kluwer.

Carson, R.T., R.C. Mitchell, W.M. Hanemann, R.J. Kopp, S. Presser, and P.A. Ruud. 1992. A Contingent Valuation Study of Lost Passive Use Values Resulting from the *Exxon Valdez* Oil Spill. Report to the Attorney General of the State of Alaska.

Carson, R.T., N.E. Flores, K.M. Martin, and J.L. Wright. 1996. Contingent valuation and revealed preference methodologies: Comparing the estimates for quasi-public goods. Land Economics 72(1):80-99.

Champ, P., K. J. Boyle, and T.C. Brown (eds.). 2003. A Primer on Nonmarket Valuation. Boston, Mass.: Kluwer.

Chattopadhyay, S. 1999. Estimating the demand for air quality: New evidence based on the Chicago housing markets. Land Economics 75(1):22-38.

Clark, C. 1976. Mathematical Bioeconomics. New York: John Wiley and Sons.

Courant, P.N., and R.C. Porter. 1981. Averting expenditure and the cost of pollution. Journal of Environmental Economics and Management 8(4):321-329.

Coursey, D.L., J.L. Hovis, and W.D. Schulze. 1987. The disparity between willingness to accept and willingness to pay measures of value. Quarterly Journal of Economics 102:679-690.

Costanza, R., and S.C. Farber. 1987. The economic value of wetlands systems. Journal of Environmental Management 24:41-51.

Costanza, R., S.C. Farber, and J. Maxwell. 1989. Valuation and management of wetland ecosystems. Ecological Economics 1:335-361.

Creel, M.D., and J.B. Loomis. 1990. Theoretical and empirical advantages of truncated count data estimators for analysis of deer hunting in California. American Journal of Agricultural Economics 72(2):434-441.

Cropper, M.L. 1981. Measuring the benefits from reduced morbidity. American Economic Review 71:235-40.

Cummings, R.G., and L.O. Taylor. 1998. Does realism matter in contingent valuation surveys? Land Economics 74(2):203-215.

Cummings, R.G., and L.O. Taylor. 1999. Unbiased value estimates for environmental goods: A cheap talk design for the contingent valuation method. American Economic Review 89 (3):649-665.

Danielson, L.E., and J.A. Leitch. 1986. Private vs. public economics of prairie wetland allocation. Journal of Environmental Economics and Management 13(1):81-92.

Davis, R.K. 1964. The value of big game hunting in a private forest. Transactions of the 29th North American Wildlife and Natural Resources Conference. Washington, D.C.: Wildlife Management Institute.

Desvousges, W.H., F.R. Johnson, and H.S. Banzhaf. 1998. Environmental Policy Analysis with Limited Information: Principles and Applications to the Transfer Method. Cheltenham, U.K.: Edward Elgar.

Diamond, P.A., and J.A. Hausman. 1994. Contingent valuation: Is some number better than no number. Economic Perspectives 8(4):45-64.

Dickie, M. 2003. Defensive behavior and damage cost methods. Chapter 11 in A Primer on Nonmarket Valuation, P. Champ, K.J. Boyle, and T.C. Brown (eds.). Boston, Mass.: Kluwer.

Doss, C.R., and S.J. Taff. 1996. The influence of wetland type and wetland proximity on residential property values. Journal of Agricultural and Resource Economics 21(1):120-129.

Ellis, G.M., and A.C. Fisher. 1987. Valuing the environment as input. Journal of Environmental Management 25:149-156.

EPA (U.S. Environmental Protection Agency). 2000a. Guidelines for Preparing Economic Analyses. EPA 240-R-00-003. Washington, D.C.: U.S. Environmental Protection Agency.

EPA. 2000b. Technologies and Costs for Removal of Arsenic from Drinking Water. Washington, D.C.: Office of Groundwater and Drinking Water, Standards and Risk Management Division, Targeting and Analysis Branch.

EPA. 2003. Health Effects of PCBs, Available on-line at *http://www.epa.gov/opptintr/pcb/effects.html*. Accessed August 2003.

Epp, D., and K.S. Al-Ani. 1979. The effect of water quality on rural nonfarm residential property values. American Journal of Agricultural Economics 61(3):529-534.

Epple, D. 1987. Hedonic prices and implicit markets: Estimating demand and supply functions for differentiated products. Journal of Political Economy 95:59-80.

Farber, S., and R. Costanza. 1987. The economic value of wetlands systems. Journal of Environmental Management 24:41-51.

Feitelson, E. 1992. Consumer preferences and willingness-to-pay for water-related residences in non-urban settings: A vignette analysis. Regional Studies 26(1):49-68.

Freeman, A.M., III. 1985. Supply uncertainty, option price and option value. Land Economics 61(2):176-81.

Freeman, A.M., III. 1991. Valuing environmental resources under alternative management regimes. Ecological Economics 3:247-256.

Freeman, A.M. III. 1993a. The Measurement of Environmental and Resource Values: Theory and Methods. Washington, D.C.: Resources for the Future.

Freeman, A.M. III. 1993b. Nonuse values in natural resource damage assessment. In Valuing Natural Assets: The Economics of Natural Resource Damage Assessment, R. J. Kopp, and V.K. Smith (eds.). Washington, D.C.: Resources for the Future.

Gaden, M. 1997. Lake Huron Committee Endorses Plan to Control St. Marys River Sea Lampreys. Ann Arbor, Mich.: Lake Huron Committee, Great Lakes Fisheries Commission.

Gan, C., and E.J. Luzar. 1993. A conjoint analysis of waterfowl hunting in Louisiana. Journal of Agricultural and Applied Economics 25:36-45.

Great Lakes Fishery Commission. 2002. Sea Lampreys: A Great Lakes Invader 2002. Available on-line at *http://www.glfc.org/lampcon.asp.* Accessed on October 16, 2002.

Green, P.E., and V. Srinivasan. 1978. Conjoint analysis in consumer research: Issues and outlook. Journal of Consumer Research 5:103-123.

Greene, G., C.B. Moss, and T.H. Spreen. 1997. Demand for recreational fishing in Tampa Bay, Florida: A random utility approach. Marine Resource Economics 12:293-305.

Haab, T.C., and R.L. Hicks. 1997. Accounting for choice set endogeneity in random utility models of recreation demand. Journal of Environmental Economics and Management 34(2): 127-47.

Hammack, J., and G.M. Brown, Jr. 1974. Waterfowl and Wetlands: Towards Bioeconomic Analysis. Washington, D.C.: Resources for the Future.

Hanemann, W.M. 1991. Willingness to pay and willingness to accept: How much can they differ? American Economic Review 91:635-647.

Hanemann, W.M. 1994. Valuing the environment through contingent valuation. Economic Perspectives 8(4):19-44.

Harrington, W., A.J. Krupnick, and W.O. Spofford, Jr. 1989. The economic losses of a waterborne disease outbreak. Journal of Urban Economics 25(1):116-137.

Hausman, J.A. 1993. Contingent Valuation: A Critical Assessment. Amsterdam: North Holland.

Heinz (H. John Heinz III Center for Science, Economics, and the Environment). 2000. Evaluation of Erosion Hazards. EMC-97-CO-0375. Washington, D.C.: Federal Emergency Management Agency.

Herriges, J.A., and C.L. Kling (eds.). 1999. Valuing Recreation and the Environment: Revealed Preference Methods in Theory and Practice. Aldershot, U.K.: Edward Elgar.

Hoehn, J.P., and J.B. Loomis. 1993. Substitution effects in the valuation of multiple environmental programs. Journal of Environmental Economics and Management 25(1, Part 1):5-75.

Hoehn, J.P., T. Tomasi, F. Lupi, and H.Z. Chen. 1996. An Economic Model for Valuing Recreational Angling Resources in Michigan. Report to the Michigan Department of Natural Resources and the Michigan Department of Environmental Quality. East Lansing: Michigan State University.

Holmes, T., and W. Adamowicz. 2003. Attribbute-based methods. Chapter 6 in A Primer on Nonmarket Valuation, P. Champ, K.J. Boyle, and T.C. Brown (eds.). Boston, Mass.: Kluwer.

International Joint Commission. 2003. Status of Restoration Activities in Great Lakes

Areas of Concern. Washington, D.C.: International Joint Commission.

Jakus, P.M., D. Dadakas, and J.M. Fly. 1998. Fish consumption advisories: Incorporating angler-specific knowledge, habits, and catch rates in a site choice model. American Journal of Agricultural Economics 80(5):1019-1024.

Jakus, P.M., M. Downing, M.S. Bevelhimer, and J. Mark Fly. 1997. Do fish consumption advisories affect reservoir anglers' site choice? Agricultural and Resource Economics Review 26(2):198-204.

Johnston, R.J., T.F. Weaver, L.A. Smith, and S.K. Swallow. 1995. Contingent valuation focus groups: Insights from ethnographic interview techniques. Agricultural and Resource Economics Review 24(1):56-69.

Kahn, J.R. 1987. Measuring the economic damages associated with terrestrial pollution of marine ecosystems. Marine Resource Economics 4(3):193-209.

Kahn, J.R., and W.M. Kemp. 1985. Economic losses associated with the degradation of an ecosystem: The case of submerged aquatic vegetation in Chesapeake Bay. Journal of Environmental Economics and Management 12:246-263.

Kaplowitz, M.D., and J.P. Hoehn. 2001. Do focus groups and personal interviews reveal the same information for natural resource valuation? Ecological Economics 36:137-147.

Kaplowitz, M.D., F. Lupi, and J.P. Hoehn. 2003. Multiple-methods for developing and evaluating a stated preference survey for valuing wetland ecosystems. In Questionnaire Development, Evaluation and Testing Methods, S. Presser, J. Rothgeb, J. Martin, E. Singer, and M. Couper (eds.). New York: John Wiley and Sons.

Kiel, K.A. 1995. Measuring the impact of the discovery and cleaning of identified hazardous waste sites on housing values. Land Economics 71(4):428-35.

Kirchhoff, S., B.G. Colby, and J.T. LaFrance. 1997. Evaluating the performance of benefit transfer: An empirical inquiry. Journal of Environmental Economics and Management 33 (1):75-93.

Kriesel, W., A. Randall, and F. Lichtkoppler. 1993. Estimating the benefits of shore erosion protection in Ohio's Lake Erie housing market. Water Resources Research 29(4):795-801.

Krutilla, J.V. 1967. Conservation reconsidered. American Economic Review 57:777-786.

Larson, D.M., and P.R. Flacco. 1992. Measuring option prices from market behavior. Journal of Environmental Economics and Management 22(2):178-198.

Leggett, C.G., and N.E. Bockstael. 2000. Evidence of the effects of water quality on residential land prices. Journal of Environmental Economics and Management 39:121-144.

Lewis, C.F.M. 1999. Holocene lake levels and climate, lakes Winnipeg, Erie, and Ontario. Pp. 6-19 in Proceedings of the Great Lakes Paleo-Levels Workshop: The Last 4000 Years in C.E. Sellinger and F. H. Quinn (eds.). Ann Arbor, Mich.: National Oceanic and Atmospheric Administration, Great Lakes Environmental Research Laboratory.

Louviere, J.B. 1988. Analyzing Individual Decision Making: Metric Conjoint Analysis. Sage University Series on Quantitative Applications in the Social Sciences, Series No. 67. Newbury Park, Calif.: Sage Publications, Inc.

Louviere, J.B., D. Hensher, and J. Swait. 2000. Stated Choice Methods—Analysis and Application. Cambridge, U.K.: Cambridge University Press.

Loomis, J.B. 1988. The bioeconomic effects of timber harvesting on recreational and commercial salmon and steelhead fishing: A case study of the Siuslaw National Forest. Marine Resource Economics 5(1):43-60.

Lupi, F., and J.P. Hoehn. 1998. A Partial Benefit-Cost Analysis of Sea Lamprey Treatment Options on the St. Marys River. East Lansing, Mich.: Michigan State University for the Great Lakes Fishery Commission.

Lupi, F., M.D. Kaplowitz, and J.P. Hoehn. 2002. The economic equivalency of drained and restored wetlands in Michigan. American Journal of Agricultural Economics 84(5): 1355-1361.

Lynne, G.D., P.Conroy, and F.J. Prochaska. 1981. Economic value of marsh areas for marine production processes. Journal of Environmental Economics and Management 8: 175-186.

Madriaga, B., and K.E. McConnell. 1987. Some issues in measuring existence value. Water Resources Research 23:936-942.

Mahan, B., S. Polasky, and R. Adams. 2000. Valuing urban wetlands: A property price approach. Land Economics 76(1):100-113.

Mäler, K-G. 1974. Environmental Economics: A Theoretical Inquiry. Baltimore, Md.: Johns Hopkins University Press.

Mäler, K-G., and J. R Vincent (eds.). 2003. Handbook of Environmental Economics, Volume 2—Valuing Environmental Changes. North-Holland: Amsterdam.

McConnell, K.E., and I.E. Strand. 1989. Benefits from commercial fisheries when demand and supply depend on water quality. Journal of Environmental Economics and Management 17:284-292.

Michael, H.J., K.J. Boyle, and R. Bouchard. 2000. Does the measurement of environmental quality affect implicit prices estimated from hedonic models? Land Economics 76(2): 283-298.

Mitchell, R.C., and R.T. Carson. 1981. An Experiment in Determining Willingness to Pay for National Water Quality Improvements. Unpublished report. Washington, D.C.: Resources for the Future.

Mitchell, R.C., and R.T. Carson. 1989. Using Surveys to Value Public Goods: The Contingent Valuation Method. Washington, D.C.: Resources for the Future.

Montgomery, M., and M.S. Needelman. 1997. The welfare effects of toxic contamination in freshwater fish. Land Economics 73(2):211-223.

Morey, E. 1981. The demand for site-specific recreational activities: A characteristics approach. Journal of Environmental Economics and Management 8:345-371.

Morey, E.R., and D.M. Waldman. 1998. Measurement error in recreation demand models: The joint estimation of participation, site choice and site characteristics. Journal of Environmental Economics and Management 35(3):262-276.

Morey, E., R.D. Rowe, and M. Watson. 1993. A repeated nested-logit model of atlantic salmon fishing. American Journal of Agricultural Economics 75:578-592.

Morrison, M., and K.J. Boyle. 2001. Comparative reliability of rank and choice data in stated preference models. Paper presented at the European Association of Environmental and Resource Economics 2001 Conference, Southampton, U.K., June 2001.

Morrison, M., J. Bennett, and R. Blamey. 1999. Valuing improved wetland quality using choice modelling. Water Resources Research 35:2805-2814.

Needleman, M., and M.J. Kealy. 1995. Recreational swimming benefits of New Hampshire Lake water quality policies: An application of a repeated discrete choice model. Agricultural and Resource Economics Review 24:78-87.

NRC (National Research Council). 2004. Indicators for Waterborne Pathogens. Washington, D.C.: The National Academies Press.

NOAA (NOAA Panel on Contingent Valuation). 1993. Natural Resource Damage Assessment under the Oil Pollution Act of 1990. Federal Register 58(10): 4601-4614.

Palmquist, R.B. 1984. Estimating the demand for the characteristics of housing. The Review of Economics and Statistics 66 (3): 394-404.

Palmquist, R.B. 1991. Hedonic methods. Pp. 77-120 in Measuring the Demand for Environmental Quality, J.B. Braden and C.D. Kolstad (eds.). Amsterdam: North Holland.

Palmquist, R.B. 2003. Property value models. In Handbook of Environmental Economics, K-G Mäler, and J.R. Vincent (eds.). North Holland: Elsevier.

Parsons, G.R. 1992. The effect of coastal land use restrictions on housing prices: A repeat sales analysis. Journal of Environmental Economics and Management 22:25-37.

Parsons, G.R. 2003a. The travel cost model. Chapter 9 in A Primer on Nonmarket Valuation, P. Champ, K.J. Boyle, and T.C. Brown (eds.). Boston, Mass.: Kluwer.

Parsons, G.R. 2003b. A Bibliography of Revealed Preference Random Utility Models in Recreation Demand. Available on-line at *http://www.ocean.udel.edu/cms/gparsons/index.html*. Accessed on June 14, 2004.

Paterson, R.W., and K.J. Boyle. 2002. Out of sight, out of mind: Using GIS to incorporate topography in hedonic property value models. Land Economics 78(3):417-425.

Peeters, P. 1998. Into Lake Michigan's Waters. Wisconsin Natural Resources Magazine.

Pendleton, L.H., and R. Mendlesohn. 1998. Estimating the economic impact of climate change on the freshwater sport fisheries of the northeastern U.S. Land Economics 74(4):483-496.

Phaneuf, G.J., C.L. Kling, and J.A. Herriges. 1998. Valuing water quality improvements using revealed preference methods when corner solutions are present. American Journal of Agricultural Economics 80(5):1025-1031.

Poe, G.L., and R.C. Bishop. 1999. Valuing the incremental benefits of groundwater protection when exposure levels are known. Environmental and Resource Economics 13(3):341-367.

Poor, J., K.J. Boyle, L.O. Taylor, and R. Bouchard. 2001. Objective versus subjective measure of environmental quality in hedonic property-value models. Land Economics 77(4):482-493.

Portney, P.R. 1994. The contingent valuation debate: Why economists should care. The Journal of Economic Perspectives 8(4): 3-18.

Rae, D.A. 1983. The value to visitors of improving visibility at Mesa Verde and Great Smoky National parks. In Managing Air Quality and Scenic Resources at National Parks and Wilderness Areas, R.D. Rowe and L.G. Chestnut (eds.). Boulder, Colo.: Westview Press.

Randall, A. 1991. Total and nonuse values. Chapter 10 in Measuring the Demand for Environmental Quality, Braden, J.B., and C.D. Kolstad (eds.). Amsterdam: North Holland.

Randall, A., and J.P. Hoehn. 1996. Embedding in market demand systems. Journal of Environmental Economics and Management 30(3):369-380.

Roe, B., K.J. Boyle, and M.F. Teisl. 1996. Using conjoint analysis to derive estimates of compensating variation. Journal of Environmental Economics and Management 31:145-159.

Sargent-Michaud, J., and K.J. Boyle. 2002. Cost comparisons for arsenic contamination avoidance alternatives for Maine households on private wells. Staff Paper REP 506. Orono, Me.: Maine Agriculture and Forest Experiment Station, University of Maine.

Sathirathai, S., and E.B. Barbier. 2001. Valuing mangrove conservation, southern Thailand. Contemporary Economic Policy 19(2):109-122.

Settle, C., and J.F. Shogren. 2002. Modeling native-exotic species within Yellowstone

Lake. American Journal of Agricultural Economics 84(5):1323-1328.
Shabman, L.A., and S.S. Batie. 1978. Economic value of natural coastal wetlands: A critique. Coastal Zone Management Journal 4(3):231-247.
Siderelis, C., G. Brothers, and P. Rea. 1995. A boating choice model for the valuation of lake access. Journal of Leisure Research 27(3):264-282.
Smith, V.K. 1983. Option value: A conceptual overview. Southern Economics Journal 49(3): 654-668.
Smith, V.K. 1987. Nonuse values in benefit cost analysis. Southern Economic Journal 54(1):19-26.
Smith, V.K. 1991. Household production and environmental benefit estimation. Chapter 3 in Measuring the Demand for Environmental Quality, J. B. Braden and C. Kolstad (eds.). Amsterdam: North Holland.
Smith, V.K. 1997. Pricing what is priceless: A status report on non-market valuation of environmental resources. Pp. 156-204 in The International Yearbook of Environmental and Resource Economics 1997/1998, H. Folmer and T. Tietenberg (eds.). Cheltenham, U.K.: Edward Elgar.
Spalatro, F., and B. Provencher. 2001. An analysis of minimum frontage zoning to preserve lakefront amenities. Land Economics 77(4):469-481.
Steinnes, D.N. 1992. Measuring the economic value of water quality: The case of lakeshore land. Annals of Regional Science 26:171-176.
Stratus (Stratus Consulting Inc). 1997. Assessment Plan: Lower Fox River/Green Bay NRDA. Prepared for the U.S. Fish and Wildlife Service, Fort Snelling, Minn. Boulder, Colo.
Sudman, S., N.M. Bradburn, and N. Schwartz. 1996. Thinking about Answers: The Application of Cognitive Processes to Survey Methodology. San Francisco, Calif.: Jossey-Bass Publishers.
Swallow, S.K. 1994. Renewable and nonrenewable resource theory applied to coastal agriculture, forest, wetland and fishery Linkages. Marine Resource Economics 9:291-310.
Taylor, L. 2003. The hedonic method. Chapter 10 A Primer on Nonmarket Valuation, P. Champ, K.J. Boyle, and T.C. Brown (eds.). Boston, Mass.: Kluwer.
Tourangeau, R., L.J. Ripps, and K. Rasinski. 2000. The Psychology of Survey Response. Cambridge, U.K.: Cambridge University Press.
Tschirhart, J. 2000. General equilibrium of an ecosystem. Journal of Theoretical Biology 203:13-32.
Tschirhart, J., and D. P. Finhoff. 2001. Harvesting in an eight-species ecosystem. Journal of Environmental Economics and Management. Available on-line at: *http://uwadmnweb.uwyo.edu/econfinance/Tschirhart/hvstall.pdf*. Accessed October 6, 2004.
USACE (U.S. Army Corps of Engineers). 2000. Shore Protection Structures, Lake Michigan Potential Damages Study. Detroit, Mich.: Detroit District Office.
Vandenberg, T.P., G.L. Poe, and J.R. Powell. 2001. Assessing the accuracy of benefits transfers: Evidence from a multi-site contingent valuation study of ground water quality. Chapter 6 in The Economic Value of Water Quality, J.C. Bergstrom, K.J. Boyle, and G.L. Poe (eds.). Cheltenham, U.K: Edward Elger.
Ward, F.A., and D. Beal. 2000. Valuing Nature with Travel Cost Models. Cheltenham, U.K.: Edward Elger.
Weisbrod, B.A. 1964. Collective-consumption services of individual-consumption goods. Quarterly Journal of Economics 78:471-77.
Welsh, M.P., and G.L. Poe. 1998. Elicitation effects in contingent valuation: Compari-

sons to a multiple bounded discrete choice approach. Journal of Environmental Economics and Management 36:170-185.

Wilman, E. 1984. The External Costs of Coastal Beach Pollution: A Hedonic Approach. Washington, D.C.: Resources for the Future.

Wilson, M.A., and S.R. Carpenter. 1999. Economic valuation of freshwater ecosystems services in the United States: 1971-1997. Ecological Applications 9:772-783.

Wisconsin DNR (Wisconsin Department of Natural Resources). 2002. Lower Fox River State of Wisconsin, April 30 2001. Available on-line at *http://www.dnr.state.wi.us/org/water/wm/lowerfox/background.html.* Accessed on December 2003.

Woodward, R.T., and Y.-S. Wui. 2001. The economic value of wetland services: A meta-analysis. Ecological Economics 37:257-270.

Wu, J., R.M. Adams, and W.G. Boggess. 2000. Cumulative effects and optimal targeting of conservation efforts: Steelhead trout habitat enhancement in Oregon. American Journal of Agricultural Economics 82(2):400-413.

Wu, J., K. Skelton-Groth, W.G. Boggess, and R.M. Adams. 2003. Pacific salmon restoration: Trade-offs between economic efficiency and political acceptance. Contemporary Economic Policy 21(1):78-89.

5
Translating Ecosystem Functions to the Value of Ecosystem Services: Case Studies

INTRODUCTION

Valuing ecosystem services requires the integration of ecology and economics. Ecology is needed to comprehend ecosystem structure and functions and how these functions change with different conditions. Both ecology and economics are required to translate ecosystem functions into the production of ecosystem goods and services. Economics is needed to comprehend how ecosystem goods and services translate into value (i.e., benefits for people; see also Figure 1-3). The two preceding chapters discuss much of the relevant ecological and economic literature. Chapter 3 focuses on the relevant ecological literature on aquatic and related terrestrial ecosystem functions and services, while Chapter 4 focuses on the economic literature on nonmarket valuation methods useful for valuing ecosystem goods and services. In this chapter, the focus is on the integration of ecology and economics necessary for valuing ecosystem services for aquatic and related terrestrial ecosystems. More specifically, a series of case studies is reviewed (including those taken from the eastern and western United States; see Chapter 1 and Box ES-1 for further information), ranging from studies of the value of single ecosystem services, to multiple ecosystem services, to ambitious studies that attempt to value all services provided by ecosystems. An extensive discussion of implications and lessons learned from these case studies is provided and precedes the chapter summary.

Development of the concept of ecosystem services is relatively recent. Only in the last decade have ecologists and economists begun to define ecosystem services and attempted to measure the value of these services (see for example, Balvanera et al., 2001; Chichilnisky and Heal, 1998; Constanza et al., 1997; Daily, 1997; Daily et al., 2000; Heal, 2000a,b; Pritchard et al., 2000; Wilson and Carpenter, 1999). There is a much longer history of natural resource managers and economists evaluating "goods" produced by ecosystems (e.g., forest products, fish production, agricultural production). For example, in 1926, Percy Viosca, Jr., a fisheries biologist, estimated that the value of conserving wetlands in Louisiana for fishing, trapping, and collecting activities was $20 million annually (Vileisis, 1997). In the 1960s and early 1970s, pioneering work by Krutilla (1967), Hammack and Brown (1974), and Krutilla and Fisher (1975), among others, greatly expanded the set of "goods and services" generated by natural systems considered by economists to be of value to humans (e.g., clean air, clean water, recreation, ecotourism). Economic geographers and regional scientists (e.g., Isard et al., 1969) examined spatial relationships among natural and socioeconomic systems. Recent work on ecosystem services has broadened the set of goods and services studied to include water purification, nutrient retention, and flood control, among other things. It has also emphasized the importance of understanding natural processes within ecosystems (e.g., primary and secondary productivity, carbon and nutrient cycling, energy flow) in order to understand the production of ecosystem services. Yet, as discussed throughout this report, for the most part, the importance of these natural processes in producing ecosystem services on which people depend has remained largely invisible to decision-makers and the general public. For most ecosystem services, there are no markets and no readily observable prices, and most people are unaware of their economic value. All too often it is the case that the value of ecosystem services becomes apparent only after such services are diminished or lost, which occurs once the natural processes supporting the production of these services have been sufficiently degraded. For example, the economic importance of protecting coastal marshes that serve as breeding grounds for fish may become apparent only after commercial fish harvests decline. By then, it may be difficult or impossible to repair the damage and restore the production of such services.

Although there has been great progress in ecology in understanding ecosystem processes and functions, and in economics in developing and applying nonmarket valuation techniques for their subsequent valuation, at present there often remains a gap between the two. There has been mutual recognition among at least some ecologists and some economists that addressing issues such as conserving ecosystems and biodiversity requires the input of both disciplines to be successful (Daily et al., 2000; Holmes et al., 2004; Kinzig et al., 2000; Loomis et al., 2000; Turner et al., 2003). Yet there are few existing examples of studies that have successfully translated knowledge of ecosystems into a form in which economic valuation can be applied in a meaningful way (Polasky, 2002). Several factors contribute to this ongoing lack of integration. First, some ecologists and economists have held vastly different views on the current state of the world and the direction in which it is headed (see, for example, Tierney, 1990, who chronicles the debates between a noted ecologist and economist [Paul Ehrlich and Julian Simon]). Second, ecology and economics are separate disciplines, one in natural science and the other in social science. Traditionally, the academic organization and reward structure for scientists make collaboration across disciplinary boundaries difficult even when the desire to do so exists. Third, as noted previously, the concept of ecosystem services and attempts to value them are still relatively new. Building the necessary working relationships and integrating methods across disciplines will take time.

Some useful integrated studies of the value of aquatic and related terrestrial ecosystem goods and services are starting to emerge. The following section reviews several such studies and the types of evaluation methods used. This review begins with situations in which the focus is on valuing a single ecosystem service. Typically in these cases, the service is well defined, there is reasonably good ecological understanding of how the service is produced, and there is reasonably good economic understanding of how to value the service. Even when valuing a single ecosystem service however, there can be significant uncertainty about either the production of the ecosystem service, the value of the ecosystem service, or both. Next reviewed are attempts to value multiple ecosystem services. Because ecosystems produce a range of services that are frequently closely connected, it is often difficult to discuss the valuation of a single service in isolation. However, valuing multiple ecosystem services typically multiplies the difficulty of valuing a single ecosystem service. Last to be reviewed are analyses that attempt to encompass all services produced by an ecosystem. Such cases can arise with natural resource damage assessment, where a dollar value estimate of total damages is required, or with ecosystem restoration efforts. Such efforts will typically face large gaps in understanding and information in both ecology and economics.

Proceeding from single services to entire ecosystems illustrates the range of circumstances and methods for valuing ecosystem goods and services. In some cases, it may be possible to generate relatively precise estimates of value. In other cases, all that may be possible is a rough categorization (e.g., "a lot" versus "a little"). Whether there is sufficient information for the valuation of ecosystem services to be of use in environmental decision-making depends on the circumstances and the policy question or decision at hand (see Chapters 2 and 6 for further information). In a few instances, a rough estimate may be sufficient to decide that one option is preferable to another. Tougher decisions will typically require more refined understanding of the issues at stake. This progression from situations with relatively complete to relatively incomplete information also demonstrates what gaps in knowledge may exist and the consequences of those gaps. Part of the value of going through an ecosystem services evaluation is to identify the gaps in existing information to show what types of research are needed.

MAPPING ECOSYSTEM FUNCTIONS TO THE VALUE OF ECOSYSTEM SERVICES: CASE STUDIES

Despite recent efforts of ecologists and economists to resolve many types of challenges to successfully estimating the value of ecosystem services, the number of well-studied and quantified cases studies remains relatively low. The following section reviews cases studies that have attempted to value ecosystem services in the context of aquatic ecosystems. These examples illustrate different levels of information and insights that have been gained thus far from the combined approaches of ecology and economics.

Valuing a Single Ecosystem Service

This review begins with studies of the value of ecosystem services using examples that attempt to value a single ecosystem service. These cases provide the best examples of both well-defined and quantifiable ecosystem services and of services that are amenable to application of economic valuation methodologies. The best-known example of a policy decision hinging on the value of a single ecosystem service involves the provision of clean drinking water for New York City, which is reviewed first. Other examples include cases where ecosystems provide habitat for harvested fish or game species and cases where they provide flood control.

In all of the cases reviewed in this section, the ecosystem service is well-defined, although there may be some scientific uncertainty surrounding quantification of the amount of the service provided. In some cases, adequate methods for valuing the single ecosystem service exist. Further, for some cases, such as the New York City example below, information about a single ecosystem service may prove sufficient to support rational environmental decision-making. In other cases, this will not be so, and further work to assess a more complete set of ecosystem services will be necessary. Under no circumstances, however, should the value of a single ecosystem service be confused with the value of the entire ecosystem, which has far more than a single dimension. Unless it is kept clearly in mind that valuing a single ecosystem service represents only a partial valuation of the natural processes in an ecosystem, such single service valuation exercises may provide a false signal of the total economic value of the natural processes in an ecosystem.

Providing Clean Drinking Water: The Catskill Mountains and New York City's Watershed

One of the best-studied water supply systems in the world is the one that provides drinking water for more than 9 million people in the New York City metropolitan area (Ashendorff et al., 1997; NRC, 2000a; Schneiderman, 2000). New York City's water supply includes three large reservoir systems (Croton, Catskill, and Delaware) that contain 19 reservoirs and 3 controlled lakes. This system, including all tributaries, encompasses a total area of 5,000 km^2 with a reservoir capacity of 2.2×10^9 m^3. This complex array of natural watersheds requires a wide range of management to sustain the water quality supplied to the reservoirs and aqueducts. Historically, these watersheds have supplied high-quality water with little contamination. However, increased housing developments with onsite septic systems, combined with nonpoint sources of pollution such as runoff from roads and agriculture, have posed threats to water quality. Further significant deterioration of water quality would force the U.S. Environmental Protection Agency (EPA) to require New York City build a water filtration system[1] to ensure that drinking water delivered to consumers would meet federal drinking water standards. By 1996, New York City faced a choice: it could either build water filtration system or protect its watersheds to ensure high-quality drinking water.

The cost of building a new, larger filtration system necessary to meet water quality standards was estimated to lie in the range of $2 billion to $6 billion. Moreover, the city estimated that it would spend $300 million annually to operate the new filtration plant. Together, the costs of building and operating the filtration system were estimated to be in the range of $6 billion to $8 billion (Chichilnisky and Heal, 1998).

Instead of investing in a water filtration facility, New York City opted to invest more in protecting watersheds. Maintaining water quality in the face of increased human population densities in the watershed required increased protection of riparian buffer zones along rivers and around reservoirs. These zones help to regulate nonpoint sources of nutrients and pesticides from stormwater runoff, septic tanks, and agricultural sources. In 1997 the city received "filtration avoidance status" from the EPA by promising to upgrade watershed protection. The 1997 Watershed Memorandum Agreement resulted from negotiations among the State of New York, New York City, the EPA, municipalities within the watershed, and five regional environmental groups. The agreement provided a framework for compliance with water quality standards and contained plans for land acquisition through mutual consent, watershed regulations, environmental education workshops, and partnership programs with community groups. For example, a farmer-led Watershed Agricultural Council provides programs for the approximately 350 dairy and livestock farms in the watershed to minimize nutrient input from agricultural runoff (Ashendorff et al., 1997).

Under this agreement, New York City is obligated to spend $250 million during a 10-year period to purchase lands within the watershed (up to 141,645 hectares). In this part of the overall response, the New York City Department of Environmental Protection land acquisition program purchases undeveloped land from willing sellers rather than relying on condemnation and the power of eminent domain. Property rights to develop land in the watershed rests in the hands of local landowners. In some cases these rights are regulated by local ordinances. New York City's 1953 Watershed Rules and Regulations give the city some authority over watershed development to limit water pollution. Decades-old resentment remains among some residents of upstate watersheds because earlier land acquisitions to build the reservoirs displaced entire communities. Moreover, recent concerns about security of the reservoirs have also polarized residents whose road access has been limited. Exactly what legal rights New York City has and what legal rights local municipalities, and local landowners

[1] In the late 1990s, the plan was to build one centralized plant for the Catskill/Delaware portion of the larger watershed (see NRC, 2000a for further information). However, it has since been determined that the Croton portion of the watershed also has to build a separate filtration plant.

have to make decisions is not fully resolved. The long-term costs of riverbank protection, upkeep of sewage treatment plants by municipalities and overall maintenance costs of this approach remain uncertain.

On the other hand, a series of regulations prohibiting certain types of development in certain places (e.g., areas in close proximity to watercourses, reservoirs, reservoir stems, controlled lakes, wetlands) was agreed upon. The city together with the Catskill Watershed Corporation developed a comprehensive geographical information system to track land uses and to analyze runoff and storm flows resulting from precipitation. Runoff is sensitive to connections among stream network, and to the amount of impervious surface in the watershed (e.g., roads, buildings, driveways, parking lots), which results in increased peak flows that can cause flooding and bank erosion (Arnold and Gibbons, 1996; Gergel et al., 2002). To minimize these effects, new construction of impervious surfaces within 300 feet of a reservoir, rivers, or wetland is prohibited. Road construction within 100 feet of a perennial stream and 50 feet of an intermittent stream is also prohibited. Septic system fields cannot be located within 100 ft of a wetland or watercourse or 300 feet of a reservoir because these on-site sewage treatment and disposal systems do not work effectively in saturated soil. Septic fields also interfere with the natural nutrient processing in floodplains, wetlands, and riparian buffer zones along streams. Funds are available to subsidize upgrades of local wastewater treatment plants and septic systems throughout the watershed. There are 38 wastewater treatment plants in the watershed that are not owned by New York City. Overall, New York City projected that it would invest $1 billion to $1.5 billion in protecting and restoring natural ecosystem processes in the watershed (Ashendorff et al., 1998; Chichilnisky and Heal, 1998; Foran et al., 2000; NRC, 2000a). Incentives for landowners to improve riparian protection through conservation easements and educational outreach efforts were combined with management of state-owned lands to minimize erosion and protect riparian buffers.

In this case, it was not necessary to value all or part of the services of the Catskill watershed; it was merely necessary to establish that protecting and restoring the ecological integrity of the watershed to provide clean drinking water was less costly than replacing this ecosystem service with a new water filtration plant. As discussed in Chapter 4, Shabman and Batie (1978) suggest that a replacement cost approach can provide a "proxy" valuation estimation for an ecological service if the alternative considered provides the same service, the alternative compared is the least-cost alternative, and there is substantial evidence that the service would be demanded by society if it were provided by that least-cost alternative. In the Catskill case the proposed filtration plant would provide very similar services (more on this below). Of course, the city will have to provide clean water somehow. So these conditions are met and the cost of replacing the provision of clean drinking provided by the watershed with a filtration plant, less the cost of protecting and restoring the watershed, can be thought of as a measure of the ecosystem service value to New York City as a water purification tool. If, however, demand side management can reduce demand for water at less cost than it costs to provide the water via the filtration plant, then demand side management costs would provide the relevant avoided costs. Both methods—natural processes in watersheds and a water filtration plant—are capable of providing clean water that meets drinking water standards.

This case also appears to provide clear environmental policy direction. For New York City, it is likely to be far less costly to provide safe drinking water by protecting watersheds, thereby maintaining natural processes, than to build and operate a filtration plant. Further, protecting watersheds to provide clean water also enhances provision of other ecosystem services (e.g., open space for recreation, habitat for aquatic and terrestrial species, aesthetics). As discussed throughout this report, such ecosystem services are arguably far harder to value economically. Since these values add to the value of protecting watersheds for the provision of clean water, which is the preferred option even without consideration of these additional values, it is not necessary to establish a value for these services for policy purposes. Thus, protecting watersheds can be justified on the basis of the provision of clean drinking water alone.

Despite the appearance of being a textbook case for valuing a single ecosystem service, several issues make the answer to ecosystem valuation less obvious than at first glance. The replacement cost approach assumes that the same service will be provided under either alternative. In reality, it is unlikely that watershed protection and filtration will provide identical levels of water quality and reliability over time because engineered systems can fail—especially during storms when heavy flows overwhelm the system. Likewise, natural watersheds can also vary in their effectiveness in response to severe storm flows or other disturbances (Ashendorff et al., 1997). Managed watersheds can require some maintenance costs to sustain ecosystem services such as clean up of accidental spills or fish kills to prevent pollution or control of invasive species such as zebra mussels (Covich et al., 2004; Giller et al., 2004). Both engineered and ecosystem approaches are vulnerable but they differ in the types of uncertainty associated with each investment.

New York City's watershed investment plan includes several maintenance costs such as thorough, multistaged monitoring of water quality and disease surveillance that triggers active management and localized water treatment. Baseline data on water quality and biodiversity of stream organisms in the watershed (e.g., aquatic insects) are being collected by the Stroud Water Research Center (2001) annually to determine if the city's recent management efforts are effective. By reducing the risk of contaminants from various sources, the city can minimize use of disinfectants at the final water treatment stages. Reducing chemical use saves money directly and may also have health benefits since chlorination can produce halogenated disinfection by-products (e.g., chloroform, trihalomethane) in drinking water, especially in ecosystems with high levels of organic matter (Symanski et al., 2004; Villanueva et al., 2001; Zhang and Minear, 2002). Some of these by-products may be carcinogens. On the other hand, filtration may provide higher-quality drinking water because chlorination is not completely effective in killing pathogens, particularly when there are high levels of suspended materials (Schoenen, 2002).

Despite the regulations and the comprehensive framework contained in the city's watershed protection plan, considerable uncertainties exist about whether the plan can sustain high quality water supplies over the longer-term. Enforcement of the regulations and monitoring the rapid rate of suburban growth constitute a major challenge, and these development pressures in the area may increase the opportunity costs of watershed protection. Construction in the headwaters of streams, permitted under the plan, may result in increased runoff rates and erosion. Filling tributary channels with sediments can take place incrementally, with each step occurring at a small scale. In addition, numerous small-scale changes may transform the watershed in detrimental ways over time without sufficient oversight and long-term planning. The U.S. Army Corps of Engineers (USACE) has authority under Section 404 of the Clean Water Act to review permits. However, without site-by-site reviews of small projects (less than four hectares), allowable incremental alterations can have significant cumulative effects on small streams. Decreased stream density (stream length per drainage basin area) would occur if natural stream channels were replaced by pipes and paved over for development, resulting in loss of the essential ecological processes of organic matter breakdown and sediment retention (Meyer and Wallace, 2001; Paul and Meyer, 2001).

Additional uncertainties might impact decision-making, besides the adequacy of protection in the watersheds. Model uncertainty that arises from imperfect understanding of ecosystem function and the translation to ecosystem services is a major issue for most ecosystem valuation studies. In this case, there is model uncertainty because the hydrologic modeling used for determining water supplies is affected by the definition of spatial and temporal boundaries. For example, other municipalities in New York and New Jersey use water from the Catskills. Changes in water diversions from the Catskill Mountains can affect outflows to the Delaware River and modify salinities in the lower sections of the river used by Philadelphia (Frei et al., 2002). Given the additional uncertainties of future regional droughts, floods, and extreme temperatures, as well as acid rain and nitrogen deposition from atmospheric sources, planners must consider the range of intrinsic natural variability in decision-making. Planners can cope with aspects of model and parameter uncertainty by carefully monitoring land uses in the basin and incorporating environmental data into any new regulations that might be required. A long series of studies on nutrient budgets and acid deposition provides some essential baseline information for the Catskills (e.g., Frei et al., 2002; Lovett et al., 2000; Murdoch and Stoddard, 1992, 1993; Stoddard, 1994). Other locations may lack sufficient information, and thus, considerable sources of uncertainty will limit the analysis of complete replacement costs.

In this case, the provision of clean drinking water supplies through the protection of natural processes in watersheds rather than through the human-engineered solution of building and operating a water filtration system offers an estimate of the value of restoring an ecosystem service that provides clear advice to a policy decision. Replacement costs for natural processes in watersheds providing clean drinking water are estimated to be in the neighborhood of $6 billion to $8 billion, which is far higher than estimates of the cost necessary to protect the watersheds. Because the policy question is relatively specific (i.e., whether to build a filtration plant or to protect watersheds), currently available economic methods of ecosystem service valuation are sufficient.

Even in this example however, obtaining a precise estimate of the value of the provision of clean water through watershed conservation is probably not possible given existing knowledge. First, it is not clear that the two approaches, filtration and watershed protection, provide the same level of water quality and reliability. There are numerous dimensions to the provision of clean drinking water, such as the concentrations of various trace chemicals, carcinogens, and suspended solids, natural variance of water quality, and the adequacy of supply. It is unlikely that the two approaches will deliver water that is identical in all of these dimensions under all conditions. Second, there is no guarantee that protecting watersheds will continue to be successful. Increased development pressure on lands outside the riparian buffer zones or inadequate enforcement may require building a filtration system at some point in the future. If the watershed protection plans prove to be insufficient in the future, the investments in protection will still likely reduce future costs of building filtration plants because the quality of the water to be treated will be enhanced through these land-use programs.

Finally, it should be emphasized that (1) the value of providing clean drinking water is only a partial measure of the value of ecosystem services provided by the watershed, and (2) replacement cost is rarely a good measure of the value of an ecosystem service. Even if water quality benefits alone did not justify watershed protection, such a finding would not justify abandoning efforts at watershed protection. To make that decision would require a broader effort to measure the value of the wider set of ecosystem services produced by Catskill watersheds. It is less clear that estimates to answer this broader question are sufficiently precise to provide policy-relevant answers (see Chapters 2 and 6 for more on framing). Replacement cost methods can be used as a measure of the value of ecosystem services only when there are alternative ways to provide the same service and when the service will be demanded if provided by the least cost alternative. Replacement cost does not constitute an estimate of value of the service to society; it represents the value of having the ability to produce the service through an ecosystem rather than through an alternative method.

Other Surface Water Examples

Other cities have used similar strategies to invest in maintaining the ecological integrity of their watersheds as a means of providing high quality drinking water that meets all federal, state, and local standards. Boston, Seattle, San

Francisco, and Greenville, South Carolina, are other examples where the value of ecosystem services could be estimated using a replacement cost approach for building and operating water treatment plants that are roughly equivalent in the quality of drinking water supplied (NRC, 2000a). The costs of producing safe drinking water were traditionally derived from production cost estimates associated with engineering treatments. Filtration plants were built to remove organic materials, and then some form of chemical purification was used to control microorganisms. Engineers generally considered natural ecosystems such as rivers and lakes mostly from the viewpoint of volumes, transport systems, resident times, dilution, and natural "reoxygenation." In other words, they viewed many natural ecosystems as large pipes rather than as complex habitats for a diverse biota. Yet even viewed strictly through the lens of water supply systems, protecting natural processes within ecosystems may be superior to engineering solutions, and such a result may be sufficient for decision-making purposes. Replacement cost estimates for provision of clean drinking water, however, provide an estimate of just one source of value and should not be confused with the complete value of ecosystem services provided by watersheds. Further, as discussed in Chapter 4, replacement cost is a valid approach to economic valuation only in highly restricted circumstances—namely, that there are multiple ways to achieve the same end and the benefits exceed the costs of providing this end.

Provision of Drinking Water from Groundwater: San Antonio, Texas

In contrast with the Catskill case, there has been a lack of studies to date on the economic value of the Edwards Aquifer (see also Box 3-5) that supplies drinking water to San Antonio as well as water for irrigation and other uses. Groundwater supplies approximately half of America's drinking water (EPA, 1999). It is relied on heavily in some parts of the arid West where surface waters are scarce. The long-term supply of groundwater is a concern in some of these areas (Howe, 2002; Winter, 2001). For example, depletion of the Ogallala Aquifer is creating great uncertainties about future water supplies throughout a large region of the central United States (Glennon, 2002; Opie, 1993). Similarly, depletion of groundwater aquifers in the Middle Rio Grande Basin is creating uncertainty about the future supply of drinking water for Albuquerque, New Mexico (NRC, 1997, 2000b). Aquifers generally provide high quality drinking water, but pollution lowers water quality in some areas, such as the Cape Cod Aquifer where there are threats from sewage and toxic substances leaching into groundwater from the Massachusetts Military Reservation (Barber, 1994; Morganwalp and Buxton, 1999).

The long-term sustainability of groundwater depends on matching extraction with recharge (Sanford, 2002). It is often difficult to predict the timing and rate of recharge because of complications of local geology, time lags, and climate uncertainties. Recharge of the porous karstic limestone that characterizes the Edward Aquifer occurs primarily during wet years when precipitation infiltrates deeply into the soils and underlying rock (Abbott, 1975). Drought conditions have complex effects on lowering recharge rates while simultaneously tending to increase the demand for water. The greatest source of uncertainty about groundwater recharge is the range of natural interannual variability in precipitation and land-use changes. Increasing demands from a growing population and the difficulty in predicting climate change raise questions about the adequacy of groundwater supplies in arid regions (Grimm et al., 1997; Hurd et al., 1999; Meyer et al., 1999; Murdoch et al., 2000).

Aquifer depletion has both economic and ecological consequences. The costs for deeper drilling and pumping increase as groundwater is depleted. Removal of water in the underground area may cause collapse of the overlying substrata. These collapses decrease future storage capacity below ground and may cause damage on the surface as areas subside, buckle, or collapse. In some areas, depleted groundwater may cause the intrusion of low-quality water from other aquifers or from marine-derived salt or brackish waters that could not readily be restored for freshwater storage and use.

Depletion of groundwater supplies creates uncertainty and generally is offset by supplies from surface waters. An interesting exception is San Antonio (the ninth largest city in the United States) that relies primarily on groundwater for its source of municipal water. An outbreak of cholera in 1866 from polluted surface waters prompted the City of San Antonio to switch to groundwater from the Edwards Aquifer. The aquifer is estimated to contain up to 250 million acre-feet of water with a drainage area covering approximately 8,000 square miles. The average annual recharge is estimated at approximately 600,000 acre-feet of water (Merrifield , 2000). Given this large supply, the Edwards Aquifer plays a major role in the economy of San Antonio and south-central Texas (Glennon, 2002). In some parts of this region, clean, free-flowing springs and artesian wells provide drinking water without the cost of pumping and with minimal treatment. San Antonio built its first pumping station in 1878. The U.S. Geological Survey (USGS) has monitored aquifer recharge rates since 1915 and water quality monitoring began in 1930. In 1970 the Edwards Aquifer was designated a "sole source aquifer" by the EPA under the Safe Drinking Water Act. Currently, more than 1.7 million people rely on the Edwards Aquifer for water. Industrial and agricultural demands on the Edwards Aquifer have increased, and the city has planned for new reservoir storage as part of its water supply several times over the last two decades. As the demand for water in the area has grown, concerns have arisen over both the quantity and the quality of groundwater available (Wimberley, 2001).

Depletion also raises the specter that adequate supply will not be available for future demand at any price. The $3.5 billion-a-year tourist industry in San Antonio is centered on the city's River Walk, which relies primarily on recycled groundwater (Glennon, 2002). Uncertainties over the long-term availability of water make long-term planning problematic and threaten long-term investments. For example, aquaculture companies (e.g., Living Waters Artesian Springs,

Ltd.) expanded their catfish operations in March 1991, but subsequently closed in November 1991 because of concerns over pumping rates and the impaired water quality of return flows (i.e., high concentrations of dissolved nutrients) to surface- and groundwaters associated with the Edwards Aquifer.

Groundwater storage is critical in most aquatic ecosystems to provide persistence spring and stream habitats during dry seasons or during drought. Several springs (Comal, San Antonio, San Pedro) in the area began to dry up following a seven-year drought in the 1950s. Chen et al. (2001) used a climate change model to estimate the regional loss of welfare at $2.2 million to $6.8 million per year from prolonged drought. They estimated groundwater recharge based on historic data for recharge rates as influenced by precipitation and temperature. These researchers forecasted municipal and irrigation demand for five scenarios, including current condition and four different levels of climate change. Estimates of demand elasticity were based on models and methods used in other studies of arid regions. Given the projected reductions in available water, it would be necessary to protect endangered species in springs and groundwater at an additional reduction of 9 to 20 percent in pumping that would add $0.5 million to $2 million in costs.

The economic value of organisms living in groundwater and in springs, wetlands, and downstream surface flows supplied by groundwater is difficult to estimate. However, their value is generally assumed to be high because of their many functional roles in maintaining clean water as well as their existence values. For example, many diverse microbial communities and a wide range of invertebrate and vertebrate species live in groundwater, springs, and streams (Covich, 1993; Gibert et al., 1994; Jones and Mulholland, 2000). Their main functions are breaking down and recycling organic matter that forms the base of a complex food web (Covich et al., 1999, 2004). Depletion of groundwater aquifers results in possible loss of habitat for endemic species protected by state and federal regulations. For example, the Edwards Aquifer-Comal Springs ecosystem provides critical habitat for several endangered and threatened species, including salamanders (the Texas blind salamander and San Marcos Spring salamander), fish (the San Marcos gambusia and fountain darter), and Texas wild rice (Glennon, 2002; Sharp and Banner, 2000). In all, 91 species and subspecies of other organisms are endemic in this aquifer and its associated springs (Bowles and Arsuffi, 1993; Culver et al., 2000, 2003; Longley, 1986).

Most studies predicting groundwater supply focus on usable water quantities given drought frequencies and recharge. Land use is also important because it influences demand as well as runoff and recharge. As a result of water shortages in San Antonio, regulations controlling development were issued beginning in 1970. These regulations included rules for limiting economic development within the recharge zone. As noted previously, economic development often increases the extent of impervious surfaces that, in turn, cause more rapid runoff and loss of infiltration during and after precipitation events. Studies indicate that when impervious cover exceeds 15 percent of the surface of a watershed, there are adverse impacts on surface water quality and subsurface water recharge (e.g., Veni, 1999).

The quality of groundwater is also an issue. Increasing concerns about water pollution of the Edwards Aquifer led former (now deceased) Congressman Henry B. Gonzalez of San Antonio to propose the Gonzalez Amendment to the Safe Drinking Water Act of 1974. The amendment dealt with protection of sole source aquifers used for water supplies (Wimberley, 2001). Leachate from landfills, leaking petroleum storage tanks, and pesticides all pose contamination threats that could render groundwater unusable. In 1987, a regional committee was formed to determine how the aquifer could be further protected. Henry Cisneros, then mayor of San Antonio, chaired the committee and proposed a plan that limited total withdrawals and called for a reservoir construction program (the Applewhite Reservoir was proposed but ultimately not approved).

A severe drought in 1990 and above-average pumping combined to dry up two of the aquifer's major springs (Merrifield, 2000). In 1993, the Sierra Club sued the state under the Endangered Species Act for failure to guarantee a minimum flow of 100 cubic feet per second to Comal and San Marcos Springs (*Sierra Club vs. Lujan*, 1993 W.L. 151353). The State and the U.S. Fish and Wildlife Service entered into an agreement to resolve this conflict. The Texas legislature created the Edwards Aquifer Authority to control pumping and reallocate water through market mechanisms (McCarl et al., 1999; Schiable et al., 1999). This approach reallocated water from lower economic uses (such as agricultural irrigation) to higher-valued uses (such as domestic and industrial water supplies and environmental and recreational uses). In 2000, the Edwards Aquifer Authority decided to ban the use of any type of sprinkler in the eight-county region whenever flow at Comal Springs declined to 150 cubic feet per second (cfs) or less. In September 2002, the USGS reported that the flow had declined to 145 cfs and the ban went into effect.

Groundwater is a renewable resource that provides both extractive use value and in situ value. In situ value refers to the value created by having a stock of groundwater in the aquifer. Extraction of groundwater generates current extractive use value but can result in lower in situ value if extractions rates exceed aquifer recharge rates. Efficient use of groundwater requires extraction only when extractive use value per unit exceeds in situ value per unit of groundwater. Most economic analyses, such as those discussed above, have focused on extractive use values because these are most readily quantified. Extractive use values include the value of water for municipal and agricultural uses as well as recreation.

Characterizing the in situ value of groundwater is more difficult. Aquifer depletion imposes direct economic costs on water users by increasing pumping costs. Depletion can also impose costs through a loss of ecosystem services, such as processing of organic matter by diverse microbes and invertebrates, providing possible dilution of some types of surface-originating contaminants, and sustaining populations of rare and endangered species that are often restricted to very local habitats (Culver et al., 2000). Further, depleting the stock of groundwater means that less water is available for use, or for maintenance of ecosystem services in the future. With uncertain recharge because future precipitation is uncertain, there is an insurance value from maintaining adequate groundwater stocks. Maintaining adequate stocks helps avoid shortages during drought years, prevents land subsidence, and provides late summer supplies of water to springs and streams for sustaining fisheries and wildlife and for recreational uses (NRC, 1997). Estimating in situ values of groundwater requires a dynamic model that incorporates expected recharge rates, pumping costs, and demand through time. Dynamic renewable resource models of groundwater with uncertain recharge exist and could provide a basis upon which to estimate in situ values (Burt, 1964; Provencher, 1993; Provencher and Burt, 1994; Rubio and Casino, 1993; Tsur and Zemel, 1994), though uncertainties about local hydrology would make it difficult to know the correct model specification (model uncertainty).

The construction of water transfer pipelines and additional surface storage reservoirs in San Antonio is under consideration along with conjunctive storage (pumping water into subsurface storage associated with aquifers). Although surface water can substitute for groundwater for extractive uses, surface water and groundwater do not contribute to the same ecosystem functions nor do they provide the same set of ecosystem services. At present, alternatives to continued reliance on groundwater are on hold because city voters rejected development of the proposed Applewhite Reservoir as an alternative water source.

Dependence on a sole source aquifer leaves communities subject to the risk that they will not have adequate water supply if it is depleted or polluted. As population and economic activity continue to increase in the San Antonio area, it seems unlikely that the Edwards Aquifer will be sufficient to meet future demand for water. Attempts to purchase water from surrounding counties and to build more storage have been under consideration for decades but have not yet materialized. While the establishment of a water market will help reallocate a fixed amount of water to high-value uses, it does not guarantee that adequate supply will be available (Merrifield and Collinge, 1999). Weighing the benefits of extractive use of groundwater versus the value of water in situ for insurance against future drought and for maintaining natural ecosystem functions and the survival of endangered species poses difficult questions. Uncertainties about potential climate change, local hydrology, and the likely future value of ecosystem services, such as provision of drinking water and habitat necessary for the survival of endangered species, complicate the task of informing decision-makers about trade-offs between current extractive use value and in situ value of groundwater. Predictions about likely future aquifer recharge and water demand, as well as evidence about the value of other ecosystem services, such as habitat provision for endangered species, all would help in guiding decisions.

Valuation of Fish Production Provided by Coastal Wetlands and Estuaries

Coastal wetlands (e.g., seagrass meadows, marshes, mangrove forests) are increasingly recognized as providing economically valuable ecosystem services. One of the most important services provided by coastal wetlands is the provision of important habitat for many species of commercially harvested fish, crustaceans, and mollusks (Beck et al., 2001). Given their high diversity and productivity, coastal wetlands are often referred to as nurseries (Boesch and Turner, 1984; NRC, 1995).

The economic value of coastal wetlands as breeding and nursery grounds can be estimated using a production function approach (see Chapter 4 and Appendix C). In economic terms, a coastal wetland is like a production facility or factory that transforms inputs (nutrients, energy) into valuable outputs (fish, crustaceans, and mollusks). The production function approach applied to fisheries requires being able to estimate the increased quantities of various marketable species produced when coastal wetlands are preserved. Then, the value of the coastal wetland as breeding and nursery grounds can be estimated by calculating the increase in consumer and producer surplus due to the increased production. Barbier (2000) provides a review of production function approaches to economically valuing the ecological function of coastal wetlands as breeding and nursery grounds.

Estimates of value of coastal wetlands for fisheries production have ranged widely. For example, Barbier and Strand (1998) estimated that conversion of one square kilometer of mangrove in Campeche, Mexico, to other than natural uses reduced the value of annual shrimp harvest by more than $150,000 for 1980 to 1981. Such a large value argues for protecting the mangroves even when ignoring the value of other ecosystem services. On the other hand, Swallow (1994) found that loss of normal-quality wetlands reduced fishery values by an estimated $2.77 per hectare, or $277 per square kilometer. Swallow concluded that protecting normal-quality wetlands is not justified because the economic value of increased value of shrimp production is less than the value of agricultural development. Basing such a conclusion on the economic value of a single ecosystem service, however, is premature; only when the value of all ecosystem services provided by the wetland is less than the value of agricultural development can such a conclusion be justified.

A major difficulty with the production function approach in the context of coastal wetlands and fisheries is the complex nature of the ecological relationships involved. Subtle changes in nutrient cycles, water temperatures and currents, and fluctuations in the populations of predators and prey, all can have a large influence on the number of fish that reach adulthood. Large variations in fish populations occur even with no apparent change in physical conditions.

The production function models of wetlands and fisheries employed by economists to date have assumed simple ecological relationships that ignore most of this complexity. Starting with Lynne et al. (1981), these models assume that the productivity of the systems is a simple nonlinear function of the area of coastal wetlands. Static production function models assume that productivity increases with the natural logarithm of area (Bell, 1989, 1997; Farber and Costanza, 1987; Lynne et al., 1981), or that the natural logarithm of productivity increases with the natural logarithm of area (Ellis and Fisher, 1987; Freeman, 1991). Dynamic production function models (Barbier and Strand, 1998) include effects of population stock size as well as area of coastal wetlands. Increasing coastal wetland area shifts the natural population growth function up (stock-recruit function) that defines population in one period as a function of the population in the previous period. However, both the static and the dynamic production function models do not account for other important environmental factors such as the aforementioned nutrient cycling, temperature, or currents, nor do they attempt to account for stochasticity in ecological conditions or in species populations. While these models are suggestive of increased fisheries productivity from wetlands, more work is needed before quantitative estimates of the value of increased productivity can stand up to critical review. An ongoing challenge will be to discern realistic ecological relationships between structure and function of coastal wetland ecosystems and fisheries productivity amid the complex and seemingly chaotic fluctuations in fishery stocks.

How fisheries are managed also influences estimates of value (Freeman, 1991). An optimally managed fishery typically generates far higher economic returns than does an open-access fishery. For example and as noted previously, Barbier and Strand (1998) estimated that the annual value of a square kilometer of mangrove was more than $150,000 in 1980 to 1981, but dropped to less than $90,000 in 1989 to 1990 when overfishing had depleted stocks, resulting in lower harvests. In addition, market prices, which depend on consumer preferences as well as production from other ecosystems, will affect estimates of value.

For commercially marketed outputs, well understood methods can be used to estimate the change in consumer plus producer surplus from a change in available resource stock. The major difficulty in applying the production function approach is the great uncertainty typically present in understanding the link between structure and function of coastal wetlands and productivity of fisheries. Complexity of ecosystems, chance events, and natural variability of populations all make it difficult to discern the input-output relationships that are necessary for estimating a production function. Assumptions about fisheries management and market conditions will also influence estimates of economic value.

Provision of Flood Control Services by Floodplain Wetlands

Flood control is an important ecosystem service provided by riverine and coastal floodplains. Floodplains absorb excess water during floods that otherwise might inundate and damage developed areas. In addition to providing flood control, floodplain ecosystems provide critical resources for plant and animal communities. Despite their importance, humans have attempted to replace or supplement natural flood control services provided by floodplains by building flood control structures (e.g., dams, reservoirs, levees, floodwalls). The magnitude of flood control infrastructure development is evidenced by the fact that as a result of the Mississippi River flood of 1927—which inundated 5.26 million hectares and forced 700,000 persons to relocate—Congress authorized $325 million for flood control works on the Lower Mississippi River, which at that time was the largest public works expenditure in U.S. history (Hey and Philippi, 1995; Wright, 2000). In fact, during the height of the flood control movement spanning 1936 to 1951, Congress spent more than $11 billion for flood control projects (Wright, 2000). Although development of this regionally engineered infrastructure has protected some areas of the United States from flood damage, it has also served to promote floodplain development. Such development ultimately exacerbates levels of flood damage during large precipitation events. Furthermore, flood control structures have often given farmers and city dwellers a false sense of protection.

In principle, flood control services provided by floodplain ecosystems can be clearly defined and quantified. They are an input into production of a valuable service, namely reducing the probability of damage from floods. In this sense, floodplain ecosystems perform a role in of flood control similar to that of coastal wetlands in fishery production—one valuation method is to estimate how changes in the ecosystem lead to changes in production of the service in question and then to value the change in the service. The simplest method for economically valuing floodplain ecosystems in providing flood control is to multiply estimates of the change in probability of floods of various magnitudes with and without floodplain conservation by the estimate of damage that floods of various magnitudes would cause. This method is essentially what insurance companies routinely do in assessing risks.

A complication in assessing flood control is that measures to prevent floods or ameliorate the damage may cause changes in human behavior. For example, if the risk of building in a floodplain is lowered, there is less reason to avoid floodplain development. Further, if those building in the floodplain do not have to pay full costs for damages from floods (e.g., they are provided with subsidized flood insurance or with disaster payments that reimburse damages from floods), then one might expect excessive development in floodplains. Insurance companies are no stranger to this phenomenon, which has been referred to as a "moral hazard." Conducting an assessment of the value of flood control services depends on assumptions about patterns of development and infrastructure. Assuming that existing buildings and infrastructure are fixed and immovable will result in a different answer than an approach that factors in a behavioral response. While doing the latter is more realistic, it is also more difficult.

Another complication in evaluating wetlands and floodplains in providing flood control is that the value of this service also depends on human-engineered infrastructure in the form of dikes, levees, or flood control dams. Floodplain ecosystems and dams are alternative ways to prevent floods, similar to watersheds as alternatives to filtration plants to produce clean water. Information relevant to the value of floodplains in providing flood control is given by *avoided costs* of human engineered flood control through dikes, levees, or flood control dams. For example, the USACE opted to purchase 3,440 hectares of floodplain wetlands in the upper portion of the Charles River watershed in Massachusetts. By protecting this land, the Corps estimated that almost 62 million cubic meters of water could be stored on the floodplain—similar to the capacity of a proposed dam. Purchase of the development rights to these floodplain wetlands cost $10 million, which was one-tenth of the $100 million estimated for the dam and levee project originally proposed (American Rivers, 1997; Faber, 1996). This natural wetlands flood control system was able to deal with large floods during 1979 and 1982. For a discussion of replacement cost as a method to estimate the economic value of an ecosystem service see the discussion of the Catskill watershed above.

The Napa River Flood Protection Project in California provides another example that includes both structural and nonstructural flood protection approaches. These range from residential and commercial development relocation, to road reconstruction and bridge removal, along with floodplain reconstruction of 80 hectares of seasonal wetlands, intertidal mudflats, and emergent marshlands. The $155 million cost of the project is a fraction of the estimated $1.6 billion that would have to be spent by Napa County to repair flood damage over the next 100 years if the project is not implemented. The project is projected to save the community $20 million annually (USACE, 1999).

Although much anecdotal information exists regarding how flood damage is related to alterations of natural floodplains and subsequent development in high flood risk areas, determining what percentage of total flood damage costs can be attributed to wetland drainage and floodplain alterations is difficult. For example, in the Upper Mississippi River basin, a strong relationship was found between flood damage and wetland destruction; areas having fewer wetlands due to drainage generally suffered greater flood damages. Likewise, in the Puget Lowlands in Washington State, water discharge events (with a recurrence interval of 10 years prior to urbanization) increased in frequency (to a recurrence interval of 1 to 4 years) after urbanization, with the increase in probability of flooding proportional to the degree of urbanization (Moscrip and Montgomery, 1997).

Wetlands and floodplains generate other services that benefit the public, such as wastewater reclamation and reuse, pollution abatement, aquifer recharge, and recreation. One study that attempted to estimate values for a range of ecosystem services in monetary terms is a study of the multipurpose Salt Creek Greenway in Illinois (Illinois Department of Conservation, 1993; USACE, 1978). The sum of the natural values of floodplain land, other than for flood control, was estimated at $8,177 per acre. The estimated value of regional floodwater storage was $52,340 per acre (Forest Preserve District of Cook County Illinois, 1988). Combining these estimates provides an estimated total value of preserved floodplain land of $60,517 per acre. Such high values indicate that preserving floodplain ecosystems was the best use of such land, far outstripping its value in agriculture or development. Demonstrating the magnitude of these values in a clear and convincing fashion would encourage sensible land use decisions that include the preservation of floodplains where their value is high (Scheaffer et al., 2002).

In general, the value of an ecosystem service will vary with its level of provision. For example, the preservation of an additional acre of floodplain wetlands will tend to be quite valuable when only a few acres of wetlands have been similarly preserved and the probability of flooding is high. In contrast, the value of preserving an additional acre of wetlands will tend to be smaller when many acres of wetlands have already been preserved and the probability of flooding is low. Estimates such as those provided in the preceding paragraph are stated in a way that makes it seem as if the value of an additional acre of floodplain wetlands is constant. Indeed, estimates of marginal changes are sometimes derived by equating them with the average value per unit over a large change. When marginal values are not constant however, this will result in biased estimates of marginal value.

Reasonably good information to estimating the value of floodplain ecosystems in providing flood control, at least in some cases. Hydrologic models can be used to estimate the amount of water that a floodplain ecosystem can absorb during a flood. Economic values from lowering the risk of damages from floods can be estimated with reasonable precision and, in fact, are calculated by government agencies and private insurance companies on a regular basis. Trying to incorporate changes in human behavior or investments in flood control infrastructure are complications that can affect valuation estimates. As with the other cases of estimating the value of single ecosystem services, such estimates should not be confused with estimates of the value of the ecosystem itself, which would require estimates of a range of ecosystem services.

Summary

Studies that focus on economically valuing a single ecosystem service show promise of delivering results that can inform important environmental policy decisions. In some cases, the valuation exercise is clearly defined, there is sufficient natural science understanding and information available, and well-supported economic valuation methods can be applied to generate reliable estimates of value. The provision of drinking water for New York City by protecting watersheds in the Catskills is an example in which evidence of the cost of replacing an ecosystem service informed decision-making. In other cases, the valuation of ecosystem services has not advanced far enough to provide clear and compelling evidence for formulating policies that are likely to be accepted by competing interests. Although some information is available, more work is necessary before reasonably precise estimates of the value of in situ groundwater can be made in the case of the Edwards Aquifer. The impacts of drought and legal issues regarding endangered species and rights to groundwater make such economic valuation efforts quite complex. Similarly, while providing useful information, studies on the value of coastal wetlands for fishery production are in need of further refinement before a high degree of confidence can be attached to estimates of economic value. Even where there is reasonably good information and valuation methods are available, details about ecological functions, the dynamics of ecosystems, human institutions, and human behavior can make estimation of economic value a difficult task. However, the limited scope of valuing a single ecosystem service allows researchers to address many of these complications.

One danger inherent in the economic valuation of a single ecosystem service is mistaking this value for the value of the entire ecosystem. Ecosystems produce a wide range of services and the value of a single service will necessarily represent only a partial valuation of the entire ecosystem. Sometimes this partial valuation is enough for purposes of decision-making, as in the New York City example. Other times, as in the case of Swallow's (1994) integrated ecological-economic analysis of the impacts of wetlands conversion on coastal shrimp nursery habitat in North Carolina, it will not be enough. Although that particular study provides a reliable estimate of the economic costs of wetlands conversion in terms of loss of key hydrological function and consequent effects on shrimp nursery habitat, other important ecosystem services provided by wetlands were not considered or addressed. Thus, there is a danger that the study could be used to advocate too much conversion of wetlands with the concomitant loss of a multitude of ecosystem services.

Valuing Multiple Ecosystem Services

This section reviews three examples that estimate the economic value of multiple services from an ecosystem. As discussed throughout this report, ecosystems provide a wide range of services. Because of the interconnection of processes within an ecosystem, it may be difficult to isolate and study the production of one ecosystem service without simultaneously considering other services. Further, production of some ecosystem services may be in conflict with provision of others. In such cases, providing clear policy advice requires the simultaneous estimation of multiple ecosystem values. Expanding the range of ecosystem services covered brings the resulting estimates of economic value closer to providing an accurate estimate of the value of all ecosystem services. Nevertheless, these studies, although more comprehensive than single ecosystem service studies, still represent only partial estimates of the complete economic value of services generated by an ecosystem.

Fish Production, Irrigation Waters, Navigation, Flood Control, and Clean Drinking Water: The Columbia River Basin

The Columbia River basin is the fourth largest in North America, covering large portions of the States of Idaho, Oregon, and Washington and the Canadian province of British Columbia. The Columbia River provides a wide range of ecosystem services including hydroelectric power, water supply for municipalities and industries, irrigation for agriculture, transportation, recreation, fish production, and diverse aesthetic values. The basin is highly developed and contains a large number of dams, including 18 on the mainstem of the Columbia and Snake Rivers; most of the large dams are multipurpose (i.e., hydroelectric power generation, flood control, irrigation, recreation, municipal and industrial water supply). Besides hydroelectric power generation, a major economic benefit of the dams is storage of snowmelt runoff and diversion of water for irrigated crops during the growing season. Navigation is also enhanced by maintenance of sufficient river depths. The dams along the Lower Columbia and Snake Rivers allow barge transportation to Lewiston, Idaho, making it a port with access to the ocean despite being located 465 river miles inland.

However, the dams along the Snake River and the mainstem of the Columbia River have been at the center of a major controversy. On the one hand, dams provide a range of economic benefits as listed above; on the other hand, dams are blamed, at least in part, for declines of Columbia and Snake River salmon stocks. One study estimated that the number of wild adult salmon returning to the Columbia River was less than 10 percent of the presettlement numbers of 8 million to 10 million (NRC, 1996). Several fish stocks are listed on the federal threatened and endangered species list including: spring- and summer-run chinook, fall-run chinook, sockeye, steelhead, and bull trout in the Snake River; spring-run chinook, steelhead, and bull trout in the Upper Columbia; steelhead and bull trout in the Mid-Columbia; and chinook, chum, steelhead, and bull trout in the Lower Columbia. The dams have fundamentally changed the ecology of the river, altering it from free-flowing to a chain of reservoirs linked by rivers that impact both downstream migration of juvenile fish and upstream migration of spawning adults (Deriso et al., 2001; NRC, 1996; Schaller et al., 1999). These dams have also closed-off access to 55 percent of the drainage area and 31 percent of the stream miles of original salmon and steelhead habitat in the Columbia River basin (NRC, 1996).

However, dams are not thought to be the only reason for the decline in the wild salmon population in the Columbia River basin. Urban development, industry, agriculture, grazing, mining, forestry, the large-scale introduction of hatchery fish, fish harvesting, ocean conditions, and climate change are also implicated. Forestry and grazing practices that result in reduced streamside vegetation can increase water temperatures above beneficial levels for salmon (Beschta, 1997; Beschta et al., 1987; Platts, 1991; Rishel, 1982). In fact, failure to attain stream temperature standards is the most prevalent water quality violation in the Pacific Northwest (Wu et al., 2003). Water withdrawals for irrigation reduce instream flow and water diversions without screens lead to loss of juvenile fish (Jaeger and Mikesell, 2002; NRC, 1996). Removal of woody debris, changes in water velocity, and erosion causing increased siltation of streams also negatively impact salmon populations (Hicks et al., 1991; NRC, 1996). Furthermore, ocean and climate conditions influence salmon populations, including decade-long changes in ocean conditions that affect currents and upwelling in the Pacific Northwest (Hare et al., 1999; Nickelson, 1986); interannual variability in precipitation influenced by El Niño-Southern Oscillation and other periodic climate shifts (Hamlet and Lettenmaier, 1999a,b; Miles et al., 2000); and long-term climate change (Beamish and Mahnken, 2001; Beamish et al., 1999; Pulwarty and Redmond, 1997).

Decision-making about fisheries management, land management, and the operation of the hydroelectric dams involves calculations of the effect on salmon populations and on other valued ecosystem services. The effects of various alternative management actions on salmon stocks and on electricity generation, irrigated agriculture, navigation, and other economic activities have been analyzed in a number of ecological and economic studies (NRC, 2004). Debates on whether to remove hydroelectric dams on the Lower Snake River focused attention on the costs and benefits of dam removal. Several recent ecological and economic studies analyze the effects of the removal of dams (Budy et al., 2002; Grant, 2001; Gregory et al., 2002; Kareiva et al., 2000; Levin and Tolimieri, 2001; Poff and Hart, 2002; Schaller et al., 1999). The benefits of restoring migratory routes for fish to upper headwaters are widely appreciated. The costs of removing sediments that accumulate in reservoirs by dredging or by allowing sediments to be washed downstream and alter spawning substrates (by infilling gravels with fine mud) are difficult to quantify but often significant. Furthermore, elimination of some dams that currently form barriers to fish migration (preventing non-native species from moving upstream and displacing native fish species) may be important costs, not benefits, in some rivers. The USACE estimated that forgone economic benefits that would occur with the removal of four dams on the Lower Snake River would be $267 million annually (USACE, 2002), though Pernin et al. (2002) derived far lower estimates of forgone benefits from dam removal. At present, there is no consensus on how costly dam removal would be or on how effective such actions would be for salmon recovery throughout the Columbia River Basin.

Studies have been undertaken of the costs and benefits of enhancing river flows or restoring more natural patterns of flow, such as allowing more spring flooding to remove fine sediments to enhance spawning conditions (Adams et al., 1993; Fisher et al., 1991; Jaeger and Mikesell, 2002; Johnson and Adams, 1988; Moore et al., 1994, 2000; Naiman et al., 2002; Paulsen and Hinrichsen, 2002; Paulsen and Wernstedt, 1994; Wernstedt and Paulsen, 1995). Some of these studies include integrated ecological and economic models that build from biological models of fish populations to economic models of the valuation (Adams et al., 1993; Johnson and Adams, 1988; Paulsen and Wernstedt, 1995; Wernstedt and Paulsen, 1995). Studies by Johnson and Adams (1988) and Adams et al. (1993) estimated the value of increased flows in the John Day River in Oregon for recreational steelhead fishing. These researchers estimated changes in fish population by combining a hydrologic and a biological model. They then combined this estimate using contingent valuation methods to derive an estimate of value for an increased fish population.

Economic studies that focus strictly on valuing recreational or sportfishing in the Pacific Northwest include Olsen et al. (1991) and Cameron et al. (1996); though other studies have valued salmon fishing in Alaska (Layman et al., 1996) and central California (Huppert, 1989). Valuation estimates vary depending on the location of the study and the methodology employed. Other studies have focused on costs of providing increased streamflows (Aillery et al., 1999; Jaeger and Mikesell, 2002; Moore et al., 1994, 2000). Jaeger and Mikesell (2002) noted that the costs of augmenting streamflows to increase the survival of native fish in the Pacific Northwest are likely to be "modest" (between $1 and $10 per capita per year within the region). Studies have also evaluated the costs and benefits of modifying habitat condition (Loomis, 1988; Wu et al., 2000) and decreasing stream temperatures (Wu et al., 2003). Another area of research is on the cost-effectiveness of fish hatcheries that were initially built to offset losses of migratory fish after dam construction (Congleton et al., 2000; Levin et al., 2001; Lichatowich, 1999; Meffe, 1992). Populations of hatchery-reared fish are known to have different genetic composition and behaviors than wild populations of the same species, and in some cases, these hatchery-reared fish may compete with or breed with wild populations thereby diminishing the stocks of those populations best adapted for long-term survival in the wild (Fisher et al., 1991).

Efforts to rebuild salmon stocks have been going on for several decades. The Pacific Northwest Electric Power Planning and Conservation Act of 1980 created the Northwest Power Planning Council to create a plan "to protect, mitigate and enhance fish and wildlife, including related spawning ground and habitat, on the Columbia River and its tributaries while assuring the Pacific Northwest an adequate, efficient, economical and reliable power supply." Despite legal authority and expenditures of more than $3 billion to date (Northwest Power Planning Council, 2001), salmon populations have not recovered.

In part, this failure is due to the lack of scientific understanding about what measures are likely to be effective in restoring salmon: "The list of central topics that we know too little about is surprisingly long. The topics include, for example, the survival of young fish between dams compared with their survival as they pass through and over dams; the relationship of survival of young fish to the flow rates of water in rivers; the effects on survival of various management practices including logging, grazing, irrigation, agriculture, and use of hatcheries, the influence of ocean conditions. . ." (NRC, 1996). Such pervasive uncertainty has led to calls for increased research effort to reduce critical uncertainties (NRC, 1996) and for adaptive management (Lee, 1993, 1999; Walters, 1986). Several studies have analyzed the value of reducing uncertainty by learning or better forecasting ability (Costello et al., 1998; Hamlet et al., 2002; Paulsen and

Hinrichsen, 2002). At present, managers face a difficult challenge in making decisions under uncertainty (see also Chapter 6). Sometimes decisions cannot wait for science to provide clear evidence, but decision-making without clear evidence allows the management policies to be attacked as excessively risky. Such policies impose potentially high costs on certain sectors of society while lacking an adequate basis of scientific support to show that they will be either biologically effective or efficient (cost-effective). The fact that some consequences are irreversible (e.g., extinction) raises the stakes further.

Questions such as how to recover salmon populations and how to protect or restore other ecosystem services in the Columbia River basin have been, and likely will continue to be, contentious issues. The costs of recovery efforts for salmon are high, already topping several billion dollars (Northwest Power Planning Council, 2001). Changing the fisheries management, regional land use, or operation of dams could lead to fundamental changes in the functioning of the ecosystem, with consequent effects on the production of multiple ecosystem services, ranging from hydroelectric power generation to the existence value of salmon. At present, there are large gaps in the scientific understanding of how such changes would impact important elements of the ecosystem, particularly salmon populations. Even if those scientific controversies were resolved, difficult valuation questions would remain. Furthermore, estimating the existence value and spiritual value of salmon with currently available economic valuation methods is controversial (some would argue economic methods cannot fully capture such values; see Chapter 2 for further information). The large and uncertain costs and benefits of alternative proposals, which will fall disproportionately on different groups within society, amplify the difficulty of decision-making. The political nature of this controversy will make it a difficult arena for ecosystem valuation to be viewed as rational, objective, and conclusive. Despite these challenges, it is important to try to impart good information to such debates.

Upstream Versus Downstream Water Use: Losses in Downstream Economic Benefits as a Result of Upstream Diversion from Dams

The development of the Hadejia-Jama'are floodplain in northern Nigeria is one of many examples worldwide where water diversion upstream (associated with dams) is negatively affecting economic activities downstream. Supporters of dams and water diversion projects typically point to the economic benefits created by such projects but often fail to consider costs imposed elsewhere. In this particular case, economists and hydrologists worked together to estimate both upstream benefits and downstream costs (Acharya and Barbier, 2000, 2001; Barbier, 2003; Barbier and Thompson, 1998). These studies are among the few integrated case studies to assess the impact of upstream water allocation on water availability and groundwater recharge downstream and to value the effects on irrigated agriculture and potable water supplies downstream.

Barbier and Thompson (1998) combined economic and hydrological analysis to compare the benefits of upstream diversion with losses of downstream floodplain benefits in terms of agriculture, fishing, and fuel wood. They found that fully implementing all existing and planned upstream irrigation projects results in losses of approximately $20 million (1989-1990 U.S. dollars) versus the case with no irrigation upstream. Full implementation of upstream irrigation project generated estimated benefits of approximately $3 million, while floodplain losses were estimated to be around $23 million. Acharya and Barbier (2000, 2001) analyzed impacts of a one meter drop in groundwater from lower water recharge in the floodplain on dry season agriculture and rural domestic water use in villages. They estimated annual losses of $1.2 million in irrigated dry season agriculture and $4.8 million in domestic water consumption for rural households. These analyses strongly suggest that expansion of existing irrigation schemes within the river basin is not economically desirable (Barbier, 2003).

In a very different setting, Berrens et al. (1998) reported similar conclusions about upstream diversions of water. The purpose of this study was to analyze the costs of imposing minimum instream flow regulations in the Colorado River to protect endangered fish species. However, instead of costs they found that imposing instream flow restrictions generated overall positive net benefits because it allowed more water to be used further downstream where it would be put to higher-valued uses.

Cumulative alterations in hydrologic connections in the landscape exert major environmental and economic effects at different spatial scales (e.g., Pringle, 2001). In the last decade, ecologists have begun to identify and quantify the substantial environmental consequences of dams on local, regional, and even global scales (e.g., McCully, 2002; Pringle et al., 2000). However, relatively few integrated studies have evaluated economic consequences from hydrologic modifications and the resultant changes in provision of ecosystem services. Even at local scales, studies are conspicuously lacking that attempt to quantify the economic costs to downstream human activities from upstream water diversions such as those associated with dams. In many cases, damage assessments are attempted decades after a dam is completed so research is dependent on historical records to recall or reconstruct wetland environments and associated economic activities that once existed. For example, researchers are dependent on midden piles (i.e., a collection of biotic materials that can provide a paleoenvironmental history of an area) to assess the extent of shellfish production near the mouth of the Colorado River before dams diverted virtually all of its flow.

Fully evaluating the consequences of many projects, such as dams and water diversions, requires assessment of the change in value of ecosystem services that may play out at different spatial scales. Some of the consequences may occur far removed from the site of the project, such as consequences to downstream environments (floodplains, deltas, etc.). As the case studies of the

Hadejia-Jama'are floodplain illustrate, a full accounting of downstream consequences can generate a different perspective of whether a project generates positive or negative net benefits.

Other well-known examples, such as water use in the Colorado River, the hypoxic zone in the Gulf of Mexico caused by high nitrogen runoff from Mississippi River drainage, and the drying of the Aral Sea due to upstream diversion of water, further illustrate the importance of considerations of downstream consequences. Ecosystem processes are often spatially linked, especially in aquatic ecosystems (see Chapter 3 for further information). Full accounting for the consequences of these actions on the value of ecosystem services requires understanding these spatial links and undertaking integrated studies at suitably large spatial scales to fully address important effects.

Food Production, Recreational Fishing, and Provision of Drinking Water from Lakes: Lake Mendota, Wisconsin

In many ecosystems it is difficult to isolate the economic value of a single good or service because of the complex connections among species and ecosystem functions. For example, food production such as a largemouth bass may seem obvious as an economic "good" derived from a lake ecosystem. Similarly, the recreational value of fishing may be measured by economic analysis as another good. However, much of an ecosystem's productivity may not produce a harvestable yield of interest to human consumers (algae or other aquatic plants). Furthermore, the type of fish (largemouth bass, lake trout, or carp) may also vary in value as products for either food or recreation. Although productivity is a fundamental measure of ecosystem functioning (see Box 3-1), it is different from what economists would typically use to evaluate human uses of ecosystem function. Generally, ecologists measure units of energy required for a species maintenance (respiration) and the energy converted to live matter (biomass) per unit area per unit time as the total productivity, whereas economists focus on harvestable amounts of certain desirable species as the valuable yield or one type of good produced by the ecosystem. Breakdown of dead organic matter through decomposition by microorganisms might be deemed an ecosystem service that maintains clean water in the lake, but its economic value is difficult to isolate from the recycling of nutrients needed for the productivity of plants and animals. Clean drinking water, food production, and recreation are all products of a lake ecosystem, but it is not easy to measure each one separately or to resolve conflicting views on which one is more or less important if trade-offs in management decisions are required. Removing excessive nutrients from a lake will improve drinking water quality (up to some point), but the resulting effect on fish production requires careful study of the entire food web.

Lake Mendota, located on the edge of the campus of the University of Wisconsin, Madison, is probably the most thoroughly studied medium-sized lake (>4,000 hectares) in the world (e.g., Brock, 1985; Kitchell, 1992; Lathrop et al.,

1998, 2002). In the early 1980s, the combined decline of walleye populations and recreational fishing, together with concerns over unpredictable outbreaks of noxious and sometimes toxic Cyanobacteria (blue-green algae) in the lake, led to a joint research effort that demonstrated that water quality and food web management could be successfully integrated. This research effort focused on the following issues: (1) trade-offs between increased stocking for walleye and northern pike fishing or managing for bass or perch (distinctly different goods for different groups of people); (2) effects of increased water clarity following removal of algae by grazing zooplankton on deep light penetration that can result in increased growth of submerged aquatic plants[2]; and (3) effects of improved water quality (clear water with lower concentrations of dissolved nutrients) that may reduce fish productivity and result in lower recreational fishing harvest levels. Finding the right balance of the production of various ecosystem goods and services is challenging, especially since what happens in the lake ecosystem depends on management decisions for the surrounding land as well. Inflowing waters from agricultural sources and municipal sewage treatment plants can provide excessive nutrients without appropriate land and municipal wastewater management. Conventional management approaches often focus on one sector at a time. However, management to address the problems of one sector may increase problems in other sectors if important interconnections are ignored. Successful management requires understanding the linkages between sectors and may require interdisciplinary teams to address complex multisector issues.

Economic analyses of ecosystem services of Lake Mendota (Stumborg et al., 2001) and similar lake ecosystems have considered costs and benefits of managing eutrophication relative to recreation, real estate values, drinking water quality, and other site-specific attributes (Boyle et al., 1999; Brock and de Zeeuw, 2002; Carpenter et al., 1999; D'Arge and Shogren, 1998; Wilson and Carpenter, 1999). These studies illustrate the unique aspects of Lake Mendota that constrain benefits transfer of results to other lakes. They also highlight the considerable uncertainties in lake management. Significant sources of uncertainties are related to high levels of temporal variability in lake ecosystem dynamics, surrounding land-use changes, and hydrological variables. For example, regional droughts greatly reduce inflows, increase residence times of nutrients, and often decrease transport of suspended sediments that affect water quality by altering turbidity and light regimes, as well as influencing nutrient input, transport, and cycling (Kitchell, 1992). Land clearing for development generally increases peak flows of runoff, increases bank erosion of tributaries that drain into lakes, and greatly increases turbidity. Thus, despite intensive programs to remove nutrients from point sources such as sewage treatment plants, continued input of nutrients from diffuse, nonpoint sources (e.g., fertilizers from

---

[2] Macrophytes provide critical habitat structure used by juvenile fish to avoid predators, but some can become weedy and reduce dissolved oxygen in shallow, nearshore lake regions during late summer and winter months when the dead plants decay

agricultural runoff, soil erosion, septic tanks) remains a major challenge in many watersheds (NRC, 2000).

Aquatic ecologists manipulated fish and zooplankton species to regulate algal production and restore clear water to lakes. Some lakes were covered with green scum and characterized by fish kills resulting from deoxygenation during warm-water periods in late summer. Ecologists learned that successive, small increments of phosphorus additions to lakes were critical to eutrophication in many situations. The ratio of phosphorus to nitrogen was also found to alter the species composition of the planktonic algae. Low values of phosphorus led to the dominance of lake waters by green algae that were readily consumed by grazing zooplankton and fish. Incremental nutrient additions caused lakes to flip from one state (clean water) to another (green, turbid water) that altered ecosystem services and lowered real estate values of surrounding property (Carpenter et al., 1999; Wilson and Carpenter, 1999).

Although harvesting fish was known to remove nutrients, especially phosphorus, and to alter pathways of food webs to minimize algal blooms, the effects of large-scale applications of this approach to managing water quality in Lake Mendota and other lakes remained unknown until a number of field experiments and models were completed (DeMelo et al., 1992; Gulati et al., 1990; Kitchell, 1992; Reed-Anderson et al., 2000). The concept of removing some dissolved nutrients from the open waters by optimizing their incorporation into green algae that is later consumed by zooplankton, and then by juvenile fish, was widely understood to work in small ponds but was not often tested in lake ecosystems. Excretion of nutrients by grazers and predators can increase nutrient turnover and productivity, but understanding and stabilizing the balance of different consumer species in food webs remains complex. Lake management efforts use a combination of biomanipulation of food webs (Shapiro, 1990), diversions of some tributaries that have high nutrient loadings, and nutrient removal technologies that focuses on point sources. This combined management approach provides an opportunity to examine trade-offs between alternative investments in water pollution control and recreational fisheries management.

Summary

As the case studies in this section illustrate, aquatic ecosystems produce multiple services, many of which are closely interconnected. These interconnections often make it difficult to analyze one service in isolation. For example, a dam that diverts water from a river or increases nutrient input to a lake may alter ecosystem structure and function in fundamental ways, thereby causing changes in the production of a range of ecosystem goods and services. Thus, increasing the number of services to be economically valued necessarily increases the complexity of the valuation exercise and will likely increase the set of specialized skills and experience needed. Deriving a unified assessment of economic value requires integrating disciplinary skills. This integration becomes increasingly difficult both on an intellectual level and on a practical level as the number of services is increased. The interconnection of ecosystem services may take place on a spatial or temporal scale, as well. As the Hadejia-Jama'are floodplain example illustrates, there are links between the provision of ecosystem services at upstream and downstream sites. Finally, it will often be the case that there are trade-offs among the production of different services. For example, reduced nutrient input into a lake may increase recreational values by decreasing algal blooms and turbidity, but it may also lower total fish productivity. Building a dam will change a section of free-flowing river into a lake, which may result in a decrease in the population of some fish species (e.g., salmon) and in opportunities for river recreation (e.g., canoeing, kayaking, whitewater rafting) while increasing populations of lake-adapted fish species and lake-based recreation (e.g., sailing, waterskiing). Trade-offs among ecosystem services increase the likelihood of sociopolitical debates because different groups are likely to place different relative values on different services. Natural variation, such as interannual differences in flood and drought frequencies and intensities, further complicates issues associated with reaching agreement on trade-offs among different ecosystem services. Although economic valuation of multiple ecosystem services is more difficult than valuation of a single ecosystem service, interconnections among services may make it necessary to expand the scope of the analysis.

Valuing Ecosystems

This section reviews three cases that in some sense attempt to cover the economic value of all ecosystem services either for a single ecosystem or, more ambitiously, for the entire planet. The policy context of these three sets of studies is quite different. The first case study in this section reviews valuation studies done for the purpose of natural resource damage assessment for the *Exxon Valdez* oil spill. The second case, concerning the Florida Everglades, reviews studies that support what is probably the most expensive attempt at ecosystem restoration undertaken to date. The final case study by Costanza et al. (1997) represents the most ambitious attempt at valuation of ecosystem services to date. Its scope is nothing less than the value of ecosystem services for the entire planet (i.e., "the value of everything").

Exxon Valdez *Oil Spill*

In March 1989, the *Exxon Valdez* oil tanker spilled 38,000 metric tons of crude oil (about one-fifth of its total cargo) into Prince William Sound in south-central Alaska. This accident inflicted large-scale environmental damage. Approximately 2,100 km of shoreline were impacted, with 300 km heavily or moderately impacted and 1,800 km lightly or very lightly oiled. Much of this coastline consists of gravel beaches into which oil penetrated to depths as great as one meter. The carcasses of more than 35,000 birds and 1,000 sea otters were found after the spill, but this is considered to be a small fraction of the actual death toll since most carcasses sink. The best estimates are that the spill caused the deaths of 250,000 seabirds, 2,800 sea otters, 300 harbor seals, 250 bald eagles, up to 22 killer whales, and billions of salmon and herring eggs. While lingering injuries continue to plague some species, others appear to have recovered. Knowledge of the fate of the 38,000 metric tons of oil lost by the *Exxon Valdez* is imprecise; however, it is estimated that 30-40 percent evaporated, 10-25 percent was recovered, and the rest remained in the marine environment for some period of time (Shaw, 1992).

Following the accident, both private groups and governments sued Exxon for damages caused by the oil spill. Commercial fish interests pursued their own damages under federal and state law because they had a direct economic stake in the resource. Federal, state, and tribal governments serve as the legal trustees for public resources. The State of Alaska and the federal government sued for damages to public natural resources. Damage to public resources included lost recreational opportunities, diminished passive use values, and diminished use by Native peoples.

To prepare for possible trial in these cases, private parties, the State of Alaska, the federal government, and Exxon commissioned research bearing on the question of damages caused by the oil spill. Recognized researchers in a number of fields were recruited to undertake this research. The research was conducted for the purposes of litigation and took place in a highly charged atmosphere with billions of dollars of potential liability on the line. It was subject to intense scrutiny and generated heated debates over methods and results, particularly about validity and reliability of nonuse values estimated using contingent valuation methods. Although the State of Alaska and the federal government settled with Exxon over damages to public resources in 1991, debates about the validity and reliability of contingent valuation estimates of nonuse values raised by the affair continued. Some analysts extended these critiques to applications of contingent valuation to estimate use values. A conference sponsored by Exxon held in 1992 presented research papers that were quite critical of contingent valuation estimates of nonuse values (these papers were subsequently published in Hausman, 1993). In response to the ongoing controversy over the use of contingent valuation in natural resource damage assessment, the National Oceanic and Atmospheric Administration (NOAA) convened a blue-ribbon panel to assess the validity of contingent valuation applications to nonuse values, resulting in a widely cited NOAA panel report (NOAA, 1993).

Researchers used a variety of valuation techniques to assess the dollar value of damage from the *Exxon Valdez* oil spill to an array of public resources. Economic studies were conducted on recreational fishing losses (using a travel-cost model), impacts on tourism, replacement costs of birds and mammals, and a contingent valuation study of lost passive nonuse values. Studies of sportfishing activity and tourism indicators (i.e., vacation planning, visitor spending, canceled bookings) all indicated decreases in recreation and tourism activity. A major study using contingent valuation was undertaken to estimate losses in (nonuse) values from the oil spill for people who did not visit or directly use the resources of Prince William Sound. There were also studies of lost value from commercial fishing. Commercial fishing losses, although part of the economic measure of damage to the ecosystem, were not part of the public resource injuries. Recreational fishing losses were counted as part of the public resource injuries.

Recreational fishing losses were estimated by two different teams, one representing Exxon and one representing the State of Alaska. Both teams used a random utility travel-cost model to estimate forgone use values but they arrived at estimates that differed by an order of magnitude. Hausman et al. (1992, 1993) estimated losses at $2.6 million to $3.2 million in the first year after the oil spill (1989) depending on the specific model used. This damage estimate would be expected to decline in future years as salmon stocks recovered from the spill. Carson and Hanemann (1992) estimated losses as high as $50 million per year. These differences occurred largely because Hausman et al. (1992, 1993) assumed 16,000 fewer recreational trips per year while Carson and Hanemann assumed 180,000 fewer trips. Hausman et al. (1992, 1993) also estimated lost recreational use values for hunting and hiking or viewing as well as a gain in recreational use value for pleasure boating (due to more trips taken to observe the aftermath of the spill). In total, they estimated "lost interim use values" due to the oil spill of $3.8 million in 1989.

An extensive contingent valuation study (Carson et al., 1994) estimated a loss of $2.8 billion in passive nonuse values by people who did not use or anticipated using Prince William Sound in the future. That estimate was derived from a national in-person survey that asked respondents about their willingness to pay to prevent the ecological harm of an oil spill of the magnitude of the *Exxon Valdez*. The survey found that median household willingness to pay to avoid similar injury to the marine ecosystem of the Prince William Sound region was $31 per household—which results in a value of $2.8 billion when summed across all households in the United States. However, it can be argued that this estimate was conservative and that the value of the ecological damage was far higher. For example, the persons surveyed were informed that ecological damages included 75,000 to 150,000 seabirds, 580 sea otters, and 100 harbor seals, compared to best estimates of 250,000 seabirds, 2,800 sea otters, and 300 harbor seals. Survey respondents were also told that no long-term damage would occur to the ecosystem and that wildlife populations would return to previous densities within three to five years. In addition, willingness to pay was used as the measure of damages, rather than willingness to accept (compensation) estimates, which typically are higher (Hanemann, 1991; see also Chapters 2 and 4). On the other hand, Hausman et al. (1993) were quite skeptical of estimates of nonuse values of several billion dollars when their estimate of use value was only several million dollars.

The replacement costs study identified a per-unit replacement cost of various seabirds and mammals, as well as eagles (Brown, 1992). For example, the market price or the costs of relocating otters vary from $1,500 to $50,000 per otter. Replacement costs cannot be added to the public and private losses noted above, however, because these are expenditures to restore both the ecological services of the ecosystem and the aspects of these services enjoyed by humans (e.g., viewing wildlife and fishing).

A market model was used to evaluate private economic losses to commercial fisheries. Cohen (1995) estimated that the upper bound of the accident's first-year social costs was $108 million. Second-year effects may have been as high as $47 million. Although estimates of economic losses to commercial fisheries are typically far less controversial than estimates of nonmarket values, there remain a number of sources of uncertainty. Cohen (1995) was not able to fully consider the numerous sources of variability inherent in the marine environment that may have contributed to harvest volume impacts but were provisionally attributed to the oil spill. In addition, efforts to distinguish effects of the oil spill on the value of harvest from other potential influences were hindered by inadequacies in economic data on supply responses of other U.S. commercial fisheries and the Japanese commercial fish market (Cohen, 1995). The analysis did not attempt to analyze economic harm to other components of south-central Alaska's regional economy (e.g., fish processing and service sectors) or the extent to which the oil spill contributed to changes in the overall economic climate in south-central Alaska (Cohen, 1995).

Natural resource damage assessments require accurate assessment of the dollar value of damages to ecological resources. However, difficulties in understanding ecosystems, the production of services, and the values of those services are likely to lead to imprecise estimates. A precise determination of the damages caused by the *Exxon Valdez* oil spill is constrained by the dynamic interaction of numerous biological and economic variables (Cohen, 1995; Paine et al., 1996; Shaw, 1992). It is difficult to measure the full impact of the oil spill, to predict the time path of ecosystem recovery, and the extent of recovery that will ultimately occur. Furthermore, it is difficult to disentangle the effects of the oil spill from other environmental changes. Therefore, some unavoidable uncertainty will remain in attempts to quantify the link between the oil spill and changes in the provision of ecosystem services valued by humans. On top of this, valuing changes in the ecosystem involves both use values and passive nonuse values, the latter being notoriously difficult to estimate with much precision. However, even valuing damages to marketed commodities (e.g., the value of lost commercial fishing), where traditional uncontroversial market methods were used, proved difficult and a source of disagreement. Although studies of the value of ecosystem services can generate useful information, the degree of imprecision of the resulting estimates of values leaves plenty of room for arguments in court in natural resource damage assessment cases.

Restoration of the Florida Everglades

The Comprehensive Everglades Restoration Plan (CERP) is a framework (see also Box 3-6) and guide to restore the water resources of central and south Florida including the Everglades. This plan covers an area of 18,000 square miles and is predicted to take more than 30 years to implement. It is designed to regulate the quality, quantity, and distribution of water flows (CERP, 2001). The Florida Everglades ecosystem is one of the most endangered wetland complexes in the United States. More than one-half of the original marshes contained in this highly productive and diverse ecosystem have been drained. The remaining area is dissected by 2,253 kilometers of canals that transport water loaded heavily with nutrients from fertilizer and waste runoff from urban and agricultural lands. The Everglades provides habitat for 14 endangered or threatened species including the Florida panther (*Felis concolor coryi*), wood stork (*Mycteria americana*), and Florida Everglades snail kite (*Rostrhamus sociabilis plumbeus*).

The hydrologic connectivity (Pringle, 2003) between many different ecosystems within the Everglades makes quantifying the changes in ecosystem services due to restoration an extremely complex issue. The Everglades provide recharge water for aquifers across the state. Water flow through the Everglades also affects the salinity and biological integrity of connecting marine waters of Florida Bay. The effects of hydrologic alterations on these interconnected ecosystems are still subject to dispute. These and related issues have served as the basis of several previous National Research Council reports (e.g., NRC, 2002a,b). For example, the effectiveness of regional aquifer storage and recovery[3] as a component of the CERP Plan is limited (NRC, 2002a). While aquifer storage and recovery have many advantages, disadvantages include low recharge and recovery rates relative to surface storage. Likewise, ecological impacts of altered hydrologic flow scenarios into Florida Bay also require more study (NRC, 2002b).

The Florida Everglades includes 4 national parks and preserves, 13 national wildlife refuges, 2 national marine sanctuaries, 17 state parks, 10 state aquatic preserves, and 5 wildlife management areas. Everglades National Park was created in 1947 to protect the approximately 20 percent of the remaining wetlands and is thus a vestige of the original Everglades ecosystem (which once included what is presently the Everglades Agricultural Area, the Water Conservation Area, and western portions of coastal urban areas). Large-scale drainage efforts over the last several decades have led to rapid agricultural, commercial, and residential growth (Englehardt, 1998) to the extent that native flora and fauna of the Everglades and adjacent interconnecting systems are imperiled. Efforts to

---

[3] Pyne (1995) defines aquifer storage and recovery as "the storage of water in a suitable aquifer through a well during times when water is available, and recovery of the water from the same well during times when it is needed."

restore hydrologic function (i.e., flows) to the region are complicated by the magnitude and extent of human modification of the landscape.

Waters of the Kissimmee River flow south into Lake Okeechobee (the second-largest freshwater lake in the United States) and then into agricultural fields through an extensive system of flood control canals and reservoirs. Eventually the waters flow into the Everglades and into mangrove forests and estuaries on the Atlantic and Gulf Coasts. The Kissimmee was once a broad (1-2 miles wide), 103-mile-long river that meandered through an extensive network of floodplain wetlands (20,000 hectares). The ecosystem provided habitat for more than 300 fish and wildlife species, including resident and over-wintering waterfowl, a diverse wading bird community, and 13 game fish species. Channelization of the Kissimmee and drainage of approximately two-thirds of the floodplain wetlands were undertaken in the 1960s by the U.S. Army Corps of Engineers to improve flood protection and to provide drainage for agriculture. This has damaged the river-floodplain ecosystem, resulting in a 92 percent reduction in over-wintering waterfowl and negative effects on the native fish community (Englehardt, 1998). Moreover, agricultural drainage waters contain elevated phosphorus concentrations and have caused enrichment of Lake Okeechobee and the Everglades. Algal blooms have resulted in dramatic reductions in dissolved oxygen which has led to the death of many aquatic species; for example, nesting bird populations have decreased by 90 percent over the past 60 years.

One aspect of the CERP is to reestablish historic geomorphic and hydrologic conditions so that the Kissimmee River will once again be connected with its floodplain. This is being accomplished by back-filling the central portion of the dredged flood control canal (mainstem Kissimmee) and reestablishing side channels and backwaters (Toth, 1996). The restoration effort is also attempting to reduce phosphorus levels in the ecosystem by constructing stormwater treatment areas (large constructed wetlands). Other efforts to restore the Everglades include increasing water flows through the region, mimicking historic flow patterns, cleaning up polluted waters (e.g., Guardo et al., 1995), and purchasing private lands to protect them from development.

The economic valuation of restoration alternatives for the Everglades involves many challenges, primarily due to the complexity of the ecological systems (Davis and Ogden, 1994; Englehardt, 1998; Toth, 1996). Although restoration efforts promise to increase habitat for a wide variety of species, it is difficult to predict *how* different species will respond to changes in water quantity and quality. For example, ongoing restoration of the Everglades is dependent on numerous computer models to understand ecosystem processes, test alternatives, and evaluate restoration performance (Sklar et al., 2001). Landscape models used for restoration include hydrologic models, transition probability models, gradient models, distributional mosaic models, and individual-based models. When several landscape models are combined, they have the potential to contribute to water management and policymaking for Everglades restoration (Sklar et al. 2001); however, they have shortcomings based on their inherent assumptions and lack of important information. Although this is one of the most studied ecosystems in the world, much additional ecological knowledge is necessary (Kiker et al., 2001) to improve existing models and develop new ones. Curnutt et al. (2000) developed spatially-explicit species index models to predict how a number of species and species groups (e.g., cape seaside sparrow, snail kite, a species group model of long-legged wading birds) would respond to different hydrological restoration management alternatives. While no one scenario was beneficial to all species, the model allowed assessment of relative species responses to alternative water management scenarios.

Englehardt (1998) evaluated ecological benefits and impacts of proposed and alternative restoration plans in monetary terms. Current plans for restoration involve discharge of phosphorus-enriched water from artificial wetlands (stormwater treatment areas) to relatively pristine Everglades marshes for 3-10 years, risking conversion of the ecosystem to a eutrophic cattail marsh. Uncertain benefits and impacts were analyzed probabilistically, following principles of net present value analysis. This analysis indicated that alternative "bypass plans" would avoid the loss of up to 1,200 hectares of sawgrass marsh at a cost that is probabilistically justified by the value of the ecosystem preserved. This type of analysis can help clarify trade-offs but is complicated by the reality that restoration alternatives may have competing ecological benefits and losses over time. Again, there is also often a lack of scientific understanding and agreement (Englehardt, 1988).

Aillery et al. (2001) provide an analysis of trade-offs between restoration and agricultural economic returns to the Everglades Agricultural Area under alternative water retention targets. They developed a model linking economic and physical systems (including agricultural production, soil loss, and water retention). Effects of water retention scenarios, such as groundwater retention and surface water storage development, on production returns and agricultural resource use were estimated. Not surprisingly, the results suggest that small increases in water retention can be achieved with minimal losses in agricultural income, while agricultural returns decline more significantly with higher water retention targets.

To date there have been no attempts at a comprehensive economic valuation of the Everglades restoration efforts. Given the hydrological, ecological, and economic complexities of South Florida, a complete accounting of values is unlikely anytime in the near future. However, advances in our understanding of hydrological, ecological, and economic relationships could be of great help in guiding future restoration efforts. Such data can be useful in comparing the net benefits of alternative management policies even if an overall estimate of ecosystem values remains elusive.

The Value of Everything: Multiple Services in Multiple Ecosystems

In an ambitious and controversial paper, Costanza et al. (1997) attempted to estimate the total economic value of the services provided by all ecosystems on earth. The paper received a great deal of attention, not all of it favorable. A follow-up briefing article in *Nature* the following year stated that "The paper was a box-office success but was panned by the critics" (Nature, 1998).

In the paper, Costanza et al. (1997) estimated values for 17 ecosystem services[4] from 16 ecosystem types including wetlands, forests, grasslands, estuaries, and other marine and terrestrial ecosystems. To derive estimates of the economic value of ecosystem services, Costanza et al. (1997) began with existing estimates of the productivity of a hectare for each ecosystem type for each service and a willingness to pay estimate for the service. Multiplying these estimates generated a per hectare value of the ecosystem service for each ecosystem type. They then aggregated across all services to establish a value per hectare for each ecosystem type. Finally, they multiplied this per-hectare value by the number of hectares of each ecosystem type and summed across ecosystem types to derive the total value of ecosystem services. For the bottom line, they estimated that the annual value of ecosystem services for the earth ranged from $16 trillion to $54 trillion, with a mean estimate of $33 trillion. This value was notably higher than the value of global GDP (gross domestic product) at the time ($18 trillion).

Critics have pointed out a number of serious flaws that lead to conclusions that the estimate has little scientific merit (e.g., Bockstael et al., 2000; Toman, 1998) while some attacked the approach as a meaningless exercise. If the question is the value of the life support system of the planet, there can be only one of two answers depending upon whether a willingness to pay or a willingness to accept approach is used. Willingness to pay should be bounded by global ability to pay (i.e., global GDP, or $18 trillion). If willingness to accept is used, then as Toman (1998) concludes, $33 trillion is "a serious underestimate of infinity."

Other criticisms focused on problems with the methods and assumptions used in the paper. The paper itself has a long list of "sources of error, limitations and caveat" (Costanza et al., 1997). Obviously, there will be large data gaps in any such exercise. In addition, aggregation issues pose particular trouble in this study. According to Bockstael et al. (2000),

> ...Simple multiplication of a physical quantity by 'unit value' (derived from a case study that estimated the economic value for a specific resource) is a serious error. Small changes in an ecosystem's services do not adequately characterize, with simple multipliers, the loss of a global ecosystem service.

---

[4] These 17 services, in order of importance, were nutrient cycling (accounting for more than 50 percent of the total value), cultural values, waste treatment, water supply, disturbance regulation, food production, gas regulation, water regulation, recreation, raw materials, climate regulation, erosion control, biological control, habitat or refugia, pollination, genetic resources, and soil formation.

> Values estimated at one scale cannot be expanded by a convenient physical index of area, such as hectares, to another scale; nor can two separate value estimates, derived in different contexts, simply be added together.

A similar aggregation problem occurs in ecology, "A linear aggregation rule treats each change as if it could be made independent of the other constituent elements. In doing so, it assumes independence within and across the ecosystems being considered, and it ignores the possible effects of feedback cycles" (Bockstael et al., 2000). The approach used by Costanza et al. (1997) also assumes that ecosystem service production is "scale-free" in the sense that provision per unit area is constant no matter how big or small the ecosystem under consideration. Other papers (see also Chapter 3) have since stressed the importance of more focused analysis that matches the scale of analysis for ecosystem valuation to the scale of management questions (Balmford et al., 2002; Daily et al., 2000).

However, even some harsh critics of the paper have concluded that it served a useful role in getting more attention on the values of ecosystem services. One prominent economist said the paper was "a recklessly heroic attempt to do something futile" but that it was "very useful—it stirred things up a lot." (Nature, 1998)

Summary

In one sense, attempting to economically value all ecosystem services can be viewed as the correct approach to take because it offers a complete accounting. It would certainly be advantageous to have evidence on *all* benefits and costs prior to decision-making because anything less will be partial and incomplete and risks giving incorrect advice to decision-makers. Yet trying to attain the "value of everything" through a complete and reliable accounting of all ecosystem services cannot be done with current understanding and methods and is unlikely to be accomplished anytime soon. Problems arise because knowledge of the translation from ecosystem function to ecosystem services is often incomplete as is the translation from services to values. For studies of the value of a single ecosystem service, and to some extent for studies of the value of multiple ecosystem services, attention can be directed toward services that are easier and relatively straightforward to value, such as the economic value of reducing the likelihood of flood damage or providing clean drinking water without filtration. In the case of the *Exxon Valdez* and the Florida Everglades restoration however, many of the important values are linked to the existence of species or the existence of the ecosystem itself in something akin to its original (pre-human-altered) condition. Valuing such services presents difficult challenges even when ecological knowledge is relatively complete. In addition, aggregation issues can cause problems in comprehensive approaches to ecosystem service valuation, particularly when scaling up the valuation exercise to cover multiple ecosystems.

IMPLICATIONS AND LESSONS LEARNED

This chapter has reviewed a number of applications of ecosystem valuation ranging from economic valuation of a single ecosystem service to attempts to value all services for an ecosystem and even for the entire planet. The valuation of ecosystem services is still relatively new and requires the integration of ecology and other natural sciences with economics. Such integration is not easy to accomplish. Still, examples of approaches and interdisciplinary studies that provide such integration indicate successful beginnings. Some of the lessons emerging from the case studies reviewed in the previous sections are discussed below.

Extent of Ecological and Economic Information for Valuing Ecosystem Services

As examples in this chapter have shown, the ability to generate useful information about the value of ecosystem services varies widely across cases. For some policy questions, enough is known about ecosystem service valuation to help in decision-making. A good example is the value of providing drinking water for New York City by protecting watersheds in the Catskills rather than building a more costly filtration system. As other examples make clear, knowledge and information may not yet be sufficient at present to estimate the value of ecosystem services with enough precision to answer policy-relevant questions.

The inability to generate sufficiently precise and reliable estimates of ecosystem values for purposes of informing decision-making may arise from any combination of the following three reasons: (1) there may be insufficient ecological knowledge or information to estimate the quantity of ecosystem services produced or to estimate how ecosystem service production would change under alternative scenarios; (2) existing economic methods may be unable to generate reliable and uncontroversial estimates of value for the provision of various levels of ecosystem services; and (3) there may be a lack of integration of ecological and economic analysis.

Much of the difficulty in generating reliable estimates of the value of ecosystem services derives from the fact that ecosystems are complex and dynamic and our understanding of them is typically incomplete or flawed. Learning how such ecosystems evolve and change as inputs to the system change can be a slow process (perhaps not even as fast at the system itself is changing). The example of the Everglades and the difficulty in designing a restoration plan aptly illustrate problems inherent in attempting to understand and manage aquatic ecosystems because the links from ecosystem condition and function to the production of goods and services may be hard to decipher. Other examples reviewed include fish production in coastal wetlands and salmon production in the Columbia River, where changes in ocean currents, flow of nutrient, water temperature, precipitation patterns, disease prevalence, predator and prey populations, and other factors can impact fish populations. Although an increase in fish population from one year to the next could be related to a beneficial change in management strategy, it may also be due to changes in ocean conditions or other causes. In other cases, it is not necessary to understand the entire ecosystem in order to be able to estimate the production of an ecosystem service of interest with reasonable precision, such as the degree of flood control provided by wetlands. However, without adequate ecological understanding of ecosystem structure and function, it will not be possible to predict the level of some ecosystem services provided or the way provision levels may change under alternative management options.

Other difficulties arise because some ecosystem services are notoriously difficult to value. As stated previously, it is clear that people place value on such things as the continued existence of species, wilderness, beautiful scenery, and restoring ecosystems to a pre-human-altered condition. Ignoring such values, essentially assigning a value of zero to them, is clearly incorrect. What value should be assigned, however, is often far from clear and subject to debate. Estimating existence values and other nonconsumptive or nonuse values is among the most difficult challenges in environmental economics. For entire ecosystem valuation efforts, such as the *Exxon Valdez* case or the Everglades restoration, estimating such values cannot be avoided because they may account for a significant fraction of total economic value. Although the development and application of nonmarket valuation approaches have advanced significantly over the past two decades (see Chapter 4), there remains controversy, both within the economics profession and outside it, regarding the reliability of economic valuation methodologies (contingent valuation in particular) for environmental goods and services. For some ecosystem services such as valuing commercial fish harvests or the reduction of flood damage, the valuation exercise is more straightforward and uncontroversial. Difficulties may remain in knowing the level of services provided (e.g., how many fish are produced by coastal wetlands) or in obtaining relevant data (e.g., costs of fish harvesting), but there is relatively little disagreement about the utility of existing valuation methodology. One method, however, deserves particular mention and caution.

Using replacement or avoided cost to value an ecosystem service is justified under a restricted set of circumstances—namely, when there are alternative ways of providing the same service and the value of the service exceeds the cost of providing it, such as the provision of drinking water for New York City by increasing the protection of watersheds in the Catskills. However, this approach is sometimes applied when these conditions do not hold, thereby generating numbers that may bear no relation to the actual economic value of ecosystem services. For example, tallying up the large sum of money necessary to restore Prince William Sound to something close to its pre-spill condition does not necessarily imply that the economic value for services provided by the ecosystem is anywhere close to this cost.

Even when ecologists understand a system reasonably well and economists can apply widely accepted valuation methods, an effort at valuing ecosystem services may still fail if ecologists and economists fail to integrate their approaches. Unless the correct questions are asked at the outset, ecological information may not be of particular use for generating estimates of the production of ecosystem services in a useful form for economists to apply valuation methods. For their part, economists may apply valuation methodologies to cases that are not built on solid ecological grounding. It is important for ecologists and economists to talk at the outset of the valuation exercise to design a unified approach. Although it is easy enough to state or even recommend that ecologists and economists need to work together on integrated studies, accomplishing such integration is often difficult because of institutional constraints and reward structures that are largely disciplinary-based. Advances in interdisciplinary efforts may be risky or professionally unrewarding, especially for junior faculty members. It is important to overcome some of the institutional barriers that prevent ready and effective collaboration between ecologists and economists. Explicitly interdisciplinary programs, such as Dynamics of Coupled Natural and Human Systems as part of the Biocomplexity in the Environment Program[5] at the National Science Foundation (NSF), represent a move in the right direction. Expanding "Schools of the Environment" at universities, where faculty from different disciplines interact routinely in addressing environmental issues, is another way to help overcome disciplinary barriers.

As discussed throughout this report, the adequacy of information in providing estimates of the economic value of ecosystem services that are policy relevant depends in large part on what policy question is asked. If the relevant policy question (or questions) can be answered by a relatively narrow evaluation of ecosystem services, the value of ecosystem services can likely be estimated with a relatively high degree of confidence with existing methods. For example, it is possible to answer questions about whether to conserve watersheds to provide clean water is worthwhile, as in the Catskills, or to conserve floodplains for flood control, as in the Salt Creek Greenway in Illinois. However, if the questions were reframed to identify the complete value of the conservation of watersheds or floodplains, there is insufficient information available on which to generate a reliable and credible answer. The issue of the effect of framing in terms of the policy context is also discussed in Chapters 2 and 6.

[5] The NSF Dynamics of Coupled Natural and Human Systems Program emphasizes quantitative understanding of short- and long-term dynamics of natural capital, including how humans value and influence ecosystem services and natural resources, and considering uncertainty, resilience, and vulnerability in complex environmental systems. Further information is available on-line at *http://www.nsf.gov/od/lpa/news/publicat/nsf0203/cross/pma.html*.

Scope of Coverage, Spatial and Temporal Scale

Aquatic ecosystems produce a broad range of ecosystem services. Typically, however, ecological and economic information suitable for estimating reasonably precise values for ecosystem services exists for only a relatively narrow range of services. Lack of natural science (often ecological) information or understanding, or imprecision of valuation estimates for certain services, limits the ability to obtain precise estimates of economic value over the entire range of services provided by an ecosystem. In addition, there is considerable variation in ecosystem structure and function across space and time. As a consequence, the value of services from a particular ecosystem at a particular time may not necessarily be a good predictor of the economic value of services for other ecosystems or even the same ecosystem at a different time. Such ecosystem idiosyncrasies make benefits transfer problematic (see Chapter 4 for a discussion of benefits transfer). For these reasons, measures of the economic value of ecosystems services will continue to be partial and incomplete, at least for the foreseeable future. Some limit on the scope and scale of analysis is inevitable, but just where to set the boundaries for analysis is an important question.

The difficulty in obtaining estimates of economic value for the full range of ecosystem services presents analysts with a problematic trade-off. While relatively precise estimates of the value of ecosystem services may be derived for a fairly narrow set of services, an ecosystem valuation study that analyzes only a partial list of services may be insufficient for policy purposes. For example, suppose a proposed development would destroy a wetland. If relatively uncontroversial estimates of ecosystem service value such as flood reduction and increased fishery production do not exceed the value of development, it may be necessary to estimate values for a wider array of ecosystem services to inform the decision. However, when there are large uncertainties associated with estimates of value of these other ecosystem services, even collecting information on a wider set of ecosystem service values may not yield a clear recommendation about whether it is better to protect the wetland or allow development.

A second difficulty with limiting the scope of coverage of an ecosystem valuation study is the interconnection of processes within an ecosystem. Changing the inflow of nutrients into a lake will change ecosystem function and result in changes in fish productivity, recreational opportunities, and other ecosystem services. When there is a conflict between the provision of different ecosystem services—for example, hydroelectric power generation and fish production—the analysis should include the potentially conflicting ecosystem services if it is to be of use in policy decisions. Further, there may be cascading effects in which changes in one part of an ecosystem can ripple through the ecosystem, causing additional effects that may be difficult to foresee. For example, removal of a top predator may cause an increase in small predators, and changes in the herbivore prey base, with consequent changes in vegetation. It may be difficult to predict a priori how ecosystem functions and services will change when a predator is removed.

The preceding paragraphs strongly favor a more complete scope of coverage and a systems approach to valuing ecosystem services. However, expanding the scope of services covered by the analysis not only increases the workload and range of expertise necessary to design and conduct the analysis, but it will also likely to force analysts to estimate values for services whose production is poorly understood or for which valuation methods may generate imprecise estimates. There are no case studies that include a broad range of ecosystem services for which the value of these services can be estimated within a narrow range with much confidence.

In addition to questions about the scope of services studied, analysts will face difficult issues about the proper spatial and temporal scales. Spatial heterogeneity also limits the utility of benefits transfer, in which the estimates of value generated for one ecosystem are applied to other ecosystems. On the other hand, analyzing every ecosystem in detail can be prohibitively expensive and time consuming. In generating estimates of the economic value of ecosystem services across larger spatial scales, some method of extrapolation may be unavoidable, but such extrapolations bear careful scrutiny.

Interconnections in the production of ecosystem services across whatever spatial boundaries are chosen are virtually inevitable. A real danger of being too narrow in spatial scale is that important linkages in the production of ecosystem services or in the value of those services will be ignored. For example, focusing on upstream benefits from dams in the case of the Hadejia-Jama'are floodplain in northern Nigeria, while ignoring downstream losses, would give an incorrect assessment of the net benefits of dams and water diversions. Besides obvious physical interconnections, other types of interconnections may create important linkages in the production of ecosystem services. One mechanism that creates important interconnections across ecosystems occurs when multiple conditions contribute to the level of service provided. For example, protecting the summer habitat for neotropical migrant birds may be for naught if their winter habitat is destroyed. Protecting coastal wetlands in Louisiana as fish breeding grounds will be more or less valuable depending on the level of nitrogen export from Mississippi River drainage and the extent of the hypoxic zone. Another interconnection may occur with the existence of ecological thresholds and cumulative effects (as discussed in Chapter 3). Stress may be tolerated with little damage to an ecosystem service until a threshold is reached, at which point system function might change drastically, giving rise to a large change in ecosystem services. A classic example is the change in a shallow lake from oligotrophic to eutrophic conditions. A study of the consequences of increased nutrient export from a single stream into a lake may show that there is no change in economic value of the ecosystem services produced by the lake. However, the cumulative effects of increasing nutrient export from all streams into the lake could be sufficient to trigger a regime shift, causing a large change in the value of ecosystem services.

There may be interconnections between ecosystem services on the valuation side even when no biophysical connections exist between ecosystems. The marginal value of an ecosystem service typically depends on the quantity of service supplied rather than being constant (e.g., demand curves generally slope downward). So, for example, a collapse in fish harvest in one ecosystem will tend to increase the economic value of fishery production from other ecosystems. In all valuation studies, some assumption must be made about the level of related ecosystem services produced elsewhere. In addition, the value of particular ecosystem services may also be a function of the level of provision of other ecosystem services or other human-produced services. In other words, there may be important complementarity or substitutability among services.

Most existing valuation techniques used by economists work well for valuing marginal changes but may be more problematic for valuing larger changes. Market price is an accurate signal of the marginal change in value for a small change in the quantity of a marketed good. However, to estimate the change in value from a nonmarginal change in quantity requires information about how price changes with quantity (i.e., the shape of the demand curve), information that may not be readily available. There are similar difficulties for nonmarketed services. For example, it is difficult obtain values for nonmarginal changes in hedonic studies (see Chapter 4). Changes in ecosystem structure and function, and hence in the provision of ecosystem services, however, may require nonmarginal valuation, such as with regime shifts (e.g., oligotrophic to eutrophic conditions in lakes) or large-scale disturbances. For nonmarginal changes, it is not valid simply to multiply the change in provision of the ecosystem service by an estimate of the marginal value of the service under current conditions to derive an estimate of the total change in economic value. Estimates of changes in total value must account for changes in marginal values as conditions change. Failure to take this fact into account can lead to serious errors—for example, in claiming that diamonds are of greater value than water, based on the fact that the price of diamonds (which are scarce) is high while the price of water (which is not scarce in many places) is low.

Because of biological or physical connections and the dependence of marginal value on conditions, great care must be exercised when estimates of value derived at one scale of analysis are applied at a different scale. Typically, there are no simple rules for aggregating values from small scales to larger scales. Some of the most pointed criticisms of the Costanza et al. (1997) study involved aggregation issues.

The temporal scale to be considered also presents challenges to the economic valuation of ecosystem services. Just as ignoring downstream effects in a spatial sense generates an incorrect assessment of net benefits, ignoring the future costs or benefits of decisions will result in an incorrect assessment of the present value of net benefits. For example, ignoring the loss of future benefits when stocks of groundwater are depleted or when the population of a commercially valuable species such as salmon declines will not provide adequate signals of the value of conserving such resources. The difficult issue of comparing present and future values arises when the consequences of a decision impact not only present but also future conditions. A common approach in economic studies is to discount future values. However, there is concern about discounting, especially for decisions having long-term consequences that will have repercussions for decades, centuries, or even longer (see Chapters 2 and 6 for further information). Assessing future consequences necessarily introduces uncertainty into the valuation of ecosystem services. Numerous events that affect ecosystems (e.g., disease outbreaks, fire patterns, weather) and human systems (e.g., innovation, changes in preferences, political change) cannot be predicted in advance. Knowing that ecosystem conditions may change or that values may shift places a premium on the ability to learn and adapt through time and to avoid outcomes with irreversible consequences (or consequences that can be reversed only at great expense). Adaptive management (see Chapter 6) and avoiding difficult-to-reverse decisions prior to reducing uncertainty arose in the context of managing salmon in the Columbia River basin.

The estimate of value of ecosystem services typically depends on a number of current conditions both in the ecosystem itself and in other interconnected systems, many of which are not explicitly stated. A change in fundamental underlying conditions, such as with climate change or an invasive species, may result in large changes in the estimated value of ecosystem services.

Finally, although there is great danger that studies will be partial and incomplete, as discussed in this section, there is also the possibility that the economic value of some ecosystem services will be counted more than once. When value is attributed to coastal wetlands as an input to fishery production, it cannot also be attributed to increased fishery production as an output. Unless studies are carefully designed and executed, such "double-counting" issues may arise.

SUMMARY: CONCLUSIONS AND RECOMMENDATIONS

This chapter has reviewed a series of case studies that value ecosystem services from aquatic and related terrestrial ecosystems, with a focus on their integration of ecology and economics. The case studies varied from those valuing a single ecosystem service, to multiple ecosystem services, to ambitious attempts to value all services from an ecosystem and even the entire planet. Many of the topics and issues addressed in this chapter directly respond to the committee's statement of task (see Box ES-1). An extensive summary of implications and lessons learned from these reviews is provided in the previous section, and no attempt is made to resummarize that section here.

Based on the case studies reviewed in this chapter and the various implications and lessons learned, the committee makes the following specific conclusions regarding efforts to improve the valuation of ecosystem services:

- Studies that focus on valuing a single ecosystem service show promise of delivering results that can inform important policy decisions. In no instance, however, should the value of a single ecosystem service be confused with the value of the entire ecosystem, which has far more than a single dimension.

Unless it is understood clearly that valuing a single ecosystem service represents only a partial valuation of the natural processes in an ecosystem, such single service valuation exercises may provide a false signal of total value.

- Even when the goal of a valuation exercise is focused on a single ecosystem service, a workable understanding of the functioning of large parts or possibly the entire ecosystem may be required.
- Although valuation of multiple ecosystem services is more difficult than valuation of a single ecosystem service, interconnections among services may make it necessary to expand the scope of the analysis.
- Ecosystem processes are often spatially linked, especially in aquatic ecosystems. Full accounting of the consequences of actions on the value of ecosystem services requires understanding these spatial links and undertaking integrated studies at suitably large spatial scales to fully cover important effects. In generating estimates of the value of ecosystem services across larger spatial scales, extrapolation may be unavoidable but should be applied with careful scrutiny.
- The value of ecosystem services depends on underlying conditions. Ecosystem valuation studies should clearly present assumptions about underlying ecosystem and market conditions and how estimates of value could change with changes in these underlying conditions.

Building on these preceding conclusions, the committee provides the following recommendations:

- There is no perfect answer to questions about the proper scale and scope of analysis in ecosystem services valuation. Decisions about the scope and scale of analysis should be dictated by a clearly defined policy question.
- Estimates of value should be placed in context. Assumptions about conditions in ecosystems outside the ecosystem of interest should be clearly specified. Assumptions about human behavior and institutions should be clearly specified.
- Concerted efforts should be made to overcome existing institutional barriers that prevent ready and effective collaboration among ecologists and economists regarding the valuation of aquatic and related terrestrial ecosystem services. Furthermore, existing and future interdisciplinary programs aimed at integrated environmental analysis should be encouraged and supported.

REFERENCES

Abbott, P.L. 1975. On the hydrology of the Edwards Limestone, south-central Texas. Journal of Hydrology 24(3/4):251-269.

Acharya, G., and E.B. Barbier. 2000. Valuing groundwater recharge through agricultural production in the Hadejia-Jama'are wetlands in Northern Nigeria. Agricultural Economics 22: 247-259.

Acharya, G., and E.B. Barbier. 2001. Using domestic water analysis to value groundwater recharge in the Hadejia'Jama'are floodplain in northern Nigeria. American Journal of Agricultural Economics 84(2):415-26.

Adams, R.M., R.P. Berrens, A. Cerda, H.W. Li, and P.C. Klingeman. 1993. Developing a bioeconomic model for riverine management: Case of the John Day River, Oregon. Rivers 4:213-226.

Aillery, M., M.R. Moore, M. Weinberg, G. Schaible, and N. Gollehan. 1999. Salmon recovery in the Columbia River basin: Analysis of measures affecting agriculture. Marine Resources Economics 14:15-40.

Aillery, M., R. Shoemaker, and M. Caswell. 2001. Agriculture and ecosystem restoration in South Florida. American Journal of Agricultural Economics 83(1):183-195.

American Rivers. 1997. Protecting wetlands along the Charles River. Available on-line at *http://www.amrivers.org/floodcase.html#protecting*. Accessed June 14, 2004.

Arnold, C.L. and C.J. Gibbons. 1996. Impervious surface coverage: The emergence of a key environmental indicator. American Planners Association Journal 62: 243-258.

Ashendorff, A., M.A. Principe, A. Seely, J. LaDuca, L. Beckhardt, W. Faber, and J. Mantus. 1997. Watershed protection for New York City's supply. Journal of American Water Works Association 89(3):75-88.

Balmford, A., A. Bruner, P. Cooper, R. Costanza, S. Farber, R.E. Green, M. Jenkins, P. Jefferiss, V. Jessamy, J. Madden, K. Munro, N. Myers, S. Naeem, J. Paavola, M. Rayment, S. Rosendo, J. Roughgarden, K. Trumper, and R.K. Turner. 2002. Economic reasons for saving wild nature. Science 297:950-953.

Balvanera, P., G.C. Daily, P.R. Ehrlich, T.H. Ricketts, S.A. Bailey, S. Kark, C. Kremen, and H. Pereira. 2001. Conserving biodiversity and ecosystem services. Science 291:2047.

Barber, L.B. II. 1994. Sorption of chlorobenzenes to Cape Cod aquifer sediments. Environmental Science and Technology 28(5):890-897.

Barbier, E.B. 2000. Valuing the environment as input: Review of applications to mangrove-fishery linkages. Ecological Economics 35:47-61.

Barbier, E.B. 2003. Upstream dams and downstream water allocation: The case of the Hadejia-Jama'are Floodplain, Northern Nigeria. Water Resources Research 39(11):1311-1319.

Barbier, E.B. and I. Strand. 1998. Valuing mangrove-fishery linkages: A case study of Campeche, Mexico. Environmental and Resource Economics 12:151-166.

Barbier, E. B., and J.R. Thompson. 1998. The value of water: Floodplain versus large-scale irrigation benefits in northern Nigeria. Ambio 27:434-440.

Beamish, R.J., and C. Mahnken. 2001. A critical size and period hypothesis to explain natural regulation of salmon abundance and the linkage to climate and climate change. Progress in Oceanography 49:423-437SI.

Beamish, R.J., D.J. Noakes, G.A. McFarlane, L. Klyashtorin, V.V. Ivanov, and V. Kurashov. 1999. The regime concept and natural trends in the production of Pacific salmon. Canadian Journal of Fisheries and Aquatic Sciences 56:516-526.

Beck, M., K.L. Heck, K.W. Able, D. L. Childers, D.B. Eggleston, B.M. Gillanders, B. Halpern, C.G. Hays, K. Hoshino, T.J. Minello, R.J. Orth, P.F. Sheridan, and M.P. Weinstein. 2001. The identification, conservation and management of estuarine and marine nurseries for fishes and invertebrates. BioScience 51:633-641.

Bell, F.W. 1989. Application of Wetland Valuation Theory to Florida Fisheries. Report No. 95, Florida Sea Grant Program. Tallahassee, Fla.: Florida State University.

Bell, F.W. 1997. The economic value of saltwater marsh supporting marine recreational fishing the southeastern United States. Ecological Economics 21:243-254.

Berrens, R., D. Brookshire, M. McKee and C. Schmidt. 1998. Implementing the safe minimum standard approach: Two case studies from the U.S. Endangered Species Act. Land Economics 74(2):147-161.

Beschta, R. 1997. Riparian shade and stream temperature: An alternative perspective. Rangelands 19(2):25-28.

Beschta, R., R.E. Bilby, G.W. Brown, L.B. Holtby, and T.D. Hofstra. 1987. Stream temperature and aquatic habitat: Fisheries and forest interactions. In Streamside Management: Forestry and Fishery Interactions E.O. Salo and T.W. Cundy (eds.). Seattle, Wash.: Institute of Forest Resources, University of Washington.

Bockstael, N.E., A.M. Freeman, R.J. Kopp, P.R. Portney, and V.K. Smith. 2000. On measuring economic values for nature. Environmental Science and Technology 34(8):1384-1389.

Boesch, D.F., and R.E. Turner. 1984. Dependence of fishery species on salt marshes: The role of food and refuge. Estuaries 7:460-68.

Bowles, D.E. and T.L. Arsuffi. 1993. Karst aquatic ecosystems of the Edwards Plateau region of Central Texas, USA—A consideration of their importance, threat to their existence, and efforts for their conservation. Aquatic Conservation 3:317-329.

Boyle, K.J., P.J. Poor, and L.O. Taylor. 1999. Estimating the demand for protecting freshwater lakes from eutrophication. American Journal of Agricultural Economics 81:1118-1122.

Brock, T.D. 1985. A Eutrophic Lake: Lake Mendota, Wisconsin. New York: Springer-Verlag.

Brock, W.A., and A. deZeeuw. 2002. The repeated lake game. Economics Letters 76:109-114.

Brown, G., Jr. 1992. Replacement Costs of Birds and Mammals. Report for the State of Alaska. Available on-line at *http://www.evostc.state.ak.us/pdf/econ4.pdf.* Accessed January 14, 2004.

Budy, P., G.P. Thiede, N. Bouwes, G.E. Petrosky, and H. Schaller. 2002. Evidence linking delayed mortality of Snake River salmon to their earlier hydrosystem experience. North American Journal of Fisheries Management 22:35-51.

Burt, O. 1964. Optimal resource use over time with an application to groundwater. Management Science 11:80-93.

Cameron, T.A., W.D. Shaw, S.E. Ragland, J.M. Callaway, and S. Keefe. 1996. Using actual and contingent behavior data with different levels of time aggregation to model recreational demand. Journal of Agricultural and Resource Economics 21(10):130-149.

Carpenter, S.R., D. Ludwig, and W.A. Brock. 1999. Management and eutrophication for lakes subject to potentially irreversible change. Ecological Applications 9:751-771.

Carson, R.T. and W.M. Hanemann. 1992. A Preliminary Economic Evaluation of Recreational Fishing Losses Related to the *Exxon Valdez* Oil Spill. A report to the Attorney General of the State of Alaska. Available on-line at *http://www.evostc.state.ak.us/pdf/econ1.pdf.* Accessed October 6, 2004 .

Carson, R.T., R.C. Mitchell, W.M. Hanemann, R.J. Kopp, S. Presser, and P.A. Ruud. 1994. Contingent valuation and lost passive use: Damages from the *Exxon Valdez.* Discussion Paper 94-18. Washington, D.C.: Resources for the Future.

CERP (Comprehensive Everglades Restoration Plan). 2001. Project Management Plan: Florida Bay and Florida Keys Feasibility Study. Available on-line at *http://www.evergladesplan.org/pm/program/program_docs/pmp_study_florida/cerp_fb_fk.pdf.* Accessed June 14, 2004.

Chen, C.C., D. Gillig, and B.A. McCarl. 2001. Effects of climatic change on a water dependent regional economy: A study of the Texas Edwards Aquifer. Climatic Change 49:397-409.

Chichilnisky, G., and G. Heal. 1998. Economic returns from the biosphere. Nature 391:629-630.

Cohen, M.J. 1995. Technological disasters and natural resource damage assessment: An evaluation of the *Exxon Valdez* oil spill. Land Economics 71:65-82.

Congleton, J.L., W.J. LaVoie, C.B. Schreck, and L.E. Davis. 2000. Stress indices in migrating juvenile Chinook salmon and steelhead of wild and hatchery origin before and after barge transportation. Transactions of the American Fisheries Society 129:946-961.

Costanza, R., R. d'Arge, R. de Groot, S. Farber, M. Grasso, B. Hannon, K. Limburg, S. Naeem, R.V. O'Neil, J. Paruelo, R.G. Raskin, P. Sutton, and M. van den Belt. 1997. The value of the world's ecosystem services and natural capital. Nature 387:253-260.

Costello, C., R. Adams, and S. Polasky. 1998. The value of El Niño forecasts in the management of salmon: A stochastic dynamic assessment. American Journal of Agricultural Economics 80:765-777.

Covich, A.P. 1993. Water and ecosystems. Pp. 40-50 in Water in Crisis: A Guide to the World's Fresh Water Resources, P.H. Gleick (ed.). Oxford, U.K.: Oxford University Press.

Covich, A.P., K.C. Ewel, R.O. Hall, P.G. Giller, D. Merritt, and W. Goedkoop. 2004. Ecosystem services provided by freshwater benthos. Pp.45-72 in Sustaining Biodiversity and Ecosystem Services in Soils and Sediments, D. Wall (ed.). Washington, D.C.: Island Press.

Covich, A.P., M.A. Palmer, and T.A. Crowl. 1999. The role of benthic invertebrate species in freshwater ecosystems. BioScience 49:119-127.

Culver, D.C., L.L. Master, M.C. Christman, and H.H. Hobbs. 2000. Obligate cave fauna of the 48 contiguous United States. Conservation Biology 14:386-401.

Culver, D.C., M.C. Christman,W.R. Elliott, H.H. Hobbs, and J.R. Reddell. 2003. The North American obligate cave fauna: Regional patterns. Biodviersity and Conservation 12: 441-468.

Curnutt, J.L., J. Comiskey, M.P. Nott, and L.J. Gross. 2000. Landscape-based spatially explicit species index models for everglades restoration. Ecological Applications 10(6):1849-1860.

Daily, G.C. (ed.) 1997. Nature's Services: Societal Dependence on Natural Ecosystems. Washington, D.C.: Island Press.

Daily, G.C., T. Soderqvist, S. Aniyar, K. Arrow, P. Dasgupta, P.R. Ehrlich, C. Folke, A. Jansson, B.O. Jansson, N. Kautsky, S. Levin, J. Lubchenco, K.G. Maler, D. Simpson, D. Starrett, D. Tilman, and B. Walker. 2000. Ecology—The value of nature and the nature of value. Science 289:395-396.

D'Arge, R.C., and J. Shogren. 1989. Okoboji experiment: Comparing non-market valuation techniques in an unusually well-defined market for water quality. Ecological Economics 1:251-259.

Davis, S., and J. Ogden. 1994. Everglades: The Ecosystem and Its Restoration. Delray Beach, Fla.: St Lucie Press.

DeMelo, R., R. Fraqnce, and D.J. McQueen. 1992. Biomanipulation: Hit or myth? Limnology and Oceanography 37:192-207.

Deriso, R.B., D.R. Marmorek, and I.J. Parnell. 2001. Retrospective patterns of differential mortality and common year-effects experienced by spring and summer Chinook salmon (*Oncorhunchus tshawytscha*) of the Columbia River. Canadian Journal of Fisheries and Aquatic Science 58:2419-2430.

Ellis, G.M., and A.C. Fisher. 1987. Valuing the environment as input. Journal of Environmental Management 25:149-156.

Englehardt, J.D. 1998. Ecological and economic risk analysis of Everglades: Phase I restoration alternatives. Risk Analysis 18:755-771.

EPA (U.S. Environmental Protection Agency). 1999. Safe Drinking Water Act, Section 1429, Ground Water Report to Congress. EPA-816-R-99-016. Available on-line at *http://www.epa.gov/safewater/gwr/finalgw.pdf.* Accessed October 20, 2004.

Faber, S. 1996. On Borrowed Land: Public Policies for Floodplains. Cambridge, Mass.: Lincoln Institute of Land Policy.

Farber, S., and R. Costanza. 1987. The economic value of wetlands systems. Journal of Environmental Management 24:41-51.

Fisher, A.C., W.M. Hanemann, and A. Keeler. 1991. Integrating fishery and water resource management: A biological model of a California salmon fishery. Journal of Environmental Economics and Management 20:234-261.

Foran, J., T. Brosnan, M. Connor, J. Delfino, J. DePinto, K. Dickson, H. Humphrey, V. Novotny, R. Smith, M. Sobsey, and S. Stehman. 2000. A framework for comprehensive, integrated, water monitoring in New York City. Environmental Monitoring and Assessment 62:147-167.

Forest Preserve District of Cook County Illinois. 1988. An evaluation of floodwater storage. River Forest, Ill.: Forest Preserve District of Cook County.

Freeman, A.M. 1991. Valuing environmental resources under alternative management regimes. Ecological Economics 3:247-256.

Frei, A., R.L. Armstrong, M.P. Clark, and M.C. Serreze. 2002. Catskill Mountain water resources: Vulnerability, hydroclimatology, and climate-change sensitivity. Annals of the Association of American Geographers 92:203-224.

Gergel, S.E., M.G. Turner, J.R. Miller, J.M. Melack, and E.H. Stanley. 2002. Landscape indicators of human impacts to riverine systems. Aquatic Sciences 64:118-128.

Gibert, J., D.L. Danielopol, and J.A. Stanford (eds.). 1994. Groundwater ecology. San Diego, Calif.: Academic Press.

Giller, P.S., A.P. Covich, K.C. Ewel, R.O. Hall, Jr., and D.M. Merritt. 2004. Vulnerability and management of ecological services in freshwater systems. Pp. 137-159 in Sustaining Biodiversity and Ecosystem Services in Soils and Sediments, D. Wall (ed.). Washington, D.C.: Island Press.

Glennon, R. 2002. Water Follies: Groundwater Pumping and the Fate of America's Fresh Waters. Washington, D.C.: Island Press.

Grant, G. 2001. Dam removal: Panacea or pandora for rivers? Hydrological Processes 15:1531-1532.

Gregory, S., H. Li, and J. Li. 2002. The conceptual basis for ecological responses to dam removal. BioScience 52:713-723.

Grimm, N.B., A. Chacon, C.N. Dahm, S.W. Hostetler, O.T. Lind, P.L. Starkweather, and W.W. Wurtsbaugh. 1997. Sensitivity of aquatic ecosystems to climatic and anthropogenic changes: The basin and range, American Southwest, and Mexico. Hydrological Processes 11:1023-1041.

Guardo, M., L. Fink, T.D. Fontaine, S. Newman, M. Chimney, R. Bearzotti, and G. Goforth. 1995. Large-scale constructed wetlands for nutrient removal from stormwater runoff: An Everglades restoration project. Environmental Management 19(6): 879-889.

Gulati, R.D., E.H.R.R. Lammens, M.L. Meijer, and E. Donk (eds.). 1990. Biomanipulation-Tool for Water Management. Belgium: Kluwer Academic Publishers.

Hamlet, A.F., and D.P. Lettenmaier. 1999a. Columbia River streamflow forecasting based on ENSO and PDO climate signals. Journal of Water Resources Planning and Management 125:333-341.

Hamlet, A.F., and D.P. Lettenmaier. 1999b. Effects of climate change on hydrology and water resources in the Columbia River Basin. Journal of the American Water Resources Association 35:1597-1623.

Hamlet, A.F., D. Huppert, and D.P. Lettenmaier. 2002. Economic value of long-lead streamflow forecasts for Columbia River hydropower. Journal of Water Resources Planning and Management 128:91-101.

Hammack, J., and G.M. Brown, Jr. 1974. Waterfowl and Wetlands: Toward Bioeconomic Analysis. Baltimore, Md.: Johns Hopkins University Press-Resources for the Future.

Hanemann, W.M. 1991. Willingness to pay and willingness to accept: How much can they differ? American Economic Review 18:635-647.

Hare, S.R., N.J. Mantua, and R.C. Francis. 1999. Inverse production regimes: Alaska and West Coast Pacific salmon. Fisheries 24:6-14.

Hausman, J.A. (ed.). 1993. Contingent Valuation: A Critical Assessment. Amsterdam: North-Holland.

Hausman, J.A., G.K. Leonard, and D. McFadden. 1995. A utility-consistent combined discrete choice and count data model: Assessing recreational use losses due to natural resource injury. Journal of Public Economics 56:130.

Hausman, J.A., G.K. Leonard, and D. McFadden. 1993. Assessing use value losses caused by natural resource injury. Pp. 341-363 in Contingent Valuation: A Critical Assessment, J.A. Hausman (ed.). Amsterdam: North-Holland.

Heal, G.M. 2000a. Valuing ecosystem services. Ecosystems 3:24-30.

Heal, G.M. 2000b. Nature and The Marketplace: Capturing the Value of Ecosystem Services. Washington, D.C.: Island Press.

Hey, D.L., and N.S. Philippi. 1995. Proceedings of the scientific assessment and strategy team workshop on hydrology, floodplain ecology and hydraulics. Pp. 47-52 in Science for Floodplain Management into the 21st Century, G.E. Freeman, and A.G. Frazier (eds.). Volume 5. Washington, D.C.: Government Printing Office.

Hicks, B.J., R.L. Beschta, and R.D. Hart. 1991. Long-term changes in streamflow following logging in Western Oregon and associated fishery implications. Water Resources Bulletin 27:217-226.

Holmes, T.P., J.C. Bergstrom, E. Huszar, S.B. Kask, and F. Orr. 2004. Contingent valuation, net marginal benefits, and the scale of riparian ecosystem restoration. Ecological Economics 49:19-30.

Howe, C.W. 2002. Policy issues and institutional impediments in the management of groundwater: Lessons from case studies. Environment and Development Economics 7:625-641.

Huppert, D. 1989. Measuring the value of fish to anglers: Application to central California anadromous species. Marine Resource Economics 6:89-107.

Hurd, B., N. Leary, R. Jones, and J. Smith. 1999. Relative regional vulnerability of water resources to climate change. Journal of the American Water Resources Association 35:1399-1409.

Illinois Department of Conservation. 1993. The Salt Creek Greenway Plan. Springfield, Ill.: Illinois Department of Conservation.

Isard, W., K. Bassett, C. Chogull, J. Furtado, R. Izumita, J. Kissin, E. Romanoff, R. Seyfarth, and R. Tatlock. 1969. Linkage of socio-economic and ecologic systems. Ekistics 28:28-34.

Jaeger, W.K., and R. Mikesell. 2002. Increasing streamflow to sustain salmon and other native fish in the Pacific Northwest. Contemporary Economic Policy 20:366-380.

Johnson, N.S., and R.M. Adams. 1988. Benefits of increased streamflow: The case of the John Day River steelhead fishery. Water Resources Research 24(11):1839-1846.

Jones, J.B. and P.J. Mulholland (eds.). 2000. Stream and Ground Waters. San Diego, Calif.: Academic Press.

Kareiva, P., M. Marvier, and M. McClure. 2000. Recovery and management options for spring/summer chinook salmon in the Columbia River basin. Science 290: 977-979.

Kiker, C.F., J.W. Milon, and A.W. Hodges. 2001. Adaptive learning for science-based policy: The Everglades restoration. Ecological Economics 37(3):403-416.

Kinzig, A. (and 50 co-authors). 2000. Nature and Society: An Imperative for Integrated Environmental Research. A report from an NSF sponsored workshop held in June 2000. Available on-line at *http://lsweb.la.asu.edu/akinzig/NSFReport.pdf.* Accessed December 10, 2003.

Kitchell, J.F. (ed.). 1992. Food Web Management. A Case Study of Lake Mendota. New York: Springer-Verlag.

Krutilla, J.V. 1967. Conservation reconsidered. American Economic Review 57:777-786

Krutilla, J.V., and A.C. Fisher. 1975. The Economics of Natural Environments: Studies in the Valuation of Commodity and Amenity Resources. Baltimore, Md.: Johns Hopkins University Press-Resources for the Future.

Lathrop, R.C., S.R. Carpenter, C.A. Stow, P.A. Soranno, and J.C. Panuska. 1998. Phosphorus loading reductions needed to control blue-green algal blooms in Lake Mendota. Canadian Journal of Fisheries and Aquatic Sciences 55:1169-1178.

Lathrop, R.C., B.M. Johnson, T.B. Johnson, M.T. Vogelsang, S.R. Carpenter, T.R. Hrabik, J.F. Kitchell, J.J. Magnuson, L.G. Rudstam, and R.S. Stewart. 2002. Stocking piscivores to improve fishing and water clarity: A synthesis of the Lake Mendota biomanipulation project. Freshwater Biology 47:2410-2424.

Layman, C., J. Boyce, and K. Criddle. 1996. Economic evaluation of chinook salmon sport fishery of the Gulkana River, Alaska, under current and alternative management plans. Land Economics 72:113-128.

Lee, K.N. 1993. Compass and Gyroscope: Integrating Science and Politics for the Environment. Covelo, Calif.: Island Press.

Lee, K.N. 1999. Appraising adaptive management. Conservation Ecology 3(2):3. Available on-line at *http://www.consecol.org/vol3/iss2/art3.* Accessed December 10, 2003.

Levin, P.S., and N. Tolimieri. 2001. Differences in the impacts of dams on the dynamics of salmon populations. Animal Conservation 4:291-299.

Levin, P.S., R.W. Zabel, and J.G. Williams. 2001. The road to extinction is paved with good intentions: Negative association of fish hatcheries with threatened salmon. Proceedings of the Royal Society of London Series B-Biological Sciences 268:1153-1158.

Lichatowich, J. 1999. Salmon Without Rivers. Washington, D.C.: Island Press.

Longley, G. 1986. The biota of the Edwards Aquifer and the implications for paleozoogeography. In The Balcones Escarpment, P.L. Abbott and C.M. Woodruff, Jr. (eds.). Available on-line at *http://www.lib.utexas.edu/geo/BalconesEscarpment/BalconesEscarpment.html.*

Loomis, J.B. 1988. The bioeconomic effects of timber harvesting on recreational and commercial salmon and steelhead fishing: A case study of the Siuslaw National Forest. Marine Resource Economics 5:43-60.

Loomis, J., P. Kent, L. Strange, K. Fausch, A. Covich. 2000. Measuring the total economic value of restoring ecosystem services in an impaired river basin: Results from a contingent valuation survey. Ecological Economics 33:103-117.

Lovett, G.M., K.C. Weathers, and W.V. Sobczak. 2000. Nitrogen saturation and retention in forested watersheds of the Catskill Mountains, New York. Ecological Applications 10:73-84.

Lynne, G.D., P. Conroy, and F.J. Prochaska. 1981. Economic value of marsh areas for marine production processes. Journal of Environmental Economics and Management 8:175-186.

McCarl, B.A., C.R. Dillon, K.O. Keplinger, and R.L. Williams. 1999. Limiting pumping form the Edwards Aquifer: An economic investigation of proposal, water markets and spring flow guarantees. Water Resources Research 35:1257-1268.

McCully, P. 2002. Silenced Rivers: The Ecology and Politics of Large Dams. 2nd edition. London: Zed.

Meffe, G.K. 1992. Techno-arrogance and halfway technologies: Salmon hatcheries on the Pacific coast of North America. Conservation Biology 6:350-354.

Merrifield, J. 2000. Goundwater resources: The transition from capture to allocation. Policy Studies Review 17(1):105-124.

Merrifield, J., and R. Collinge. 1999. Efficient water pricing policies as an appropriate municipal revenue source. Public Works Management and Policy 4(2):119-130.

Meyer, J.L. and J.B. Wallace. 2001. Lost linkages in lotic ecology: Rediscovering small streams. Pp. 295-317 in Ecology: Achievement and Challenge, M. Press, N. Huntly, S. Levin (eds.). Boston, Mass.: Blackwell Science.

Meyer, J.L., M.J. Sale, P.J. Mulholland, and N.L. Poff. 1999. Impacts of climate change on aquatic ecosystem functioning and health. Journal of the American Water Resources Association 35:1373-1386.

Miles, E.L., A.K. Snover, A.F. Hamlet, B. Callahan, and D. Fluharty. 2000. Pacific Northwest regional assessment: The impacts of climate variability and climate change on the water resources of the Columbia River Basin. Journal of the American Water Resources Association 36:399-420.

Moore, M.R., N.R. Gollehon, and M.B. Carey. 1994. Multicrop production decisions in western irrigated agriculture—The role of water price. American Journal of Agricultural Economics 76:859-874.

Moore, M.R., N.R. Gollehon, and D.M. Hellerstein. 2000. Estimating producer's surplus with the censored regression model: An application to producers affected by Columbia River Basin salmon recovery. Journal of Agricultural and Resource Economics 25:325-346.

Morganwalp, D.W., and H.T. Buxton (eds.). 1999. U.S. Geological Survey Toxic Substances Hydrology Program, March 8. Volume 3—Subsurface Contamination from Point Sources. USGS Water-Resources Investigations Report 99-4018C.

Moscript, A. L., and D. R. Montgomery. 1997. Urbanization, flood frequency, and salmon abundance in Puget Lowland streams. Journal of the American Water Resources Association 33:1289-1297.

Murdoch, P.S., and J.L. Stoddard. 1992. The role of nitrate in the acidification of streams in the Catskill Mountains of New York. Water Resources Research 28:2707-2720.

Murdoch, P.S., and J.L. Stoddard. 1993. Chemical characteristics and temporal trends in eight streams of the Catskills Mountains, New York. Water, Air, and Soil Pollution 67:257-280.

Murdoch, P.S., J.S. Baron, and T.L. Miller. 2000. Potential effects of climate change on surface-water quality in North America. Journal of the American Water Resources Association 36:347-366.

Naiman, R.J., S.E. Bunn, C. Nilsson, G.E. Petts, G. Pinay, and L. Thompson. 2002. Legitimizing fluvial ecosystems as users of water: An overview. Environmental Management 30:455-467.

Nature. 1998. Audacious bid to value the planet whips up a storm. 395:430.

Nickelson, T.E. 1986. Influence of upwelling, ocean temperature, and smolt abundance on marine survival of Coho Salmon in the Oregon production area. Canadian Journal of Fisheries and Aquatic Sciences 43:527-535.

NOAA (NOAA Panel on Contingent Valuation). 1993. Natural Resource Damage Assessment under the Oil Pollution Act of 1990. Federal Register 58(10):4601-4614.

Northwest Power Planning Council. 2001. Inaugural Annual Report of the Columbia Basin Fish and Wildlife Program, 1978-1999. Northwest Power Planning Council Document 2001-2. Portland, Ore.: Northwest Power Planning Council.

NRC (National Research Council). 1995. Understanding Marine Diversity: A Research Agenda for the Nation. Washington, D.C.: National Academy Press.

NRC. 1996. Upstream: Salmon and Society in the Pacific Northwest. Washington, D.C.: National Academy Press

NRC. 1997. Valuing Groundwater: Economic Concepts and Approaches. Washington, D.C.: National Academy Press

NRC. 2000a. Watershed Management for Potable Water Supply: Assessing the New York City Strategy. Washington, D.C.: National Academy Press.

NRC. 2000b. Investigating Groundwater Systems on Regional and National Scales. Washington, D.C.: National Academy Press.

NRC. 2002a. Regional Issues in Aquifer Storage and Recovery for Everglades Restoration. Washington, D.C.: National Academy Press.

NRC. 2002b. Florida Bay Research Programs and Their relation to the Comprehensive Everglades Restoration Plan. Washington, D.C.: National Academy Press.

NRC. 2004. Managing the Columbia River: Instream Flows, Water Withdrawals, and Salmon Survival. Washington, D.C.: The National Academies Press.

Olsen, D., J. Richards, and R.D. Scott. 1991. Existence and sport values for doubling the size of the Columbia River basin salmon and steelhead runs. Rivers 2(1):44-56.

Opie, J. 1993. Ogallala: Water for a Dry Land. Lincoln, Neb.: University of Nebraska Press.

Paine, R.T., J.L. Ruesink, A. Sun, E.L. Soulanille, M.J. Wonham, C.D.G. Harley, D.R. Brumbaugh and D.L. Secord. 1996. Trouble on oiled waters: Lessons from the *Exxon Valdez* oil spill. Annual Review of Ecology and Systematics 27:197-235.

Paul, M.J., and J.L. Meyer. 2001. Streams in the urban landscape. Annual Review of Ecology and Systematics 32:333-365.

Paulsen, C.M., and R.A. Hinrichsen. 2002. Experimental management for Snake River spring-summer chinook (*Oncorhynchus tshawytscha*): Trade-offs between conservation and learning for a threatened species. Canadian Journal of Fisheries and Aquatic Sciences 59:717-725.

Paulsen, C.M., and K. Wernstedt. 1995. Cost-effectiveness analysis for complex managed hydrosystems: An application to the Columbia River Basin. Journal of Environmental Economics and Management 28:388-400.

Pernin, C.G., M.A. Bernstein, A. Mejia, H. Shih, F. Reuter, and W. Steger. 2002. Generating Electric Power in the Pacific Northwest: Implications of Alternative Technologies. Santa Monica, Calif.: RAND.

Platts, W.S. 1991. Livestock grazing. In Influences of Forest and Rangeland Management on Salmonid Fishes and Their Habitat. American Fisheries Society Special Publiaction 19: 389-423.

Poff, N.L., and D.D. Hart. 2002. How dams vary and why it matters for the emerging science of dam removal. BioScience 52:659-668.

Polasky, S. (ed.). 2002. The Economics of Biodiversity Conservation. Aldershot, Hampshire, U.K.: Ashgate Publishing Limited.

Pringle, C.M. 2002. Hydrologic connectivity and the management of biological reserves: A global perspective. Ecological Applications 11:981-998.

Pringle, C.M. 2003. What is hydrologic connectivity and why is it ecologically important? Hydrological Processes 17:2685-2689.

Pringle, C.M., M. Freeman, and B. Freeman. 2000. Regional effects of hydrologic alterations on riverine macrobiota in the New World: Tropical-temperate comparisons. BioScience 50:807-823.

Pritchard, L., Jr., C. Folke, and L. Gunderson. 2000. Valuation of ecosystem services in institutional context. Ecosystems 3:36-40.

Provencher, B. 1993. A private property rights regime to replenish a groundwater aquifer. Land Economics 69(4):325-340.

Provencher, B., and O. Burt. 1994. A private property rights regime for the commons: The case for groundwater. American Journal of Agricultural Economics 76(4):875-888.

Pulwarty, R.S., and K.T. Redmond. 1997. Climate and salmon restoration in the Columbia River Basin: The role and usability of seasonal forecasts. Bulletin of the American Meteorological Society 78:381-397.

Pyne, R.D.G. 1995. Groundwater Recharge and Wells: A Guide to Aquifer Storage Recovery. Boca Raton, Fla.: Lewis Publishers.

Reed-Anderson, T., S.R. Carpenter, and R.C. Lathrop. 2000. Phosphorus flow in a watershed-lake ecosystem. Ecosystems 3:561-573.

Rishel, G.B., J.A. Lynch, and E.S. Corbett. 1982. Seasonal stream temperature changes following forestry harvest. Journal of Environmental Quality 11:112-116.

Rubio, S.J., and B. Casino. 1994. Competitive versus efficient extraction of a common property resource: The groundwater case. Journal of Economic Dynamics and Control 25:1117-1137.

Sanford, W. 2002. Recharge and groundwater models: An overview. Hydrogeology Journal 10:110-120.

Schaller, H.A., C.E. Petroky, and O.P. Langness. 1999. Constrasting patterns of productivity and survival rates for stream-type chinook salmon (*Oncorhynchus tsawytscha*) populations of the Snake and Columbia Rivers. Canadian Journal of Fisheries and Aquatic Sciences 56:1031-1045.

Scheaffer, J.R., J.D. Mullan, and N.B. Hinch. 2002. Encouraging wise use of floodplains with market-based incentives. Environment (January-February):33-43.

Schiable, G.D., B.A. McCarl, and R.D. Lacewell. 1999. The Edwards Aquifer water resource conflict: USDA farm programs resource use incentives. Water Resources Research 35: 3171-3183.

Schneiderman, J.S. 2000. From the Catskills to Canal Street: New York City's water supply. Pp. 166-180 in The Earth Around Us, J.S. Schneiderman (ed.). Boulder, Colo.: Westview Press.

Schoenen, D. 2002. Role of disinfection in suppressing the spread of pathogens with drinking water: Possibilities and limitations. Water Research 3:3874-3888.

Shapiro, J. 1990. Biomanipulation: The next phase—Making it stable. Pp. 13-27 in Biomanipulation-Tool for Water Management, Gulati, R.D., E.H.R.R. Lammens, M.L. Meijer, and E. Donk (eds.). Belgium: Kluwer Academic Publishers.

Sharp, J.M, Jr., and J.L. Banner. 2000. The Edwards Aquifer: Water for thirsty Texans. Pp. 154-165 in The Earth Around Usk, J.S. Schneiderman (ed.). Boulder, Colo.: Westview Press.

Shaw, D.G. 1992. The *Exxon Valdez* oil-spill: Ecological and social consequences. Environmental Conservation 19:253-258.

Sklar, F.H., H.C. Fitz, Y. Wu, R. Van Zee, and C. McVoy. 2001. The design of ecological landscape models for Everglades restoration. Ecological Economics 37(3):379-401.

Stoddard, J.L. 1994. Long-term changes in watershed retention of nitrogen. Pp. 223-284 in Environmental Chemistry of Lakes and Reservoirs, L.A. Baker (ed.). Washington, D.C.: American Chemical Society.

Stroud Water Research Center. 2001. Water Quality Monitoring in the Source Water Areas for New York City: An Integrative Watershed Approach. Contribution No. 2001007. Avondale, Pa..

Stumborg, B.E., K.A. Baerenklau, and R.C. Bishop. 2001. Nonpoint source pollution and present values: A contingent valuation study of Lake Mendota. Review of Agricultural Economics 23:120-132.

Swallow, S.K. 1994. Renewable and non-renewable resource theory applied to coastal agriculture, forest, wetland and fishery linkages. Marine Resource Economics 9:291-310.

Symanski, E., D.A. Savitz, and P.C. Singer. 2004. Assessing spatial fluctuations, temporal variability, and measurement error in estimated levels of disinfection by-products in tap water: Implications for exposure assessment. Occupational and Environmental Medicine 61:65-72.

Tierney, J. 1990. Betting the planet. New York Times Magazine. December 2.

Toman, M.A. 1998. Why not calculate the value of the world's ecosystem services and natural capital? Ecological Economics 25:57-60.

Toth, L.A. 1996. Restoring the hydrogeomorphology of the channelized Kissimmee River. Pp. 369-383 in River Channel Restoration: Guiding Principles for Sustainable Projects, A. Brookes and F. D. Shields Jr. (eds.). New York: John Wiley and Sons.

Tsur, Y., and A. Zemel. 1995. Uncertainty and irreversibility in groundwater resource management. Journal of Environmental Economics and Management 29(2):149-161.

Turner, R.K., J. Paavola, P. Cooper, S. Farber, V. Jessamy, and S. Georgiou. 2003. Valuing nature: Lessons learned and future research directions. Ecological Economics 46:493-510.

USACE (U.S. Army Corps of Engineers). 1978. Nonstructural Plan for the East Branch of the DuPage River. Chicago, Ill.: USACE.

USACE. 1999. A Citizen's Guide to the City of Napa, Napa River and Napa Creek Flood Protection Project. Guidebook prepared by the U.S. Army Corps of Engineers and Napa County Flood Control and Water Conservation District. Available on-line at *http://www.usace.army.mil*. Accessed December 10, 2003.

USACE. 2002. Lower Snake River juvenile salmon migration feasibility study. Available on-line at *http://www.nww.usace.army.mil/lsr/final_fseis/study_kit/studypage.htm*. Accessed October 20, 2004.

Vileisis, A. 1997. Discovering the Unknown Landscape: A History of America's Wetlands. Washington, D.C.: Island Press.

Villanueva, C.M., M. Kogevinas, J.O. Grimalt. 2001. Drinking water chlorination and adverse health effects: A review of epidemiological studies. Medicina Clinica 117:27-35.

Walters, C.J. 1986. Adaptive Management of Renewable Resources. New York: Macmillan.

Wernstedt, K., and C.M. Paulsen. 1995. Economic and biological analysis to aid system-planning for salmon recovery in the Columbia River Basin. Journal of Environmental Management 43:313-331.

Wilson, M.A., and S.R. Carpenter. 1999. Economic valuation of freshwater ecosystems services in the United States: 1971-1997. Ecological Applications 9:772-783.

Wimberley, L.A. 2001. Establishing "sole source" protection: The Edwards Aquifer and the Safe Drinking Water Act. Pp. 169-181 in On the Border: An Environmental History of San Antonio, C. Miller (ed.). Pittsburgh, Pa.: University of Pittsburgh Press.

Winter, T.C. 2001. The concept of hydrologic landscapes. Journal of the American Water Resources Association 37:335-349.

Wright, J.M. 2000. The Nation's Responses to Flood Disasters: A Historical Account. Association of State Floodplain Managers Inc., Madison, Wis. Available on-line at *http://www.floods.org/PDF/hist-fpm.pdf.* Accessed December 10, 2003.

Wu, J., R.M. Adams, and W.G. Boggess. 2000. Cumulative effects and optimal targeting of conservation efforts: steelhead trout habitat enhancement in Oregon. American Journal of Agricultural Economics 82:400-413.

Wu, J., K. Sketon-Groth, W.G. Boggess, and R.M. Adams. 2003. Pacific salmon restoration: Trade-offs between economic efficiency and political acceptance. Contemporary Economic Policy 21(1):78-89.

Zhang, X., and R.A. Minear. 2002. Characteristics of high molecular weight disinfection by products resulting from chlorination of aquatic humic substances. Environmental Science and Technology 36:4033-4038.

6
Judgment, Uncertainty, and Valuation

INTRODUCTION

Some aspects of the economic valuation of aquatic and related terrestrial ecosystem services inevitably involve investigator judgments, and some are unavoidably uncertain. This chapter aims to identify the needs for investigator judgments and how they arise, how such judgments should be made, and how they should be presented to environmental decision-makers. It also seeks to describe the sources and types of uncertainty, indicate which are most significant, and suggest how analysts and decision-makers can and should respond. More specifically, this chapter provides a review of issues related to framing, methodological judgments, and peer review; the sources and management of uncertainty and how these relate to valuation and policymaking considerations; and a summary of the chapter and its conclusions and recommendations. Although unavoidable, uncertainty and the need to exercise professional judgment are not debilitating to ecosystem services valuation. It is important to be clear, however, when such judgments are made, to explain why they are needed, and to indicate the alternative ways in which judgment could have been exercised. It is also important that the sources of uncertainty be minimized and accounted for in ways that ensure that one's conclusions and resulting decisions regarding ecosystem valuation are not systematically biased and do not convey a false sense of precision.

PROFESSIONAL JUDGMENTS

The following sections describe cases in which investigators had to use professional judgments in ecosystem valuation regarding issues of: (1) how to frame a valuation study; (2) how to address the methodological judgments that have to be made during the study (such as the choice of a discount rate); and (3) how to use peer review to identify and evaluate these judgments.

Framing

Perhaps the most important choice in any ecosystem services valuation study is the selection of the question to be asked and addressed. This report has previously described the importance of a careful selection of the question in several case studies including the Catskills watershed and the *Exxon Valdez* oil spill (see Chapter 5). In the Catskills study (see also NRC, 2000), a critical decision was made early on to not attempt to value the entire suite of services provided by the watershed but rather to focus on the service of water purification. More specifically, the issue was whether the restoration of the Catskills watershed would be more cost-effective than constructing a new drinking water filtration system as a way of addressing New York City's drinking water quality problems. This definition of the issue was determined by policymakers, not by the analysts.

This very specific and policy-oriented focus meant that it was not necessary to identify and attempt to value all of the services provided by the watershed, but rather to ascertain whether the cost of restoring its water purification services exceeded or was less than the known cost of a replacement for them. As discussed in Chapter 5, this focus greatly simplified the valuation task because a full economic valuation of the services of the watershed would have required the following: (1) that all sources of value be identified, such as water purification, tourism, support of biodiversity, esthetic values, recreational fishing, streamflow stabilization, and so on; (2) that each of these services be quantified; and (3) that each service be valued. It was not even necessary to establish the restoration cost exactly, but only to compare it to the cost of the alternative (i.e., construction of a drinking water filtration system). Since the outcome of this comparison was that the cost of restoration was less than that of the alternative, New York City decided to spend more than one billion dollars on increased protection and restoration of the watershed (NRC, 2000). It is worth emphasizing that no aspects of the services of the Catskills ecosystems were valued to reach this conclusion; watershed restoration costs were compared to those of an alternative source of the desired service. If this answer had been different—if, for example, the cost of restoration had exceeded the cost of a new water filtration system—it might still have been appropriate to restore the watershed. However, in that case, a complete economic justification of such a decision would have required the valuation of a sufficient number of services of the Catskills watershed to show that the total economic value exceeded the costs of restoration, and offered New York City an attractive return on its investment. Such a valuation exercise would have been an order of magnitude more complex. Thus, not only was the question framed in a way that simplified the analysis, but the existing data were conducive to supporting the simplest possible outcome. The decision tree in Figure 6-1 illustrates this point—investigation of the New York City watershed followed the upper part of this decision tree, leading to a conclusion that avoided two complex steps that would otherwise have been required.

The *Exxon Valdez* case presents a different situation (Carson et al., 2003; Hanemann, 1994; Portney, 1994) as legal liability issues required estimates of damages to natural resources. A complete economic valuation of the costs of the massive oil spill would have required the following: (1) identification of all of the categories of impacts of the spill such as loss of fish catch, loss of tourist revenues, deaths of many species of birds, fish, mammals, and invertebrates; (2) quantification of all of these types of impacts (e.g., how much revenue from

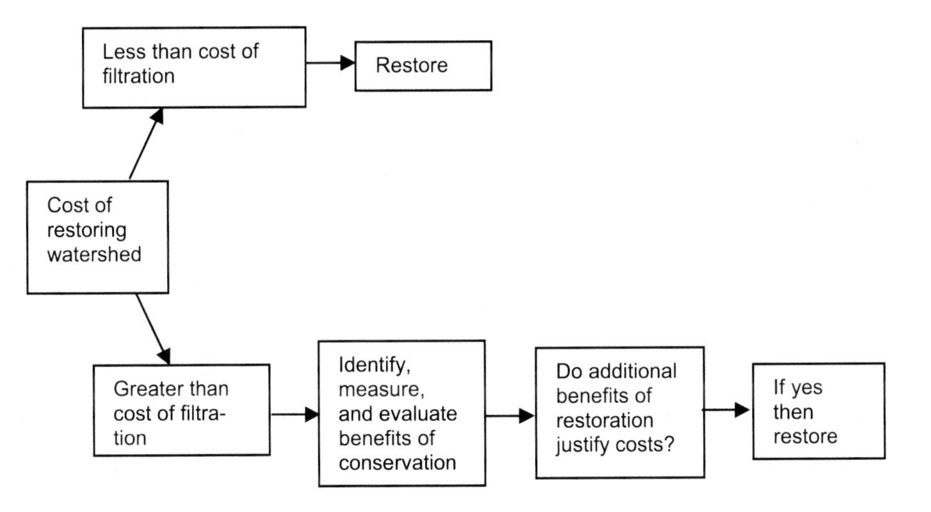

FIGURE 6-1 Decision tree for Catskills watershed study.

fishing and tourism was lost, how many animals of each type were killed); and (3) valuation of each of these losses. Clearly, completing all three stages of such an ecosystem valuation study presents a massive and challenging task.[1] Although numerous studies were commissioned by Exxon, the State of Alaska, the federal government, and other interested parties, a clear answer to the question of the dollar value of damages to ecosystem services caused by the oil spill was not produced (Portney, 1994). As noted in Chapter 5, there are difficulties in quantifying the link between the oil spill and changes in ecosystem services as well as difficulties in valuing such changes—especially when considering non-use values such as existence value. There was no obvious and simple way of framing this issue in the *Exxon Valdez* case because all aspects of the damages were relevant to disputes about compensation.

These two cases illustrate the importance of how a valuation study is framed, and how the frame used derives from the specific context within which an ecosystem valuation issue is raised. They also illustrate that the way an issue is posed may make a huge difference in the complexity of the valuation problem to be addressed.

In addition to determining the question to be asked and the complexity of the analysis required, psychologists have shown that how an issue is framed frequently affects the way in which people make judgments about that issue and

---

[1] It is important to note that under the Comprehensive Environmental Response, Compensation, and Liability Act (CERCLA) legislation the federal government was only allowed to sue for public damages, which exclude loss of tourist revenues and business profits. See Hanemann and Strand (1993) for further information.

the subsequent answers they give to questions about the issue (Kahneman and Tversky, 2000; Machina, 1987). One classic illustration concerns the difference between the way people react to a policy that can alternatively be described as either saving lives or losing lives. Suppose that 100 people are threatened by a fatal disease but a policy intervention may save half of them. This situation could be described by stating that if this policy is followed, 50 of 100 people will die. Alternatively, one could also accurately state that this policy will save the lives of 50 of the 100 people who would otherwise die. Not surprisingly, the latter description is usually found to elicit a much more positive response and a higher "willingness to pay" (see more below) that is due entirely to the differences in the way the issue is framed. In one case, the emphasis is on saving lives, while the other is on losing lives.

A similar phenomenon has been noted in the description and interpretation of event probabilities (Kunreuther et al., 2001). Suppose that a natural disaster has a 1 in 100 chance of occurring each year. One could accurately state that over a 20-year period there is a 1 in 5 chance of such an event occurring. However, the latter way of presenting the same event probability almost always produces a stronger negative reaction. For example, people are typically willing to pay more for disaster insurance if the data is presented in the second way than in the first.

In the context of valuing aquatic ecosystems and their services, framing effects could matter in the choice between whether to emphasize what will be lost or what will be preserved. If an environmental policy will result in half of an existing wetland being lost, should this be presented as half being lost or half being saved? Should an analyst emphasize the number of birds or fish saved as the result of a policy measure or the number that will die in spite (or because) of the measure? One might be tempted to answer that the correct solution is to present all relevant information and allow individuals to select based on what is important to them. Although in some cases this might be possible, in many cases the volume of relevant data will be so large that it is virtually impossible to present it all in a completely even-handed way. In such cases, some element of selection and framing will be unavoidable.

The choice between willingness to pay (WTP) and willingness to accept (WTA) as measures of the value of an ecosystem good or service (see Chapters 2 and 4 for further information) is also a choice about how an issue is framed. This choice is normally thought of as depending on where the property rights lie (Hanemann, 1991). If the recipients of an ecosystem service have a right to that service, then the loss from removing it or allowing it to be lost is what they would be willing to accept as compensation. Unlike WTP, this measure is not bounded by their wealth. If on the other hand there is no inherent right to an ecosystem good or service, then its value to people is better measured by their willingness to pay for it. Certainly, there are situations in which the underlying ownership rights are not clear and it is therefore not obvious as to which measure is the better one. For example, do polluters have a right to pollute water, or do individuals have a right to clean water? The answers to such questions determine whether clean water is most appropriately valued by WTP or WTA compensation for its loss. These are likely to result in very different valuation estimates, and unfortunately the methods of eliciting them are also rather different (see Chapter 4).

In fact, methods of eliciting WTP are better developed than those for eliciting WTA. Indeed the experience of some investigators in this area is that subjects in contingent valuation studies are more comfortable with questions about what they are WTP than with questions about WTA, as deciding what to pay for a good or service is an everyday human activity whereas one is rarely called upon to decide what to accept.[2] In such cases, the analyst should ideally report both sets of estimates in a form of sensitivity analysis. However, the committee recognizes that in some cases this may effectively double the work and in such situations a second best alterative is to carefully document the ultimate choice made and state clearly that the answer would probably have been higher or lower had the alternative measure been chosen.

The previously described Catskills watershed example (NRC, 2000) provides a good illustration of the possible ambiguity of property rights and the consequent ambivalence about whether willingness to pay or to accept is the more appropriate measure of value. Did the upstream communities have the right to pollute, at least within some limits, or did New York City have the right to clean drinking water? The answers were governed by the legislative framework, in particular the federal Clean Water Act (see footnote 1, Chapter 1), which makes a sharp distinction between point source pollution and nonpoint source pollution—the former being strictly regulated, the latter less so. It also became clear during the discussions about conserving the Catskills watershed that the answer could change as a part of the ongoing negotiations. This was made clear when the State of New York introduced the possibility of using eminent domain legislation to compulsorily allow the purchase of areas of land deemed critical. The cost to New York City of restoring the watershed was affected by these considerations because they determined how much had to be paid to landowners in the watershed to help persuade them to reduce polluting activities. These payments would obviously be higher, given better-established landowners' "rights to pollute."

There are cases in which the ability to present an environmental policy recommendation in several different frames may be important to decision-makers because it allows them to seek and obtain support from different constituencies. For example, a recommendation to use tradable air emission permits to limit emission of a pollutant can be presented as an extension of the use of market mechanisms to those who may be predisposed to support such measure because of their belief in the market mechanism. It can also be presented as a limitation on pollution to "environmentalists," who may be disposed to support such a measure because it results in a net reduction in air pollution. The fact that a particular environmental policy appeals to several different constituencies often

[2] Michael Hanemann, University of California, Berkeley, personal communication, 2004.

stems from the ability to frame it in different ways. Cross-constituency support for a measure may mean that there is widespread agreement on the measure; it may also indicate that it can be seen from several different perspectives and is framed differently to appeal to different groups.

These preceding examples suggest that framing unavoidably affects both the question that is asked in an ecosystem valuation study, and therefore the type and level of analysis needed to answer it, and the way in which people respond to any given issue. Framing in the second of these senses introduces an element of subjectivity into an ecosystem valuation analysis. Rarely, if ever, will a completely objective presentation of the issues be attainable. Analysts must be aware of this and sensitive to the different ways of presenting data and issues and make a serious attempt to address all perspectives in their presentations. Failure to do so could undermine the legitimacy of an ecosystem valuation study.

Framing in the first sense—that is, determining the question to be asked in a valuation study such as the Catskills and *Exxon Valdez* studies—represents a legitimate and appropriate attempt to fit the analysis conducted to the precise decision to be made. In the Catskills case, it was appropriate and logical to ask whether watershed restoration could meet the same needs at a lower cost. In the *Exxon Valdez* case, investigators used the information available from the impact and injury studies being conducted by the State of Alaska to present the issues to respondents and so to frame the issues. The investigators attempted to be conservative in summarizing the conclusions of these studies and were constrained by the fact that the economic and ecological studies were being conducted somewhat in parallel. Because they did not desire the survey respondents to rely on information they had individually gleaned from the media, the investigators went out of their way to describe the effects of the spill, albeit in a succinct manner. Furthermore, the investigators chose to avoid duplicating the impact and injury studies that had already been completed. Instead they relied on the presentation and discussion of these studies in the media and other public fora to have created an informed public who could use this discussion to place values on the avoidance of a similar event. Such an approach does raise questions about how informed the sample used in the *Exxon Valdez* contingent valuation study was, about the soundness of their understanding of the impact of the oil spill on the local ecosystem, and about the sensitivity of the values people placed on preventing ecosystem damage to possible further information about the issues.

Additional Methodological Judgments

In most ecosystem valuation studies, the analyst will be called on not only to frame the study but also to make additional judgments about how the study should be designed and conducted. Typically, these will address issues such as whether, and at what rate, future benefits and costs should be discounted (see

Chapter 2 for further information); whether to value goods and services by what people are willing to pay or what they would be willing to accept if these goods and services were reduced or lost; and how to account for and present distributional issues arising from possible policy measures. In many cases, different choices regarding some of these issues will make a substantial difference to the final valuation. For example, many environmental restoration projects have projected lives of a century or more, and over such long periods, even small differences in discount rates can result in order-of-magnitude differences to the present value of a stream of net benefits (Heal and Kriström, 2002). In such cases, the appropriate response is undoubtedly for the analyst to present figures on the sensitivity of the results to alternative choices.

In the case of choice of discount rate, it is a straightforward matter to present a table of results showing how valuation varies with the discount rate selected. For cases in which a measure has significant distributional impacts, it is incumbent on the analyst to identify and describe these impacts, providing details of the groups that gain and lose from the policy, and the extents of these gains and losses. The analyst may also provide an estimate for the aggregate value of an environmental policy if benefits and costs to all recipients are weighted equally and then indicate how this would change if different distributional weights were used (see Layard and Walters, 1994).

Another illustration of the importance of methodological judgments comes in the choice of an objective in an economic project evaluation. There are usually several possibilities in making this selection. The conventional approach is to follow the utilitarian route of choosing the project that generates the greatest net total benefit. In this approach, the analyst calculates all of the gains and losses to the different groups in society and then totals them, with the project having the highest total gains deemed the best. In the process of adding benefits over different groups, the analyst might apply different weights: for example, weighting gains and losses to indigent groups more than those to the affluent. Of course, in adding up gains and losses that occur at different dates, the analyst may weigh by discount factors (see Chapter 2 for further information).

An alternative approach is to follow the Rawlsian route;[3] in this case the analyst focuses exclusively on the impact of the policy measure on one social group, the poorest group in society. In such cases, the "best" policy is defined as the one that does best by this poorest group. These two different approaches, the utilitarian and the Rawlsian, often lead to significantly different outcomes (Heal, 1998). The ultimate choice depends, among other things, on which approach the analyst believes best reflects the values of the group for whom the study is being undertaken. If the client is society as a whole, are its values better reflected by utilitarian or Rawlsian goals? Similar to situations in which WTP or WTA is used in ecosystem valuation study, ideally the analyst will present the

[3] American philosopher John Rawls' chief work, *A Theory of Justice* (1971), discussed liberty and equality in the context of a social contract. Rawls stated that inequalities in the distribution of wealth and income only become just when they can work in favor of the worst-off segment of the society.

results of both approaches and explain how and why they differ. However, the reality is that this may greatly increase the complexity of the ecosystem valuation study. If time and resources allow only one approach, then it is reasonable to expect a clear explanation of how the choice was made and some discussion of alternatives.

Peer Review

The unavoidable need to make professional judgments in ecosystem valuation activities through choices of framing and methods suggests that there is a strong case for peer review to provide input on these methodological issues before study design is complete and relatively unchangeable. Although most significant ecosystem valuation studies will be reviewed by external reviewers on completion and/or publication, the committee believes that external review by peers and stakeholders could also be particularly valuable at a much earlier stage, when key judgments for the study have tentatively been chosen but there remains a legitimate opportunity for revision. Outside review at these earlier stages can make the difference between a valuation study that is widely accepted and one that is regarded as controversial or misleading (NRC, 1996).

UNCERTAINTY

The following sections discuss the major sources of uncertainty in the economic valuation of aquatic ecosystem services and how policymakers and analysts should respond.

Levels of Uncertainty: Risk and Ambiguity

The almost inevitable uncertainty facing analysts involved in ecosystem valuation can be more or less severe depending on the availability of good probabilistic information. A favorable case would be one in which, although there is uncertainty about the magnitudes of various parameters, the analyst nevertheless has good probabilistic information. That is, there is a distribution of possible magnitudes—with means, standard deviations, and other aspects of the distributions available—and these distributions are based on statistical data that are sufficiently extensive to allow some confidence in their predictions. An illustration of such a case is provided by insurance companies, which typically have many years of actuarial data on the death rates of people with different characteristics and thus can calculate the expected number of deaths in a population with some confidence.

An alternative and common scenario in ecosystem valuation is one in which there is really no good probabilistic information about the likely magnitudes of some variables and what is available is based only on expert judgment. To continue the insurance analogy, this would likely be the position of an insurance company trying to assess the risk it faces if it provides terrorist insurance for owners of prominent buildings in major cities. There is no database of events on which the company can draw, and important decisions will have to be based solely on experts' assessments of the risks. Environmental policymakers find themselves in this situation when making decisions about climate changes because there is no database that allows an estimation of the consequences of increasing concentrations of greenhouse gases. Thus, such decisions should be based on the analyses of expert groups such as the Intergovernmental Panel on Climate Change (IPCC).[4] Analysts are in a similar position when evaluating changes designed to restore functionality in complex ecosystems such as the Florida Everglades.[5]

Situations such as the first of these, where there are reliable probabilities describing the unknown magnitudes, are described as characterized by *risk*—and the word "risk" in this context refers to situations in which reliable estimates of the probabilities are available. In contrast, the term *ambiguity* describes situations in which there are no data-based probabilities. Obviously, making good decisions is harder under conditions of ambiguity than under conditions of risk (Machina, 1987).

One way in which decision-makers can attempt to bridge the gap between risk and ambiguity is to assign subjective probabilities to the different possible outcomes. A subjective probability is one that is not based on repeated trials and observed occurrence frequencies, which is the classical interpretation of a probability, but rather on strength of belief in the likelihood of an outcome. So, in situations where there are no objective frequency-based probabilities, such as the consequences of the accumulation of greenhouse gases in the atmosphere, one could ask experts to present their best judgments about the likelihood of different outcomes by probability distribution. These would be subjective probabilities. Such judgments provide probability-like numbers to use in situations in which there are no data to provide frequency-based probabilities. One might, of course, end up with as many different subjective probabilities as there are different experts (Nordhaus, 1994; Roughgarden and Schneider, 1999.)

Model Uncertainty

Model uncertainty arises for the obvious reason that in many cases the relationships between certain key variables are not known with certainty (i.e., the "true model" of an important phenomenon or process will not be known). To

---

[4] The IPCC was organized by the United Nations to provide scientific, technical, and socioeconomic data on the impacts and options for adaptation and mitigation in climate change. Further information is available on-line at *http://www.ipcc.ch*, accessed June 14, 2004.

[5] Such groups include, for example, the South Florida Ecosystem Task Force (see *http://www.sfrestore.org* for further information).

use a biogeochemical example, the relationship between the nature of riparian tree cover in a watershed and the purification of water by that watershed may never be known. How do the amount and extent of water purification depend on the types of plant communities in a watershed and the successional stage of those communities? This is an example of the relationships discussed in Chapter 3 between ecological structure and function and the provision of ecosystem goods and services to the community. This relationship is often poorly understood and inevitably a source of uncertainty in ecosystem valuation efforts. In fact, in most studies of the value of aquatic ecosystems, this will be the largest single source of uncertainty because our understanding of how the structure of an ecosystem is affected by human activities and of how these effects translate into changes in ecosystem services is often rudimentary (see, for example, the Columbia River case study in Chapter 5 for further information).

On the economic side, an analyst might not know how society's WTP for an ecosystem service depends on the way in which that service is provided. For example, how does the degree of visible cleanliness, or the degree of development and crowding, affect the value that is placed on a particular waterbody? What are the functional forms that relate the value that people place on a body of water to the parameters describing the state of that waterbody? In economic terms, what is clear is that investigators often do not know the form of the demand function for an ecosystem service. Difficulties in estimating societal values of an ecosystem's services are especially acute for nonuse values such as the existence value that individuals may have for preserving species or intact ecosystems.

As discussed in Chapter 3, a particularly important issue in evaluating environmental policies designed to change the functioning of ecosystems is the existence of thresholds at which the qualitative behavior of an ecosystem changes. There is, for example, some evidence that many streams can absorb nitrate pollution up to a certain level with little or no effect on their biochemistry, but that beyond a certain level of nitrate input, their capacity to neutralize nitrates is exhausted and their biochemistry changes sharply (Lovett et al., 2001). The discussion of Lake Mendota in Chapter 5 also illustrates this effect. In such a situation, assuming a linear or even smooth response of the behavior of the system to outside influences could lead to massive errors in forecasts of the impacts of these influences. Model uncertainties about qualitative changes in ecosystem behavior are particularly important in ecosystem valuation. These should always be of concern to analysts who should establish a range for the main sources of uncertainty whenever possible.

It is clear from the preceding examples that given the imperfect knowledge of the way people value natural ecosystems and their goods and services, and our limited understanding of the underlying ecology and biogeochemistry of aquatic ecosystems, calculations of the value of the changes resulting from a policy intervention will always be approximate.

Parameter Uncertainty

Parameter uncertainty is one level below model uncertainty in the logical hierarchy of uncertainty in the valuation of ecosystem services. Even if the mathematical form of a relationship between important variables were known, one could—and in all probability would—still be uncertain about the values of the parameters in this functional form. For example, assume that an analyst knew with certainty that the value individuals place on a lake take the form $V = A^x B^y C^z$, where A, B, and C are characteristics of the lake such as water clarity, fish populations, and cleanliness; x, y, and z are parameters; and V is the value placed on the lake. Even if the functional form were known, the exact values of the parameters x, y, and z of the function would still not be known. At best, statistical estimates of these could be obtained, giving expected values of the parameters and distributions of possible errors about these parameters.

Most commonly, an analyst seeking to value the service or services of a particular ecosystem is subject to both model, and parameter uncertainty in that he or she is not sure of the true model and, conditional on the choice of model, faces further uncertainty about the values of parameters in the model.

Reducing Uncertainty: (Quasi) Option Values and Adaptive Management

Although there is considerable uncertainty regarding the value of ecosystem services, there is often the possibility of reducing this uncertainty over time through learning. Learning can be either active (the result of actions such as research designed to generate new knowledge), or passive (the byproduct of actions taken for other purposes or simply of the passage of time). Regardless of its source, the possibility of reducing uncertainty in the future through learning can affect current decisions, particularly when the impacts of these decisions are irreversible (Arrow and Fisher, 1974; Demers, 1991; Epstein, 1980; Henry, 1974). With learning, a "quasi-option value" has to be incorporated into the analysis, beyond the inclusion of expected net benefits that reflects the value of the additional flexibility. (From now on, this is collectively referred to simply as just "option value"; see also Chapter 2.) This flexibility allows future decisions to respond to new information as it becomes available.[6]

If the destruction of a natural system is irreversible, and if its value is currently unclear but may become better known in the future, then preserving it now allows the "destroy or conserve" issue to be revisited at a time when decision-makers are better informed; whereas destroying the ecosystem forces a permanent choice without the benefit of better knowledge. It follows that with

[6] However, it is not universally true that learning in the future makes increased flexibility more desirable. For discussions of the conditions under which this holds, see Epstein (1980), Freixas and Laffont (1984), Gollier et al. (2000), and Graham-Tomasi (1995).

the possibility of learning, in a cost-benefit analysis the measurement of the benefits of ecosystem protection through ecosystem valuation should consider the possibility of learning and, in consequence, making a better decision at a later date (i.e., it should incorporate the option value; Arrow and Fisher, 1974; Hanemann, 1989; Henry, 1974).[7]

The incorporation of option value in cost-benefit analysis still entails a balancing. Although the flexibility created by preservation and by the opportunity to revisit the decision adds to the benefits of preservation, this balancing does not necessarily imply that preservation will in all cases be justified by this criterion. The benefits of ecosystem preservation (including the value of retaining the flexibility to respond to new information) will not necessarily exceed the associated costs. At present, there is little guidance about the importance of option values in ecosystem valuation. Similarly, only a limited amount of empirical work has been done to date on estimating the magnitude of option value. There is a need for further research in both of these areas in the context of ecosystem valuation.

Adaptive Management

A natural extension of the observation that better decisions can be made if one waits for additional information is the use of adaptive management, which is a relatively new paradigm for confronting the inevitable uncertainty arising among management policy alternatives for large complex ecosystems or ecosystems in which functional relationships are poorly known. Although advanced in the late 1970s and 1980s (Holling, 1978; Walters, 1986), adaptive management has recently only been applied by natural resource managers.[8] A key component of adaptive management is active learning by introducing new management policies to learn more about the system's behavior and thereby reduce uncertainty. Typically, there may be an effort to implement environmental management actions as "experiments" in order to "learn by doing," with the experiments designed to reduce critical uncertainties about the ecosystem's behavior. The usual goal of ecosystem management is to manage for resiliency (i.e., capacity for self-renewal) while optimizing benefits to society. Possible economic

---

[7] See Fisher and Hanemann (1986) for an empirical application of the concept of option value in the extinction of species.

[8] Adaptive management is an integrated, multidisciplinary approach for confronting uncertainty in natural resource issues. It is adaptive because it acknowledges that managed resources will change as a result of human intervention, surprises are inevitable, and uncertainties will emerge. Active learning is the way in which the uncertainty is winnowed. Adaptive management acknowledges that policies must satisfy social objectives, but also must be continually modified and flexible for adaptation to these surprises. Adaptive management therefore views policy as hypotheses; that is, most policies are really questions masquerading as answers, and management actions become treatments in an experimental sense. For more information on adaptive management, see Gunderson et al. (1995), Holling (1978), Lee (1993), NRC (2002, 2004), and Walters (1986).

benefits are often a part of the mix of information that stakeholders or government officials use to select management actions. Actually implementing potentially beneficial policies thus winnows the uncertainty in system response, albeit in a reversible and experimental sense. Adaptive management therefore provides a mechanism for learning systematically about the links between human societies and ecosystems. In contrast, the learning that occurs in economic models with option values is purely passive—information about the value of an environmental system is acquired with the passage of time. If one believes that additional information could be influential in selecting the best environmental policy option, then adaptive management is a natural step from the passive concept of an option value associated with gaining information to the concept of managing the ecosystem to learn and so reduce uncertainty. When an adaptive management approach is possible, which will not always be the case, the option value associated with conservation is likely to be increased because of the enhanced rate of information acquisition.

Adaptive management often uses explicit dynamic modeling or conceptual models of large complex ecosystems. These computer models are useful for two purposes. First, building an explicit numerical model requires a clear statement of what is known and what is assumed, which helps to expose broad gaps in data and understanding that are easily overlooked in verbal and qualitative assessments. Second, even crude models can help "screen" policy options and eliminate those that are simply too small in scale to be important or would be unacceptably risky given uncertainty about directions of response in key policy indicators (Walters et al., 2000). Proponents of adaptive management have long emphasized the importance of such modeling (Holling, 1978; Walters, 1986). Adaptive management is not a tool for ecosystem valuation or a method of valuation per se, nor does it require valuation. Rather, by reducing uncertainty and illuminating relationships within the ecosystem and between the ecosystem and human actions, it aids management and decision-making and may make economic valuation easier and more accurate.

DECISION-MAKING AND DECISION CRITERIA UNDER UNCERTAINTY

Decision Criteria

Just as there are different types of uncertainty, there are also different ways in which an analyst can allow for uncertainty in the support of environmental decision-making. A central issue is how to account for the range of possible outcomes (the variability of outcomes) that is an inevitable result of uncertainty. A widely used criterion for decision-making is to choose the alternative that yields the greatest expected value of benefits. This rates as equal all distributions of outcomes that have the same mean even if they have very different higher moments and so ignores information about variability. However, this approach can be adopted only if the possible values of the relevant variables are known and associated probabilities can be assigned; otherwise, expected values cannot be computed. Thus, in order to adopt the objective of maximizing expected net benefits in ecosystem valuation, one has to be able to assign probabilities, either objective probabilities from past experience or subjective probabilities (for a general discussion, see Machina, 1987).

The unpredictability of the outcome of an environmental policy under uncertainty means that while the outcome could be excellent, it also has a chance of being poor. In general, faced with the choice between policies that generate the same expected value but with different ranges of outcomes, most people would choose the policy with the lowest variability, implying that they are "risk averse." The extent of their risk aversion determines what they would be willing to pay to avoid a risk and replace it by a certain outcome. If people are very risk averse, an environmental policy that delivers a modest outcome with some certainty might be preferred to one that may deliver a truly outstanding outcome but may also deliver a very poor result. In such situations, an analyst has to decide whether to build some measure of risk aversion into the analysis and, if so, how much. There are studies of the degree of risk aversion displayed by individuals in financial markets (see Chetty, 2003, and references therein), but because risk aversion for a given person may vary with the magnitude of the risk and because it varies across people, these are not necessarily the appropriate values to use in environmental studies. In a heterogeneous population the analyst will have to make an assumption about the level of risk aversion that is appropriate for the group as a whole. In general, this is a matter in which the best solution is to state clearly that the assumption about the degree of risk aversion will affect the outcome and to conduct sensitivity analyses to indicate how this assumption impacts the outcome of the study (Heal and Kriström, 2002). If contingent valuation methods are used, it may be possible to inform subjects of the uncertainties associated with estimates presented in the study, so that their valuations reflect their own degrees of risk aversion.

A key assumption in ecosystem valuation models is that individuals seek to maximize their utility and that they will be indifferent to changes that leave their utility unchanged. Under uncertainty, the assumption is that they maximize their expected utility, which is simply the expected value of the utilities they would realize under the possible outcomes. Although widely used in economic analyses, the expected utility assumption has been controversial since in some contexts its predictions are not consistent with observed behavior (Machina, 1987). Alternative theories of behavior under uncertainty have been proposed, including prospect theory (Kahnemann and Tversky, 2000).[9] These alternatives introduce psychological responses (such as feelings of loss aversion and regret)

---

[9] Prospect theory differs in two key respects from expected utility theory, (1) the payoff is not linear in probabilities, overweighting low probabilities and underweighting large ones, and (2) outcomes are evaluated with respect to a reference point rather than with respect to their absolute value (see Kahneman and Tversky, 1979 for details; for a general review see Machina, 1987).

into models of choice. This modifies the arguments and structure of the individual's utility or payoff function, but maintains the assumption that there is a payoff function that individuals seek to maximize. Thus, these alternative theories retain the basic assumption that individual behavior is based on self-interest.

Under the assumption that individuals seek to maximize their expected utility, the value of ecosystem protection is typically defined as the amount an individual would be willing to pay to ensure that protection occurs, which is then a measure of the dollar value or benefit of protection. The ecosystem valuation process is designed to provide an estimate of this measure. In the context of uncertainty, both WTP and WTA have to be interpreted as expressing preferences over uncertain outcomes and, in particular, as reflecting individuals' aversions to the risks they perceive to be associated with the options available. To the extent that valuations reflect individuals' attitudes toward risk and those individuals are accurately informed of the uncertainties associated with a project, there is no need for the analyst to make further allowance for risk aversion.

If society is extremely risk averse, the objective of maximizing the expected value of the aggregate utility can be replaced by an objective known as "maximin." The intent in such cases is to focus on the worst possible outcome, the minimum, and then seek the policy option that makes this as favorable as possible, or maximum (hence, the name; for a discussion, see Arrow and Hurwicz, 1972; Maskin, 1979). By way of illustration, consider an aquatic ecosystem that, among other services, provides flood control to a residential area. It is possible that decision-makers believe that the loss of human life through floods is the worst possible outcome and must be prevented at all costs. Such a belief would be appropriately represented by maximin preferences, which would lead the analyst to select the project that minimizes the loss of life from flooding. Focusing exclusively on the worst possible outcome is justified only if there are good reasons to suppose that society is really risk averse and is willing to sacrifice considerable possible benefit from a policy to avoid any chance of a bad outcome. Technically, the maximin objective can be seen as a limiting case of the expected utility objective as the degree of risk aversion increases without limit. There are also arguments that suggest that the maximin may be an appropriate choice of objective in some cases of ambiguity—that is, cases in which there are no objective or subjective probabilities (Arrow and Hurwicz, 1972; Maskin, 1979). Implementing the maximin criterion does not require probabilities; it requires only that the worst possible outcome be identified, so it is particularly suited to problems for which no probabilities are available.

Recent literature on this topic (e.g., Ghirardato et al., 2002) has extended this concept to a broader analysis of decision-making with ambiguity and suggests, in outline, that under quite general conditions a decision-maker faced with ambiguity should look for the worst possible outcome, then for the best possible outcome, and then rank projects and policies by a weighted average of these. Obviously, using the maximin criterion in ecosystem services valuation is a special case because all of the weight in the weighted average is placed on the worst case. A logical extension of this line of thinking leads to concepts such as the precautionary principle and the idea of a safe minimum standard, which are discussed next.

The Precautionary Principle and Safe Minimum Standard

Another approach to environmental decision-making under uncertainty is embodied by the precautionary principle. Notably, the 1992 Rio Declaration (Article 15) (see Gollier et al., 2000) stated: "Where there are threats of serious and irreversible damage, lack of full scientific certainty shall not be used as a reason for postponing cost-effective measures to prevent environmental degradation." Alhough the precautionary principle has been attacked as a vague concept lacking a precise definition, the essence of the precautionary principle is clear and is that the burden of proof should be to demonstrate that changes do not cause irreversible environmental damage, rather than proving that a change is dangerous. Most economists, if asked to think of a justification for the precautionary principle in decision-making, would probably couch it in terms of learning, especially about the validity of a scientific model, irreversibilities, and option values. The option value linked to conserving an ecosystem whose change is irreversible is in effect a reward for cautious behavior, although it certainly does not imply that conservation is always appropriate. Gollier et al. (2000) note that the precautionary principle can also be given a formal justification in environmental decision-making without invoking irreversibilities, just by assuming that there is cumulative damage from a stock of pollutant and possible learning over time about the consequences of the pollutant.

There has been extensive discussion of irreversibility, learning, option values, and the precautionary principle in the context of policy toward climate change. Since the basic decision framework is similar to that in ecosystem conservation and valuation, it is useful to review briefly some of the more relevant conclusions from this literature. Notable references include Fisher and Narain (2002), Gollier et al. (2000), Kolstad (1996a,b), Pindyck (2000), among others.

One of the conclusions to emerge from this discussion is that while there may be an option value associated with ecosystem conservation, it is also possible that there is a value associated with not adopting conservation policy measures that require significant investments. The point is that if an environmental policy requires investment in fixed capital and there is some uncertainty about the appropriateness of the policy, and so about the value of the associated investment, there may be a benefit from delaying its adoption so as to benefit from learning about the value of the investment. Thus, if one is unsure of how effective a policy measure is and it requires a long-term and unchangeable commitment, it may be appropriate to wait to implement it until there is more information and the value is clear.

This implies that in discussions of the conservation of an ecosystem whose destruction would be irreversible and whose conservation would require an investment in fixed capital, there is an option value argument for conserving the ecosystem and also an option value argument for delaying implementation of the conservation policy until it is clear whether the associated investment in fixed capital is in fact appropriate. In such a case, there are two opposing option values and which is larger is an empirical question. An example of an effectively irreversible policy would be the construction or removal of a dam or of a system of canals, which cannot readily be undone once implemented.

One recommendation that emerges from this discussion is that under conditions of uncertainty and learning, there should be a preference for environmental policy measures that are flexible and minimize the commitments of fixed capital or that can be implemented on a small scale on a pilot or trial basis. In effect, this is adaptive management and the option value stays on one side of the equation.

In their study of Lake Mendota, Carpenter et al. (1999; see also Chapter 5) set out a quite different approach. In an intensive agricultural region, such as the Midwest of the United States, phosphorus is often applied as a fertilizer to the land and some runs off into nearby streams and lakes, including Lake Mendota. In sufficient concentrations, phosphorus can cause a change in the normal biological state of the lake that results in a potentially locally stable state of eutrophication in which the lake is unproductive for most human uses. Eutrophication of a lake can be reversed, albeit slowly. The response of a lake to phosphorus concentration is highly nonlinear and the concentration depends not only on the runoff but also on temperature and rainfall. How should the runoff of phosphorus over time be managed in order to maximize the expected discounted value of benefits net of the costs of phosphorus mitigation? In this regard, Carpenter et al. (1999) modeled the dynamics of the interacting lake and surrounding agricultural systems as a nonlinear dynamical system with several different locally stable states, one of which (eutrophication) is highly undesirable. Avoiding this state in agriculturally intensive regions is costly, so there are trade-offs to be made. Further, the stochasticity of the weather means that the problem has to be viewed in probabilistic terms. A particularly relevant conclusion that these authors (Carpenter et al., 1999) reached follows:

> An important lesson from this analysis is a precautionary principle. If phosphorus inputs are stochastic, lags occur in implementing phosphorus input policy, or decision makers are uncertain about lake response to altered phosphorus inputs, then phosphorus input targets should be reduced. In reality, all of these factors—stochasticity, lags, uncertainty—occur to some degree. Therefore, if maximum economic benefit is the goal of lake management, phosphorus input levels should be reduced below levels derived from traditional limnological models. The reduction in phosphorus input targets represents the cost a decision maker should be willing to pay as insurance against the risk that the lake will recover slowly or not at all from eutrophication. This general result resembles those derived in the case of harvest policies for living resources subject to catastrophic collapse. . . We believe that the precautionary principle that emerges from our model applies to a wide range of scenarios in which maximum benefit is sought from an ecosystem subject to hysteretic or irreversible changes.

Although Carpenter et al. (1999) mention the precautionary principle, they do not define it or state it in an operational way in the context of managing Lake Mendota. Rather, the precautionary principle is implied to be a recommendation that phosphorus levels should be below that recommended by traditional limnological models, this being a cost that decision-makers must shoulder to avoid the risk of eutrophication. Thus, this is not a concept that can be made operational without further work, and indeed it seems possible that much of what is at issue in this case is captured in economists' concepts of risk aversion and option value, which were not explicitly developed in the model of Carpenter et al.

The precautionary principle is widely cited by the environmental community as a justification for erring on the side of conservation in situations of uncertainty. However, it is not clear that the precautionary principle brings anything new to the decision criteria frameworks usually used by economists. As stated above, many of the concerns that drive people to articulate the precautionary principle are addressed by existing economic approaches to environmental decision-making but under different names. With learning and irreversibility, option values may tilt decisions in the direction of environmental conservation, more so if learning can be actively pursued through an adaptive management approach, and especially if there is a chance of a significantly negative outcome from environmental impacts. In such cases, risk aversion will normally move decisions in the same direction.

Related in some ways to the precautionary principle is the concept of a "safe minimum standard," which introduces a class of choices in which decision-makers seek to maintain populations or ecosystems at levels deemed necessary to ensure their continued existence. The most striking example in the United States is the Endangered Species Act (ESA). As originally passed, the ESA explicitly prohibited actions that would reduce the survival chances of an endangered species, whatever the economic costs of this prohibition.[10] Thus, the ESA mandated conservation irrespective of economic costs when the very existence of a species was threatened. The intent of the ESA was clearly to take species survival decisions out of the realm of economics, asserting the primacy of an ethical imperative to prevent extinction over any cost-benefit calculations. The ESA was subsequently amended to include a provision for balancing extinction against the economic costs of its prevention.[11] As amended, the ESA is consistent with the safe-minimum standard approach, under which a minimum population is protected unless it is too costly to do so. However, the consideration of costs can only be invoked in extreme cases. As a result of the ESA, when the survival of a species is at stake, one does not have to place an eco-

[10] In *Tennessee Valley Authority vs. Hill,* the Supreme Court upheld that the Endangered Species Act of 1973 was intended by Congress to ". . . halt and reverse the trend toward species extinction at whatever the cost."

[11] In 1978, the ESA was amended to "take into consideration economic impact, and other relevant impact" of listing and designation of critical habitats. See *http://endangered.fws.gov* for further information about the ESA.

nomic value on its continuation because legislators have determined that this is infinite and outweighs any possible costs. The Clean Water Act also contains provisions that explicitly set the attainment of public health-related standards outside the range of economic valuation, mandating that they be met whatever the cost.

These preceding examples illustrate situations in which U.S. society reacts to uncertainty about ecosystem services by specifying safe minimum standards (i.e., not causing conditions that would drive a species to extinction, not damaging human health) for impacts on or changes in these systems. Rather than calculate the expected costs and benefits of different levels of impacts and choosing the best, society specifies a bound on the permissible impacts. Of course, with ambiguity rather than risk, and thus no probabilities with which to work, it may be impossible to calculate expected costs and benefits so that standard cost-benefit analysis in such cases is hardly applicable.

Choosing one bound or safe minimum standard over another requires some justification and supporting analysis. One possible line of argument relates to thresholds in ecosystem behavior in response to stress (see Chapter 3). If stresses above a certain level are believed to lead to sharp deterioration in an ecosystem, this may provide a strong case for restricting impacts below this critical level. Yet even this argument relies implicitly on the idea that the costs of ecosystem stress rise sharply and are therefore likely to exceed benefits at some threshold—an argument that cannot be made plausibly without some idea of the magnitudes of the costs and benefits and of the associated margins of error. Once a safe minimum standard is chosen, however, valuation is not needed, but valuation may be needed in setting the safe minimum standard (Berrens, 1996; Berrens et al., 1998; Bishop, 1978; Ciriacy-Wantrup, 1952; Farmer and Randall, 1998; Palmini, 1999; Randall and Farmer, 1995; Ready and Bishop, 1991).

ILLUSTRATIONS OF THE TREATMENT OF UNCERTAINTY

This section briefly illustrates how uncertainty could be treated in ecosystem services valuation studies, with reference to the Catskills watershed in New York (also discussed earlier in this chapter) and the Edwards Aquifer case studies provided in Chapter 5. The section begins with an introduction to evaluating and assessing uncertainty through "Monte Carlo"[12] simulations and indicates

[12] Monte Carlo methods have been practiced for centuries, but under more generic names such as "statistical sampling." The "Monte Carlo" designation was popularized by early pioneers in the field during World War II because of the similarity of statistical simulation to games of chance and because Monte Carlo (the capital of Monaco) was a well known center for gambling and similar pursuits. For further information about the history, development, and use of Monte Carlo simulation methods, see *http://csep1.phy.ornl.gov/mc/node1.html*.

how this approach could be applied to provide a more complete description of the consequences of uncertainty regarding the inputs to the valuation process.

Monte Carlo Simulation

A sophisticated way of incorporating uncertainty in the output of an ecosystem services valuation study is to use Monte Carlo simulation. This method can provide an estimate of the probability distribution of possible values that is derived from uncertainty about the underlying parameters and relationships. A prerequisite for such an analysis is some probabilistic information about the elements of the valuation.

By way of illustration, assume that a policy intervention is being evaluated that would conserve an ecosystem at some cost in terms of forgone residential development, which was a relevant issue in the Catskills watershed in New York. Assume further that there are two elements to the benefits, (1) the quantity of clean water assured because of the policy intervention and (2) the price at which this water should be valued. Call these Q and P respectively, where both are uncertain. On the cost side there is a present cost of C_p and a continuing cost of C_f per year in the future while the benefits continue into the future. If all values were known with certainty, then the net present value of the project would be represented by the following formula if the time horizon is fifty years and the discount rate is r:

$$NV = \sum_{t=1}^{T=50} (PQ - C)(1+r)^t - C_p$$

If the parameters of this expression are known only with some degree of uncertainty, then NV is a random quantity and an analyst would desire data on its distribution. Suppose that the uncertainty is about P, Q, and C_f with r and C_p being known, and that the analyst possesses probability distributions over these uncertain variables. That is, for each of the uncertain variables there is a density function that provides the probability that the variable is within any interval. An analyst can then conduct a Monte Carlo simulation by picking a series of values for the uncertain variables as random numbers chosen according to their density function and for each set values for P, Q, and C_f computing the value of NV. This simulation is repeated many times with a different set of randomly-chosen values of P, Q, and C_f each time. The result will be a set of values for NV. As the number of repetitions of this process increases, the distribution of this set will approach that of the uncertain value of NV. An analyst can therefore obtain from this process approximations to the mean and standard deviation of the values of NV that are compatible with what is known about the uncertain parameters P, Q, and C_f.

In practice an analyst will use computer programs written for Monte Carlo simulation for this process and will need only to input information about the distributions of the uncertain parameters and a formula indicating how these are used to compute the value. Of course, and as has been emphasized previously, obtaining probabilistic information about parameter values is often not straightforward and on many occasions it will be necessary to use subjective probabilities for this purpose. A potential complication is that in some cases the distributions of the various parameters will not be independent but will be drawn from a joint distribution. For example, in the illustration above, price P and quantity Q will not be independent—high prices will tend to be associated with low quantities and vice versa. In such cases the analyst will have to specify joint rather than independent distributions, which is a somewhat more demanding task.

There is little doubt that if resources and sufficient information are available for a Monte Carlo approach, and if the analyst is able to supply the required probabilistic information, this approach provides decision-makers a better appreciation of the range of possible outcomes that are consistent with what is known or believed to be known concerning ecosystem services valuation. EPA has already applied Monte Carlo methods to some studies (EPA, 1997), and Jaffe and Stavins (2004) have reviewed these and conducted their own analyses. Although these previous applications were not in the context of ecosystem services valuation, they illustrate the feasibility of using Monte Carlo analysis to evaluate environmental policies and suggest that this approach could be applied in ecosystem valuation studies as well.

Catskills Watershed and Edwards Aquifer Cases Studies

In the Catskills case, and as noted previously, the key issue was to compare the cost of watershed restoration with the cost of the alternative to provide the service of water purification (NRC, 2000). While the costs of the alternative—construction of a drinking water filtration system) are relatively certain, the cost of increased watershed protection and restoration is uncertain, as is the effectiveness of a given level of restoration in restoring ecosystem services. The poorly understood link from ecosystem structure and function to services is again the cause of the problem. Uncertainty about the effectiveness of watershed restoration, however, can in this case be subsumed into uncertainty about costs, so that the main issue can be treated as uncertainty about the cost of restoring the ecosystem service of water purification to a level needed by New York City.

The first step in dealing with uncertainty in this case will be to obtain information about the possible costs of watershed restoration. Ideally, a probability distribution over possible costs can be obtained. It may be that the analyst feels able to provide this information without further research, but in many cases this will require modeling the restoration process and then using ecological models to link the final state of the system post-restoration to the levels of ecosystem services provided. This will provide an estimate of the cost of restoring a given level of ecosystem services. Because the parameters of the restoration process will typically be uncertain, as will those of the ecological models, it would therefore be desirable to use Monte Carlo simulation to study the distribution of restoration costs and service levels. In doing this, the uncertainty associated with the links between ecosystem structure and function on the one hand and ecosystem services on the other are central. At issue is how far one must restore the watershed, in terms of area, land use, and vegetation, in order to provide water purification services at the level required by New York City. There are no existing models that can be readily enlisted to answer this question in a routine way. Monte Carlo simulation will provide a probability distribution over the costs of restoration to an appropriate level. Then, if the decision-maker is risk neutral, the next step is to compare the expected cost of the restoration with the cost of the alternative (i.e., construction of a water filtration system). If some degree of risk aversion is appropriate, then to the expected cost of restoration should be added a risk premium that depends on the degree of risk aversion of the decision-maker and the standard deviation and higher moments of the probability distribution of possible restoration costs, and this total is to be compared with the cost of the alternative.

In the absence of a probability distribution for the restoration costs, the best approach is probably to construct three scenarios for restoration costs: a best case, worst case, and expected case. These might, for example, amount to $1 billion, $2.2 billion, and $1.6 billion. If the restoration cost is less than the replacement cost for each cost value, the choice is simple—restoration is preferable to the alternative. This would be the case provided that the worst case restoration cost is less than the cost of a new filtration system (i.e., less than about $8 billion; NRC, 2000).

A more complex case would arise when the range of restoration costs crosses the cost of replacement—for example, when the three restoration cost estimates are $1.5 billion, $9 billion, and $2.5 billion with a replacement cost of $8 billion. If probabilities were available to attach to these numbers, then an expected cost could be calculated and adjusted to allow for risk aversion, and the risk-adjusted expected restoration cost could be compared with the replacement cost.

In the case of the Edwards Aquifer, which provides water to San Antonio, Texas, uncertainty arises from several sources—one of which is our inability to forecast recharge rates for the aquifer. The dynamics of the aquifer can be written as:

$$S_t - S_{t-1} = R_t - C_t$$

Here, S_t is the stock of water in the aquifer at date t and R_t and C_t are the recharge and consumption rates, respectively. The consumption rate is relatively predictable and indeed can be controlled to some degree by limitations on water use, whereas the recharge rate depends on weather, which is inherently stochastic. There may also be a trend in the recharge rate associated with changing patterns of rainfall as a result of climate change and another resulting from land development in the intake region of the aquifer, which by increasing the amount of impervious surface can reduce the amount of water collected in the aquifer for any given level of rainfall. There are several other factors that aquifer managers have to take into account, including whether the structure of the aquifer may be damaged if water stocks are drawn down too low, and whether there are any endangered species that live in the aquifer and can be harmed by low water levels. The lowest level to which the water stock has fallen to date is an important variable because this can affect the health of aquifer-specific species. The precise ways in which the structure of the aquifer and the prospects of any endangered species depend on the minimum water level is far from clear, so this relationship is an additional source of uncertainty.

How should these considerations affect the value that resource managers place on water in the aquifer? If managers are risk averse, the recognition of uncertainty will tend to increase the value of water stocks in the aquifer. The fact that in a stochastic world there is a chance of little or no rainfall in the coming years and therefore of little or no replenishment of the water stock in the aquifer means that current stocks might possibly have to last through a long dry period, which adds to the value of having a slightly higher stock. Thus, the marginal value of a unit of water will be higher because of the risk. Likewise, the possibility of damage to endangered species or to the structure of the aquifer because of low water levels increases the value of existing water stocks, because in addition to providing more water for consumption, a higher stock will lower the risk of damage from a future low stock level.

The value of the aquifer considering uncertainty about future replenishment can be approximated by Monte Carlo simulation, using the equation for the dynamics of the aquifer with alternative future replenishment patterns that draw probabilistically from a distribution of future replenishment rates. It is also worth noting that if the structure of an aquifer can be damaged irreversibly by allowing the water level fall too low, then there may be an option value associated with the preservation of water levels above a minimum. This is the type of context in which such values are applicable—there is a possible irreversible change, as well as the opportunity to learn more about the aquifer system's responses over time.

These two cases indicate that it is conceptually straightforward to see how the analyst should allow for uncertainty in valuation studies. Application of the concepts requires that the uncertainty be characterized to some extent and that the analyst understands decision-makers' attitudes toward uncertainty. Even if a characterization of the uncertainty is not available, it will often be possible, as in the case of the Edwards Aquifer, to state clearly what the qualitative impact of uncertainty will be—whether it will raise or lower a value—even though it may not be possible to measure the extent of this change.

SUMMARY: CONCLUSIONS AND RECOMMENDATIONS

The valuation of aquatic and related terrestrial ecosystem services inevitably involves investigator judgments and some amount of uncertainty. Although unavoidable, uncertainty and the need to exercise professional judgment are not debilitating to ecosystem valuation. It is important to be clear however when such judgments are made, to explain why they are needed, and to indicate the alternative ways in which judgment could have been exercised. It is also important that the sources of uncertainty be acknowledged, minimized, and accounted for in ways that ensure that a study's results and related decisions regarding ecosystem valuation are not systematically biased and do not convey a false sense of precision.

There are several cases in which investigators have to use professional judgment in ecosystem valuation regarding how to frame a valuation study, how to address the methodological judgments that must be made during the study, and how to use peer review to identify and evaluate these judgments. Of these, perhaps the most important choice in any ecosystem services valuation study is the selection of the question to be asked and addressed (i.e., framing the valuation study). The case studies discussed in this chapter illustrate the fact that the policy context unavoidably affects the framing of an ecosystem valuation study and therefore the type and level of analysis needed to answer it. Framing also affects the way in which people respond to any given issue. Analysts need to be aware of this and sensitive to the different ways of presenting data and issues and make a serious attempt to address all perspectives in their presentations because failure to do so could undermine the legitimacy of an ecosystem services valuation study.

In most ecosystem valuation studies, an analyst will be called on to make various methodological judgments about how the study should be designed and conducted. Typically, these will address issues such as whether, and at what rate, future benefits and costs should be discounted; whether to value goods and services by what people are willing to pay or what they would be willing to accept if these goods and services were reduced or lost; and how to account for and present distributional issues arising from possible policy measures. In many cases, different choices regarding some of these issues will make a substantial difference to the final valuation.

The unavoidable need to make professional judgments in ecosystem valuation activities through choices of framing and methods suggests that there is a strong case for peer review to provide input on these issues before study design is complete and relatively unchangeable. There are several major sources of uncertainty in the valuation of aquatic ecosystem services and options for the way policymakers and analysts can and should respond. Model uncertainty arises for the obvious reason that in many cases the relationships between certain key variables are not known with certainty (i.e., the "true model" will not be known). Chapter 3 discusses the relationship between ecological structure and function and the provision of aquatic ecosystem goods and services to the community; however, this relationship is often poorly understood and will be the greatest single source of uncertainty in many studies of the value of aquatic ecosystems. On the economic side, an analyst might not know the extent to which society's willingness to pay for an ecosystem service depends on the way in which that service is provided. Parameter uncertainty is one level below model uncertainty in the logical hierarchy of uncertainty in the valuation of ecosystem services.

The almost inevitable uncertainty facing analysts involved in ecosystem valuation can be more or less severe depending on the availability of good probabilistic information and the amount of ambiguity. A favorable case would be one in which, although there is uncertainty about some key magnitudes of various parameters, the analyst nevertheless has good probabilistic information. An alternative and common scenario in ecosystem valuation is one in which there is really no good probabilistic information about the likely magnitudes of some variables, and what is available is based only on expert judgment.

Just as there are different types of uncertainty in ecosystem valuation, there are also different ways and decision criteria that an analyst can use to allow for uncertainty in the support of environmental decision-making. One of these is the use of Monte Carlo simulations as a method of estimating the range of possible outcomes and the parameters of its probability distribution. A key assumption in ecosystem valuation models is that individuals seek to maximize their utility and that they will be indifferent to changes that leave their utility unchanged. Under uncertainty, this implies they maximize their expected utility. Although widely used in economic analyses, the expected utility assumption has been controversial, since in some contexts its predictions are not consistent with observed behavior. Alternative theories of behavior under uncertainty have been proposed, including prospect theory and regret theory.

The outcome of an environmental policy choice under uncertainty is necessarily unpredictable, and risk aversion is a measure of what a person is willing to pay to avoid an uncertain outcome. In a heterogeneous population, the analyst will have to make an assumption about the level of risk aversion that is appropriate for the group as a whole. If society is extremely risk averse, then the objective of maximizing the value of the aggregate expected utility can be replaced by an objective known as maximin. Focusing exclusively on the worst possible outcome is justified, however, only if there are good reasons to suppose that society is really risk averse and is willing to sacrifice considerable potential gain from a policy to avoid any chance of a bad outcome. Implementing the maximin criterion does not require probabilities; it requires only that the worst possible outcome be identified, so it is particularly suited to valuation conditions for which no probabilities are available. A logical extension of this line of thinking leads to concepts such as the precautionary principle and the idea of a safe minimum standard, which are summarized below.

Although there is considerable uncertainty regarding the value of ecosystem services, there is often the possibility of reducing this uncertainty over time through passive and/or active learning. Regardless of its source, the possibility of reducing uncertainty in the future through learning can affect current decisions, particularly when the impacts of these decisions are (effectively) irreversible, such as the construction or removal of a dam. With learning, an option value needs to be incorporated into the analysis as part of the expected net benefits that reflects the value of the additional flexibility. This flexibility allows future decisions to respond to new information as it becomes available. It follows that with the possibility of learning, in a cost-benefit analysis the measurement of the benefits of ecosystem protection through ecosystem valuation should consider the possibility of learning (i.e., should incorporate the option value). At present, only a limited amount of empirical work has been done on estimating the magnitude of option value. A natural extension of the observation that better decisions can be made if one waits for additional information is through the use of adaptive management. Adaptive management provides a mechanism for learning systematically about the links between human societies and ecosystems, although it is not a tool for ecosystem valuation or a method of valuation per se.

Another approach to environmental decision-making under uncertainty is embodied by the precautionary principle as articulated in the 1992 Rio Declaration (Article 15). The precautionary principle is widely cited by the environmental community as a justification for erring on the side of conservation in situations of uncertainty. However, it is not clear that the precautionary principle brings anything new to the decision criteria frameworks usually used by economists. With learning and under conditions of irreversibility, option values may similarly move environmental policy decisions in the direction of environmental conservation, more so if learning can be actively pursued through an adaptive management approach and especially if there is the chance of a significantly negative outcome from environmental impacts. In such cases, risk aversion will normally move environmental decisions in the same direction. While there may be an option value associated with ecosystem conservation, there may also be an option value associated with not adopting conservation policy measures that require significant investments.

Related in some ways to the precautionary principle is the concept of a safe minimum standard, which introduces a class of choices in which decision-makers seek to maintain population or ecosystem levels sufficient for survival. Under this approach, the presumption is that the necessary population size should be maintained, unless the costs of doing so are prohibitively high. The most striking example of this in the United States is the ESA.[13] Choosing one bound or safe minimum standard over another requires some justification and supporting analysis. Once a safe minimum standard is chosen however, valuation is not needed, but valuation may be needed in setting the safe minimum standard.

---

[13] In this case there is a provision for the economic costs of conservation of endangered species to be taken into account when these costs are very high.

Based on these conclusions, the committee makes the following recommendations regarding judgment and uncertainty in ecosystem valuation activities and methods and approaches to effectively and proactively respond to them:

- Analysts must be aware of the importance of framing in designing and conducting ecosystem valuation studies so that the study is tailored to address the major questions at issue. Analysts should also be sensitive to the different ways of presenting study data, issues, and results and make a concerted attempt to address all relevant perspectives in their presentations.
- The decision to use WTP or WTA as a measure of the value of an ecosystem good or service is a choice about how an issue is framed. If the good or service being valued is unique and not easily substitutable with other goods or services, then these two measures are likely to result in very different valuation estimates. In such cases the analyst should ideally report both sets of estimates in a form of sensitivity analysis. However, the committee recognizes that in some cases this may effectively double the work and in such situations a second best alternative is to document carefully the ultimate choice made and clearly state that the answer would probably have been higher or lower had the alternative measure been selected and used.
- Because even small differences in a discount rate for a long-term environmental restoration project can result in order-of-magnitude differences to the present value of net benefits, in such cases, analysts should present figures on the sensitivity of the results to alternative choices for discount rates.
- Ecosystem valuation studies should undergo external review by peers and stakeholders early in their development when there remains a legitimate opportunity for revision of the study's key judgments.
- Analysts should establish a range for the major sources of uncertainty in an ecosystem valuation study whenever possible.
- Analysts will often need to make an assumption about the level of risk aversion that is appropriate for use in an ecosystem valuation study. In such cases, the best solution is to state clearly that the assumption about risk aversion will affect the outcome and conduct sensitivity analyses to indicate how this assumption impacts the outcome of the study.
- There is a need for further research about the relative importance of, and estimating the magnitude of, option value in ecosystem valuation.
- Under conditions of uncertainty, irreversibility, and learning, there should be a clear preference for environmental policy measures that are flexible and minimize the commitment of fixed capital or that can be implemented on a small scale on a pilot or trial basis.

REFERENCES

Arrow, K., and L. Hurwicz. 1972. An optimality criterion for decision-making under ignorance. In Uncertainty and Expectations in Economics, C.F. Carter and J.L. Ford (eds.). Oxford, U.K.: Basil Blackwell.

Arrow, K., and A. Fisher. 1974. Environmental preservation, uncertainty, and irreversibility. Quarterly Journal of Economics 98:85-106.

Berrens, R.P. 1996. The safe minimum standard approach: An alternative to measuring non-use values for environmental assets? Pp. 195-211 in Forestry, Economics and the Environment, W.L. Adamowicz, P. Boxall, M.K. Luckert, W.E. Phillips, and W.A. White (eds.). Wallingford, U.K.: CAB International.

Berrens, R., D. Brookshire, M. McKee, and C. Schmidt. 1998. Implementing the safe minimum standard. Land Economics 74:147-161.

Bingham, G., R. Bishop, M. Brody, D. Bromley, E. Clark, W. Cooper, R. Costanza, T. Hale, G. Hayden, S. Keller, R. Norgaard, B. Norton, J. Payne, C. Russell, and G. Suter. 1995. Issues in ecosystem valuation: Improving information for decision making. Ecological Economics 14(2):73-90.

Bishop, R.C. 1978. Endangered species and uncertainty: The economics of a safe minimum standard. American Journal of Agricultural Economics 60(1):10-18.

Carpenter, S.R., D. Ludwig, and W.A. Brock. 1999. Management and eutrophication for lakes subject to potentially irreversible change. Ecological Applications 9:751-771.

Carson R., R. Mitchell, M. Hanemann, R.J. Kopp, S. Presser, and P.A. Ruud. 2003. A contingent valuation and lost passive use: Damages from the *Exxon Valdez* oil spill. Environmental and Resource Economics 25:257-286.

Chetty, R. 2003. A new method of estimating risk aversion. NBER Working Paper 9988. Available on-line at *http://www.nber.org/papers/w9988*. Accessed June 15, 2004.

Chichilnisky, G. 1997. The costs and benefits of benefit-cost analysis. Environment and Development Economics 2:202-5.

Ciriacy-Wantrup, S.V. 1952. Resource Conservation: Economics and Policies. First Edition. Berkeley, Calif.: University of California Press.

Claude, H., and M. Henry. 2002. Formalization and Applications of the Precautionary Princicple. Ecole Polytechnique. Available on-line at *http://www.columbia.edu/itc/economics/henry/G6426/pdfs/Pp_aer.pdf*. Accessed January 24, 2005.

Demers, M. 1991. Investment under uncertainty, irreversibility, and the arrival of information over time. Review of Economic Studies 58:333-350.

EPA (U.S. Environmental Protection Agency). 1997. Guiding Principles for Monte Carlo Analysis. EPA/630/R-97/001. Washington, D.C.: Risk Assessment Forum.

Epstein, L. 1980. Decision making and the temporal resolution of uncertainty. International Economic Review 21:264-83.

Farmer, M.C., and A. Randall. 1998. The rationality of a safe minimum standard. Land Economics 74(3):287-302.

Fisher, A., and M.W. Hanemann. 1986. Option value and the extinction of species. Volume 4 in Advances in Applied Microeconomics, V.K. Smith (ed.). Greenwich, Conn.: JAI Press.

Fisher, A.C., and U. Narain. 2000. Global warming, endogenous risk, and irreversibility. CUDARE Working Paper Series No. 908. Berkeley, Calif.: University of California at Berkeley, Department of Agricultural and Resource Economics and Policy.

Freixas, Z., and J.J. Laffont. 1984. On the irreversibility effect. In Bayesian Models in Economic Theory, M. Boyer and R. Kihlstrom (eds.). Amsterdam: Elsevier.

Ghirardato, P., F. Maccheroni, and M. Marinacci. 2002. Ambiguity from the differential viewpoint. Working Paper no. 17/2002, April 2002, Applied Mathematics Working Paper Series, International Centre for Economic Research. Available on-line at *http://www.icer.it.* Accessed June 15, 2004.

Gollier, C., B. Jullien, and N. Treich. 2000. Scientific progress and irreversibility: An economic interpretation of the "precautionary principle." Journal of Public Economics 75:229-253.

Graham-Tomasi, T. 1995. Quasi-option value. In Handbook of Environmental Economics, D.W. Bromley (ed.). Oxford, U.K.: Blackwell.

Gunderson, L.H., C.S. Holling, and S.S. Light. 1995. Barriers and Bridges to the Renewal of Ecosystems and Institutions. New York: Columbia University Press.

Hanemann, M.W. 1991. Willingness to pay and willingness to accept: How much can they differ? American Economic Review 81(3):635-47

Hanemann, M.W. 1994. Valuing the environment through contingent valuation. Journal of Economic Perspectives 8(4):19-43.

Hanemann, M.W., and I.E. Strand. 1993. Natural resource damage assessment: Economic implications for fisheries management. American Journal of Agricultural Economics 75:1188-1193.

Hanemann, M.W. 1989. Information and the concept of option value. Journal of Environmental Economics and Management 16:23-7.

Heal, G.M. 1998. Valuing the Future. New York: Columbia University Press.

Heal, G.M, and B. Kriström. 2002. Uncertainty and climate change. Environmental Resource Economics 23:3-25.

Henry, C. 1974. Option values in the economics of irreplaceable assets. Review of Economic Studies 41:89-104.

Holling, C.S. 1973. Resilience and stability of ecological systems. Annual Review of Ecological Systems 4:1-23.

Jaffe, J., and R. Stavins. 2004. The Value of Formal Quantitative Assessment of Uncertainty in Regulatory Analysis. Available on-line at *http://www.aei-brookings.org/admin/authorpdfs/page.php?id=1045*. Accessed October 6, 2004.

Kahneman, D., amd A. Tversky. 1979. Prospect theory: An analysis of decisions under risk. Econometrica 47:313-327.

Kahneman, D, and A. Tversky (eds.). 2000. Choices, Values and Frames. New York: Cambridge University Press and the Russell Sage Foundation.

Kolstad, C.D. 1996a. Fundamental irreversibilities in stock externalities. Journal of Public Economics 60(2):221-233.

Kolstad, C.D. 1996b. Learning and stock effects in environmental regulation: The case of greenhouse gas emissions. Journal of Environmental Economics and Management 31(1):1-18.

Kunreuther, H., N. Novemsky, and D. Kahneman. 2001. Making low probabilities useful. Journal of Risk and Uncertainty 23:103-120.

Layard, R., and A.A. Walters. 1994. Income distribution. In R. Layard and S. Glaister (eds) Cost-Benefit Analysis, Second Edition. Cambridge, U.K.: Cambridge University Press.

Lee, K.N. 1993. Compass and Gyroscope: Integrating Science and Politics of the Environment. Washington, D.C.: Island Press.

Lovett, G.M., K.C. Weathers, and M.A. Arthur. 2001. Is nitrate in stream water an indicator of forest ecosystem health in the Catskills. Pp. 23-30 in Catskill Ecosystem Health. M.S. Adams (ed.). New York: Purple Mountain Press.

Machina, M.J. 1987. Choice under uncertainty: Problems solved and unsolved. Journal of Economic Perspective 1(1):121-157.

Maskin, E. 1979. Decision-making under ignorance with implications for social choice. Theory and Decision 11:319-37.

Nordhaus, W.D. 1994. Expert opinion on climatic change. American Scientist 82:45-52.

NRC (National Research Council). 1996. Understanding Risk in a Democratic Society. Washington, D.C.: National Academy Press.

NRC. 2000. Watershed Management for Potable Water Supply: Assessing the New York City Strategy. Washington, D.C.: National Academy Press.

Palmini, D. 1999. Uncertainty, risk aversion, and the game theoretic foundations of the safe minimum standard: A reassessment. Ecological Economics 29:463-472.

Pindyck, R.S. 2000. Irreversibilities and the timing of environmental policy. Resource and Energy Economics 22(3):233-259.

Portney, P.R. 1994. The contingent valuation debate: Why economists should care. Journal of Economic Perspectives 8(4):3-17.

Randall, A., and M.C. Farmer. 1995. Benefits, costs, and the safe minimum standard of conservation. In The Handbook of Environmental Economics, D.W. Bromley (ed.). Oxford, U.K.: Blackwell.

Rawls, J. 1971. A Theory of Justice. Cambridge, Mass.: The Belknap Press of Harvard University Press

Ready, R., and R.C. Bishop. 1991. Endangered species and the safe minimum standard. American Journal of Agricultural Economics 72(2):309-312.

Ready, R. 1995. Environmental valuation under uncertainty. In Handbook of Environmental Economics, D.W. Bromley (ed.). Oxford, U.K.: Blackwell.

Roughgarden, T., and S.H. Schneider. 1999. Climate change policy: Quantifying uncertainties for damages and optimal carbon taxes. Energy Policy 27(7):415-429.

Walters, C.J. 1986. Adaptive Management of Renewable Resources. New York: McGraw-Hill.

Walters, C., J. Korman, L.E. Stevens, and B. Gold. 2000. Ecosystem modeling for evaluation of adaptive management policies in the Grand Canyon. Conservation Ecology 4(2):1. Available on-line at *http://www.consecol.org/vol4/iss2/art1*. Accessed March 2, 2005.

7
Ecosystem Valuation: Synthesis and Future Directions

The committee's statement of task (see Box ES-1) identifies a number of specific questions regarding economic methods for valuing the services of aquatic and related terrestrial ecosystems. Chapter 2 sets the stage for the subsequent chapters with a general discussion of the meaning and sources of value, with a decided emphasis on the economic approach to valuation. Chapter 3 then discusses the relationship between ecosystem services and the more widely studied ecosystem functions; it addresses the types and measurement of ecosystem services and the extent of our current understanding of these services. Chapter 4 reviews the principal and currently available nonmarket economic valuation methods. These two chapters assess what is currently known about the underlying ecology (Chapter 3) and the economics (Chapter 4) necessary for conducting ecosystem services valuation. Existing efforts in ecology and economics are then discussed through an examination of multiple case studies in Chapter 5. That chapter also provides an extensive discussion of implications and lessons to be learned from past attempts to value a variety of ecosystem services. Uncertainty and judgments that arise when conducting an ecosystem valuation study and affect the measurement of values are discussed in Chapter 6.

The purpose of this final chapter is to synthesize the current knowledge regarding ecosystem valuation in a way that will be useful to resource managers and policymakers as they seek to incorporate the value of ecosystem services into their decisions. The chapter begins with a list of premises that underlie the committee's view of ecosystem valuation. This is followed by a synthesis of the major conclusions that emerge from the preceding six chapters. The committee then presents a checklist or set of guidelines for use by resource managers or policymakers when conducting or evaluating ecosystem valuation studies. This checklist identifies a number of factors to consider and questions to ask in improving the design and use of such studies. Finally, this chapter provides what the committee feels are the most pressing recommendations for improving the estimation of ecosystem values. As noted previously, although the focus throughout this report is on those services provided by aquatic and related terrestrial ecosystems, the various conclusions and recommendations provided in this report and final chapter are likely to be directly or at least indirectly applicable to valuation of the services provided by any ecosystem.

GENERAL PREMISES

There are several general premises that the committee feels accurately reflect the current state of knowledge about the value and valuation of aquatic ecosystem services. These premises frame the more detailed discussion of major conclusions that follows. The key links embodied in these premises are illustrated in Figure 7-1, which is a more detailed version of Figure 1-3.

1. **Ecosystem structure along with regulatory and habitat/production functions produce ecosystem goods and services that are valued by humans.** Examples include production of consumable resources (e.g., water, food, medicine, timber), provision of habitat for plants and animals, regulation of the environment (e.g., hydrologic and nutrient cycles, climate stabilization, waste accumulation), and support for nonconsumptive uses (e.g., recreation, aesthetics).

2. **In addition, many people value the existence of aquatic ecosystems for their own sake, or for the role they play in ensuring the preservation of plant and animal species whose existence is important to them.** This value can stem from a belief that these species or ecosystems have intrinsic value or from the benefits that humans get from their existence, even when that existence is not directly providing goods or services used by human populations. In some cases, this "nonuse" value may be the primary source of an ecosystem's value to humans.

3. **The total economic value of ecosystem services is the sum of the use values derived directly from use of the ecosystem and the nonuse value derived from its existence.** Use value can be decomposed further into consumptive uses (e.g., fish harvests) and nonconsumptive uses (e.g., recreation).

4. **Human actions affect the structure, functions, and goods and services of ecosystems.** These impacts can occur not only from the direct, intentional use of the ecosystem (e.g., for harvesting resources), but also from the unintentional, indirect impacts of other activities (e.g., upstream agriculture). Human actions are, in turn, directly affected by public policy and resource management decisions.

5. **Understanding the links between human systems and ecosystems requires the integration of economics and ecology.** Economics can be used to better understand the human behavior that impacts ecosystems, while ecology aids in understanding the physical system that is both impacted and valued by humans.

6. **Nearly all policy and management decisions imply changes relative to some baseline and most changes imply *trade-offs* (i.e., more of one good or service but less of another).** Protection of an ecosystem through a ban on or reduction of a certain type of activity implies an increase in ecosystem services but a reduction in other services provided by the restricted activity. Likewise, allowing an activity that is deemed detrimental implies a reduction in some ecosystem services but an increase in the services generated by the allowed activity.

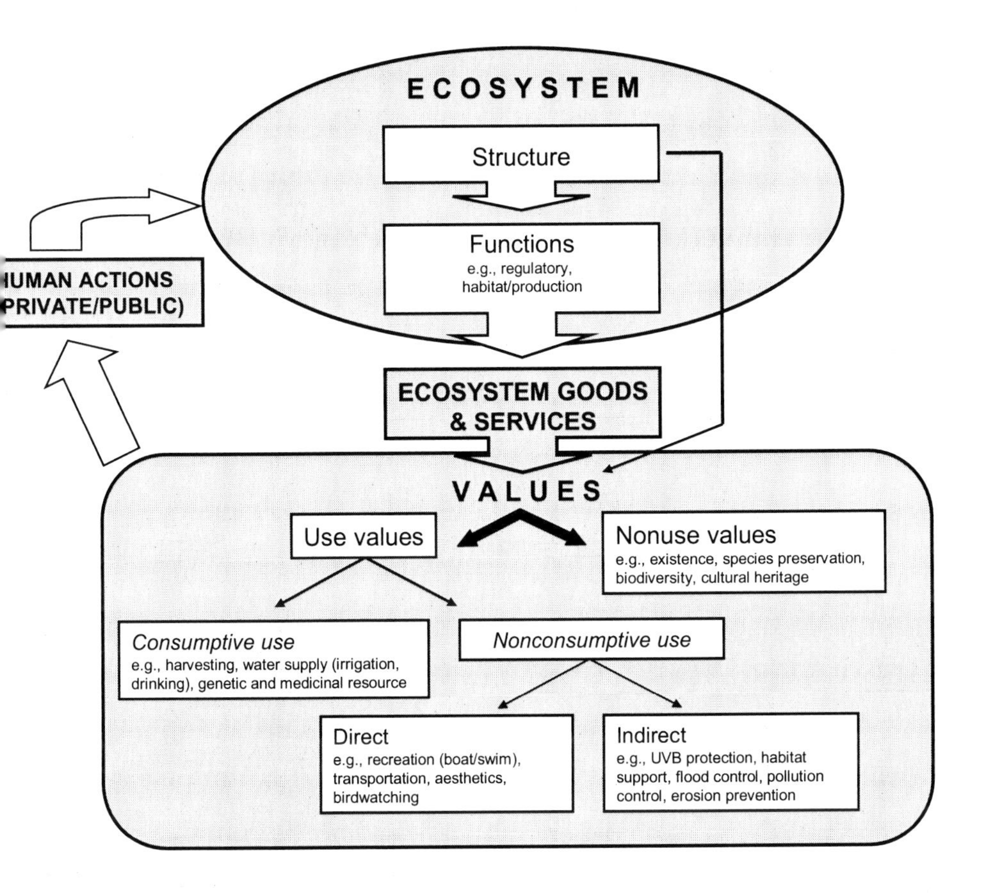

FIGURE 7-1 Connections between ecosystem structure and function, services, policies, and values.

7. Information about these trade-offs—that is, about the value of what has been increased (what is being gained) as well as the value of what has been decreased (what is being forgone or given up)—can lead to better decisions about ecosystem protection. Since decisions involve choices, whenever these choices reflect how "valuable" the alternatives are, information about those values will be an important input into the choice among alternatives.

8. Because aquatic ecosystems are complex, dynamic, variable, interconnected, and often nonlinear, our understanding of the services they provide, as well as how they are affected by human actions, is imperfect and linkages are difficult to quantify. Likewise, information about how people

value ecosystem services is imperfect. Difficulties in generating precise estimates of the value of ecosystem services may arise from insufficient ecological knowledge or data, lack of precision in economic methods or insufficient economic data, or lack of integration of ecological and economic analysis.

9. Nonetheless, the current state of both ecological and economic analysis and modeling in many cases allows for estimation of the values people place on changes in ecosystem services, particularly when focused on a single service or a small subset of total services. Use of the (imperfect) information about these values is preferable to not incorporating any information about ecosystem values into decision-making (i.e., ignoring them), since the latter effectively assigns a value of zero to all ecosystem services.

10. There is a much greater danger of underestimating the value of ecosystem goods and services than over-estimating their value. Under-estimation stems primarily from the failure to include in the value estimates all of the affected goods and services and/or all of the sources of value, or from use of a valuation method that provides only a lower bound estimate of value. In many cases, this reflects the limitations of the available economic valuation methods. Over-estimation, on the other hand, can stem from double-counting or from possible biases in valuation methods. However, it is likely that in most applications the errors from omission of relevant components of value will exceed the errors from over-estimation of the components that are included.

SYNTHESIS OF MAJOR CONCLUSIONS

The preceding general premises collectively imply that ecosystem valuation can play an important role in policy evaluation and policy and resource management decisions. The following section provides a synthesis of the major conclusions regarding ecosystem valuation that emerge from the previous chapters. **It is important to note that this is not intended to replicate or simply restate individual chapter summaries or the conclusions and recommendations of the individual chapters; rather, it is intended to integrate and summarize the broad themes that emerge from these chapters.** The synthesis is organized around these three sets of related questions:

1. What is meant by the value of ecosystem services? What components of value are being measured?
2. Why is it important to quantify the value of ecosystem services (i.e., to undertake valuation)? How will the values that are estimated (i.e., the results of the valuation exercise) be used?
3. How should these values be measured? What methods are available for quantifying values, and what are their advantages and disadvantages?

What Is Being Measured?

There is growing recognition of the crucial role that ecosystems play in supporting human, animal, plant, and microbial populations. There are several published inventories or classification schemes for the goods and services provided by aquatic ecosystems (see Chapter 3). Commonly recognized services include water purification, flood control, waste decomposition, animal and plant habitat, transportation, recreation, hydroelectricity, soil fertilization, and support of biodiversity. However, the complexity of ecosystems remains a barrier to quantifying the links from ecosystem structure and functions to the goods and services that humans value. In addition, although there is now widespread recognition that ecosystem services are "valuable," simply recognizing them as valuable may be insufficient as a guide to environmental policy choice. What is required is some way of comparing these services to other things that are also considered valuable. Without this, the value of ecosystem goods and services will not be given proper weight in policy decisions.

The concept of value, however, has many interpretations. Some notions of value are biocentric; others are anthropocentric. Some are based on usefulness (instrumental value) through contributions to human well-being (utilitarian values); others are based on inherent or intrinsic value and rights. There is a large and growing literature, much of it in the field of philosophy, devoted to defining the nature and sources of such value. To the extent that they represent dimensions that are important to people (and hence affect how they view alternative choices), all types of value can play an important role in environmental decision-making.

Given the committee's charge, this report focuses on the economic concept of value, which is generally defined in terms of the satisfaction of human wants, making it an anthropocentric and utilitarian approach. The economic definition of value postulates a potential substitutability between environmental goods or services and other goods or services that people value. It does not capture intrinsic values that stem from moral premises, although it does capture the value people place on the existence of a species or ecosystem for its own sake. For this reason, the economic concept is not an all-inclusive concept of value. Nonetheless, it is broadly defined to include not only the value derived from direct use of an ecosystem service (use value), but also nonuse values such as existence and bequest values. It thus includes the value of protection "for protection's sake," which is viewed as desirable by many humans. Economic value should not be confused with the much narrower concept of market or commercial value, which reflects only payments made or received through market transactions. In general, economic value includes many components that have no commercial or market basis, including the values individuals place on preservation of ecosystems or species, even when that preservation has no apparent use value.

Economic *valuation* is then the process of quantifying the economic value of a particular *change* in the level of a good or service. A benefit of the use of economic valuation is that it provides a process that is grounded in economic theory and information that can be used to evaluate the trade-offs that inevitably arise in environmental policy choices. By using a common metric (normally monetary) to value changes, it allows a comparison of possible changes and hence facilitates a choice among them. The use of a monetary metric (e.g., dollar equivalent) for quantifying values is based on the assumption that individuals are willing to trade the change being valued for more or less of something else that can be represented by or bought with the metric (i.e., dollars). It thus assumes that the good being valued is in principle substitutable or replaceable by other goods and services.

The economic approach to valuation does not, however, imply a unique measure of the value of a change. The economic value of a change can be defined in two alternative ways: (1) as the amount an individual or group is willing to pay to secure the change (willingness to pay) or (2) as the amount they would have to be compensated to forgo the change (willingness to accept [compensation]). These alternative measures imply different allocations of property rights and have different implications for the role of the income of those affected individuals and groups. In particular, willingness to pay is limited by ability to pay. Although contexts exist in which these two measures can be expected to yield similar values, it is nevertheless the case that without close substitutes for the service that is changing, the two can be expected to yield substantially different values. For unique ecosystems, such as the Florida Everglades, close substitutes are not available and hence the two measures can be expected to differ substantially. Usually, the willingness-to-accept measure, which is not constrained by income, yields a greater value for an improvement than does the willingness-to-pay measure. Economic theory suggests that willingness to accept is appropriate for valuing the removal of a service to which people have a right, whereas willingness to pay is appropriate for valuing the provision of a new service or more of an existing service in a situation where there is no right to receive this service, although in practice most economic valuation exercises use methodologies that measure only willingness to pay. Nonetheless, because willingness to pay provides a lower bound for willingness to accept, it is a sufficient measure for cases in which willingness-to-pay estimates exceed the value of alternatives.

Policy decisions made today and the human actions that they affect can impact an aquatic ecosystem not only now but also far into the future. The temporal dimension of policy impacts stems both from the potential effect on behavior (e.g., inducing long-term behavioral changes or irreversible decisions) and from the dynamic nature of aquatic ecosystems. As a result, the changes that result from a contemporary policy choice and the valuation of those changes must include not only current impacts but future impacts as well. In addition, aggregate value estimates require an aggregation of values over time. This is done typically through the use of discounting and the calculation of net present values. Much of the controversy surrounding the use of discounting stems from a misunderstanding of the distinction between two alternative forms of discounting:

utility discounting and consumption discounting. In particular, even when it is desirable to weigh the well-being of all generations equally (implying a zero utility discount rate), it would still be appropriate to use a positive or negative discount rate for the benefits or costs associated with changes in ecosystem services, if the general availability of these services is expected to change over time. It is important to note, however, that because they are conducted at the present time, all valuation exercises measure the values or preferences of the current generation. To the extent that the preferences of future generations differ, those differences would not be captured in the value estimates.

Why Conduct Ecosystem Valuations?

Why or when might it be important to have an estimate of the value of a change in ecosystem goods or services? As concluded above, such estimates can inform and improve environmental policy and management decisions. Again, simply stating that something has value is insufficient as a basis for policy choice. Rather, it is necessary to have a ranking of alternatives, and estimates of the values of the changes implied by different options can contribute to such a ranking. However, the specific role that valuation plays and its contribution to such processes depends on the specific way in which it will be used (i.e., on the "policy frame"). In particular, the nature of the ecosystem valuation exercise (i.e., how it is conducted and how it is used) will depend on the specific context or problem. One can distinguish between different types of valuation exercises, each of which potentially implies a different type of valuation question, different information needs, different scopes (i.e., types of ecosystem services), and different spatial and temporal scales.

One possible context in which economic valuation plays a key role is in the measurement of damages from ecosystem degradation that has already occurred as a result of some human action. This is a measure of the value of the ecosystem services that have been diminished or lost. Perhaps the most common example of this is natural resource damage assessment (NRDA), which is used to determine the amount of compensation a party responsible for the damages must pay. In this context, a point estimate of damages (rather than a distribution of possible damages) is needed. In addition, it is necessary to have a measure of *total* damages. A partial measure based on a subset of ecosystem services is not sufficient, since as noted previously, not valuing some services is equivalent to assigning those services a zero value.

Rather than valuing a change in ecosystem services that has already occurred, one might instead be interested in valuing a change that could occur. Such a change would typically be linked to a specific policy under consideration. Economic valuation has been used in an attempt to place an estimate on the value of *all* ecosystem services, not as part of a specific policy evaluation, but rather as a means to demonstrate the importance of these services. However, as noted above, economic valuation is designed to estimate the value of a *change* in the provision of services, and the techniques are normally most reliable when applied to relatively small (marginal) changes. Hence, application to very large changes (e.g., "with" and "without" scenarios) often implies an inappropriate use of the techniques.

Some valuation studies do focus on changes in ecosystem services, but still not in the context of a specific policy evaluation. For example, studies can estimate the value of a *hypothetical* change in an ecosystem services (such as a 10 percent increase in commercial fish catch rate). Most economic valuation exercises to date have been of this type. Such analyses do not require a linkage of ecological and economic models, however, because the ecological processes or responses that might generate the hypothetical change are not part of the analysis. Although greatly simplifying the analysis, the use of hypothetical scenarios makes it difficult to link the value estimates with predicted policy impacts.

Ecosystem valuation is most useful as an input into environmental decision-making when the valuation exercise is framed in the context of the specific policy question or decision under consideration; however, this presents several challenges. Such an analysis should have the following components: (1) a way of estimating the *changes* in ecosystem structure and functions that would result from implementation of the policy, (2) a way of estimating the *changes* in ecosystem services that result from the changes in structure and function, and (3) a way of estimating the value of these changes in ecosystem services (see Figure 7-1). This requires an integration of ecological and economic methods and models. The physical impacts of the policy should first be determined, and this should then be translated into a value (e.g., a willingness to pay or willingness to accept compensation for that change). Without this linkage, either it will not be possible to evaluate a specific policy (e.g., it will only be possible to consider hypothetical changes in ecosystem services) or else the subjects of the valuation exercise (e.g., the people whose values are elicited) must implicitly supply their own subjective ecological model (i.e., their own beliefs about the likely effect of the policy on the ecosystem). Thus, the values that are elicited will depend on what these individuals *think* the link between the policy and ecosystem services will or should be.

In the context of aquatic ecosystems, the impact of a given policy on ecosystem services is particularly difficult to estimate, because these ecosystems are complex, dynamic, variable, interconnected and often nonlinear. In addition, linking changes in ecosystem services to values is also difficult, because many of these services are not traded in markets and a large part of the value may stem from nonuse value. However, this task may be easier when applied on a very local scale rather than a regional or global scale, and when it is focused on a subset of services rather than trying to incorporate an exhaustive list of ecosystem services.

Whether the results of a more narrowly focused analysis are sufficient will depend on the specific environmental policy context and the decision criteria that will be used to choose among policy alternatives. Different criteria require different types of information about values. Two contexts in which valuation plays a large role are benefit-cost analysis and cost-effectiveness analysis.

Many federal statutes and regulations require benefit-cost analyses as part of regulatory policy analysis or allow a consideration (as opposed to a comparison) of benefits and costs. In either case, information about the values of changes in ecosystem services needs to be included in the measures of such benefits and costs. In some cases, a partial measure of benefits (i.e., estimating the value of changes in some subset of services) may be sufficient. If a partial measure of benefits exceeds costs, then it is not necessary to have a measure of total benefits because the additional information (i.e., values associated with the additional ecosystem services) would not change the results of the benefit-cost analysis. However, if focusing on only a subset of services yields a benefit measure that is less than cost, it is necessary to consider the value of other services not previously included to see whether inclusion of these benefits changes the results of the analysis.

Economic valuation can also be an important input into environmental policy choice when a particular service (such as water purification) must be provided and one way to provide it is through protection, preservation, or restoration of ecosystem services. In this context, the valuation exercise may simply be part of a cost-effectiveness analysis designed to determine the least-cost means of providing the required good. In such cases, the valuation exercise would only require estimation of the replacement cost—the cost of the next-best alternative means of providing the required service (e.g., the cost of a new water filtration plant instead of increased watershed protection; see also Chapters 5 and 6). In this case, the willingness to pay for the ecosystem service is the amount saved by not having to provide the good or service through alternative means. It is important to emphasize that this does *not* give a measure of the overall value of the ecosystem service, since it reflects only the costs saved by providing the service through ecosystem protection or restoration rather than through an alternative means. In such a context, the value of the ecosystem service is not the cost savings but rather the willingness to pay (or accept compensation) for the improvement in water quality resulting from the protection or restoration of the ecosystem service.

How to Value Ecosystem Services?

Given a decision on what is to be valued and why, the third and last major question to be addressed is how to conduct the economic valuation. The ability to generate useful information about the value of ecosystem services varies widely across cases for at least two reasons. First, knowledge of the link from ecosystem structure and functions to the provision of ecosystem services varies. Some ecosystems, as well as some types of aquatic services, are better understood than others. Second, some types of values (such as nonuse values) are more difficult to estimate than others. For some ecosystem services, such as commercial fish harvests or flood control, the valuation exercise is rather straightforward and uncontroversial. For others, the translation of physical changes in structure or function into values is much more difficult and, in some cases, controversial.

A variety of existing methods can be applied to measuring the economic value of ecosystem services. Some of these methods are based on observed behavior (revealed-preference measures), while others are based on survey responses (stated-preference measures).

Stated-preference methods do not seek to infer values from behavior. Rather, they seek to elicit information about values through survey responses. The two primary types of stated-preferences methods are contingent valuation and conjoint analysis. Contingent valuation was developed to estimate values for goods or services for which neither explicit nor implicit prices exist. Conjoint analysis is conceptually similar to contingent valuation, although it focuses on individual attributes and asks respondents for rankings of alternatives rather than direct statements relating to value. In either case, statistical methods are used to estimate economic values from the stated choices or ranks. Since valuation questionnaires often pose a cognitive problem for respondents, the use of focus groups, individual interviews, and pre-tests can help to ensure that the questionnaires and responses reflect the intended purpose. Although stated-preference methods have come under substantial criticism because they are not based on actual behavior, inclusion of these types of quality control mechanisms in a study design would reduce potential biases and should help in their acceptance and use in environmental decision-making.

Revealed-preference methods, on the other hand, use observed behavior to measure or infer economic values. The main revealed-preference methods that have been used to value ecosystem services are travel-cost, averting behavior, hedonic, and production function models. The travel-cost approaches can capture only the value of ecosystem services that stem from use of a particular site, for example, for recreational fishing. To the extent that an ecosystem change affects recreational fishing at one or more locations (e.g., through a change in fish quantity or quality), the value of the impact on recreational fishing can be estimated using the travel-cost approach. However, the effect of this change on other ecosystem services would not be included in the value estimates derived from the travel-cost method.

Averting behavior models are best suited for valuing ecosystem services related to human health or the provision of related services such as clean water. The premise is that people will change their behavior and invest money to avoid undesirable health outcomes. If degradation of an ecosystem leads to a reduction in the provision of a service such as clean water, the expenditure that individuals would be willing to undertake to avoid the related health impacts—for example, investing in filtration treatment technologies or purchasing alternative water sources—provides a measure of the value of what is lost as a result of the degradation. Application of this valuation approach is currently limited to cases in which the ecosystem service directly impacts individuals, they are aware of any degradation of the ecosystem and its impact on the services provided, and activities can be undertaken to avoid or reduce the negative impacts resulting from the degradation.

The basic premise of the hedonic approach to ecosystem valuation is that the ecosystem services realized by living in a particular location are one attribute that contribute to the value of a house in that location and thus affect its price. Information about how the variation in services across locations (e.g., differences in observable water quality) affects housing prices can be used to infer the value that individuals place on changes in the level of these ecosystem services. Once again, however, the resulting measure of value is only a partial measure, since it captures only the component of value realized as a result of living at a particular location.

All of the above revealed-preference methods have been applied to the valuation of some component or subset of aquatic ecosystem services. In general, however, these applications have not relied on the direct linking of ecological and economic models discussed above. In some cases, the application was to an observed environmental degradation (such as a fish consumption advisory or a water contamination episode). In others, the value of a hypothetical change in ecosystem services was estimated using information about values derived from observed variations in ecosystem services across space or time. As noted above, decoupling the economic and ecological modeling greatly simplifies the valuation exercise. However, such analyses do not provide value estimates that can readily be used directly in policy evaluation and decision-making. What is needed for this purpose is a modeling framework that links the policy to changes in ecosystem structure and functions, which in turn affects the ecosystem services that people value.

The last revealed-preference approach, the production function approach, applies integrated ecological and economic modeling in contexts in which one or more ecosystem services support or protect the production of valued final goods and services. The biological resource or ecological service is treated as an "input" into the economic activity, and like any other input, its value can be equated with the value of its marginal productivity. Although the production function approach is best illustrated in the case where the final output is marketed, as in studying the impact of habitat and water quality on commercial fisheries, it can be used equally well where the final output is not marketed—as would be the case in valuing the impact of habitat and water quality on recreational or subsistence fisheries. Most applications of the production function approach in the past have been for marketed final output. In such cases, the translation of changes in the quantities of outputs (e.g., changes in commercial harvests) into values is greatly simplified because market prices can be used as measures of value, at least for small changes. The more challenging aspect of these studies is determining policy recommendations for managing the aquatic ecosystems supporting the key ecosystem service or services of interest and, in turn, translating the change in ecosystem services into a change in the availability or cost of producing the marketed good or service. Complicating factors include threshold effects and other nonlinearities in the underlying hydrology and ecology of aquatic ecosystems, and the need to consider trade-offs between two or more environmental benefits generated by ecological services. More recent efforts have attempted to expand the integrated ecological-economic modeling underlying production function approaches to account for some of these important effects and trade-offs and to extend the approach to value "multiple" rather than "single" services provided by aquatic ecosystems.

To summarize, in many past applications to aquatic ecosystem services, revealed-preference methods have been restricted to valuing a relatively limited set of services and primarily use values. Even within the category of use values, revealed-preference approaches have been restricted to valuing certain types of ecosystem services and values, such as commercial harvests, recreation, storm protection, habitat-fishery linkages, and erosion control. In contrast, stated-preference methods have been more widely applied to all the different values listed in Figure 7-1. Furthermore, only stated-preference methods can measure certain components of value, such as existence value or other nonuse values, which may comprise a large component of the value of a change in an aquatic ecosystem. Thus, only stated-preference methods are capable of measuring the total economic value of a change (both use and nonuse values).

As noted previously, the credibility of the estimated values derived from stated-preference methods has come under greater scrutiny in academic, policy, and litigation arenas, due mainly to concerns over eliciting values from individuals' responses to surveys. In addition, although stated-preference methods have an advantage in capturing the total value of a change in the overall state of an aquatic ecosystem or in a number of interlinked ecosystem services, such methods are not concerned with how such changes arise from disturbances to the underlying regulatory functions, habitat/production functions, and structure of the ecosystem. By focusing on the values arising from single uses and services of an aquatic ecosystem, revealed-preference methods have also tended to ignore the "interconnectedness" between the functioning aquatic ecosystem and the different values that arise through ecosystem services. However, as Chapters 3-5 of this report have emphasized, this interconnectedness may matter more than previously thought in valuing the different services of aquatic ecosystems, and the challenge to economists and ecologists is to collaborate on developing more integrated ecological-economic modeling of the importance of ecosystem functioning, structure, and habitat/production functions for various ecosystem services of value to humankind.

Regardless of the methods used, there are some issues that should be considered in the design of any ecosystem valuation study. First, unless correct questions are asked at the outset, the information generated by the ecological models may not be very useful if it is not in a form suitable for the application of economic valuation methods (e.g., if it simply lists affected ecosystem services but does not quantify the resulting changes in those services). For their part, economists may apply valuation methods to ecosystem valuation scenarios not built on solid ecological foundations.

Second, as noted above, typically ecological and economic information suitable for estimating reasonably precise values for ecosystem services exists for only a relatively narrow range of services. Limiting the scope of analysis to this subset implies that valuation can be conducted with a relatively high degree of confidence with existing methods. However, limiting the scope of services considered can also lead to problems. For example, a valuation study that analyzes only a subset of ecosystem services may not be sufficient to answer some policy questions. In addition, focusing on impacts of a narrow set of services may fail to capture the interconnectedness of processes within an ecosystem and important feedback effects.

A third key issue is selection of the spatial scale for the valuation exercise. Spatial scale has two important dimensions: (1) the spatial boundaries used to define the relevant ecosystem and (2) the spatial delineation of the relevant group of people whose values will be included in the study. Being too narrow in defining the spatial scale of the ecosystem may mean ignoring important linkages and spillover effects on the production of ecosystem services or in the value of those services. In addition to the physical interconnectedness, there may also be interconnections on the valuation side due, for example, to possible complementarity or substitutability among services either within or across ecosystems.

The appropriate spatial scale for defining whose values to include in an ecosystem valuation study depends on the policy context and the decision-maker's objectives. For example, benefit-cost analysis of federal environmental policies will generally consider the values of all individuals within the United States, even though some individuals in other countries may also be affected by and value the ecosystem change. Likewise, regional analyses might include only the values of individuals within the region. However, narrowing the included population in this way could lead to policy choices (e.g., regarding land development practices) that pass a benefit-cost test at the regional or local level but not at a broader level. This situation is more likely when a substantial component of the value of ecosystem services consists of nonuse values (e.g., existence values) held by individuals outside the region.

A fourth key issue is selection of the appropriate temporal scale for the valuation exercise, which allows for consideration of future impacts of current policy choices. As noted previously, when impacts occur over time, a comparison and aggregation of present and future values is necessary, which is typically done through the use of discounting. In addition, even when present impacts can be predicted fairly accurately, it may be very difficult to predict the value of future impacts, either because the factors determining the link between policy and future ecosystem structure and function are not well understood (e.g., due to complex dynamics) or because the factors affecting the value of ecosystem services (such as income or the availability of substitutes) cannot be predicted with accuracy. Knowing that ecosystem conditions may change or that values may shift places a premium on the ability to learn and adapt through time and to avoid outcomes that cannot be reversed easily. The estimates of values associated with a particular policy change need to reflect the value of any opportunities for learning and adaptation provided by the policy.

Fifth, it is important to distinguish between the estimation of marginal and average values. Marginal values and average values can differ substantially. Evaluating changes typically requires focusing on marginal rather than average values. Most economic valuation techniques (in particular, revealed-preference methods) are well suited to valuing small changes (marginal values) but are more problematic for large changes for at least two reasons. First, marginal values reflect the level of scarcity of a particular good or service, and to the extent that large changes in ecosystems affect scarcity, they can be expected to change marginal values. These changes and the changes in implicit or explicit prices that can result are not captured by the valuation techniques. Second, in terms of ecological impacts, aquatic ecosystems can exhibit threshold effects and large changes can push the system over a threshold, causing regime shifts (e.g., from an oligotrophic to a eutrophic state). These effects would not be captured by the value of small changes that would not be sufficient to trigger such threshold effects.

The preceding discussion suggests that when valuing ecosystem services, extrapolation—across space (e.g., from one ecosystem to another), over time, or over scale (e.g., from small to large changes)—can introduce significant errors in the process and outcome. Nonetheless, some extrapolation may be necessary because of limitations in data, incomplete knowledge of underlying system structures and functions, or limits on resources for conducting the valuation study. In fact, it is likely that many valuation exercises will by necessity rely on benefit transfer methods, which take values estimated in one context and apply them in another context. Such methods should be used cautiously, with a full recognition and acknowledgement of the potential implications of the extrapolation that these methods require.

Because of limitations in data and knowledge (both ecological and economic), estimation of the value of ecosystem services will necessarily involve uncertainty. In addition, economic valuation inevitably involves some degree of subjectivity or professional judgment in framing the valuation problem.

Although unavoidable, uncertainty and the need to exercise professional judgment are not debilitating to ecosystem services valuation. Methods such as sensitivity analysis and Monte Carlo simulation allow an assessment of the *likelihood* or *probability* that the benefits of the policy will exceed its costs, or the conditions under which this would be true. However, this approach does not incorporate individual attitudes toward bearing the risks that stem from uncertainty. An approach that is more consistent with economic theory defines the benefit of a policy change (for example, the willingness to pay for the change) given that the impacts of that change are uncertain. Such a measure incorporates individuals' willingness to take or accept risks, but it is difficult to estimate and has rarely been used in practice. Possible decision criteria or management strategies that explicitly recognize the uncertainty inherent in many decisions regarding ecosystem services are maximin rules, adaptive management, the precautionary principle, and the safe minimum standard. In responding to uncertainty, it is important to recognize the possibility of learning over time and the potential value of flexibility, but not to let incomplete information bias environmental policy decisions in favor of the status quo.

GUIDELINES/CHECKLIST FOR VALUATION OF ECOSYSTEM SERVICES

The preceding synthesis of the report's major conclusions regarding ecosystem valuation suggests that a number of issues or factors enter into the appropriate design of a study of the value of a change in aquatic ecosystem services. The context of the study and the way in which the resulting values will be used play a key role in determining the type of value estimate that is needed. In addition, the type of information that is required to answer the valuation question and the amount of information that is available about key economic and ecological relationships are important considerations. This strongly suggests that the valuation exercise will be very context specific and that a single, "one-size-fits-all" or "cookbook" approach cannot be used. Instead, the resource manager or decision-maker who is conducting a study or evaluating the results of a valuation study should assess how well the study is designed in the context of the specific problem it seeks to address. The following is a checklist to aid in that assessment. It identifies questions that should be discussed openly (and in some cases debated) and satisfactorily resolved in the course of the valuation exercise.

The Policy Frame

- What is the purpose of the valuation exercise?
  - What is the policy decision to be made?
  - What decision criteria will be used and what role will the results of the valuation exercise play?
  - How will the valuation results be used?
  - What information is needed to answer the policy question?
- What is the scope of the valuation exercise?
  - What ecosystem services will be valued?
  - Is it necessary to value only one or a few ecosystem services, or is it necessary to value all services?
- What is the appropriate geographic scale of the valuation exercise?
  - Is it a local, regional, or national analysis?
  - What is the relevant population to include in the value estimates (i.e., whose values to include)?
- How is the valuation question framed?

  - Is it seeking to measure willingness to pay or willingness to accept as a measure of value? Is the question framed in terms of losses or gains?
  - What effect is framing likely to have on the valuation estimates? Is it likely to introduce systematic biases? What effect would alternative frames likely have on the value estimates?
  - What are the advantages and the limitations of the frame that is chosen?
  - Is the frame responsive to stakeholder needs and will it generate information useful to stakeholders?

The Underlying Ecology

- How well understood is the ecosystem of interest?
  - Are the important dynamics understood and reflected in the analysis?
  - Does the ecosystem exhibit important nonlinearities or threshold effects?
  - If the analysis covers multiple ecosystems (e.g., an analysis of a national wetlands policy), how similar or heterogeneous are the included ecosystems?
  - How do important sources of heterogeneity link to important variations in value?
  - Are the interlinkages between different ecological services well understood?
  - Are the complexities of the ecosystem adequately captured by the valuation method? If not, what are the implications for the valuation exercise?
- How precisely can the changes in ecological services that are likely to result from the policy be predicted?
  - Is the level of precision sufficient given the nature and purpose of the valuation exercise?
  - If not, how will the underlying ecosystem effects of the policy be characterized (e.g., as hypothetical changes in services)?

From Ecology to Economic Valuation

- Is the study designed so that the output from the ecological models can be used as an input to the economic models?
  - Does the ecological model give outputs in terms of things that people value?

- With cost-effectiveness analysis (use of replacement cost), are the alternatives providing the same goods or services with the same reliability?

- Given the services to be valued, what existing valuation methods are available?
  - Which seem most appropriate?
  - To what extent is integrated ecological-economic modeling required to capture multiple services and their values, and the "interconnectedness" between the structure and functioning of aquatic ecosystem and the services of value generated?
  - For any given method, which services are captured in the estimated values and which are not?
  - Whose values are captured by the method?
  - Is the measure a "true" measure or an underestimate (e.g., a lower bound) or overestimate of the true value?
    - Under what conditions can it serve as a reasonable proxy for true values?
    - Are those conditions met?
  - Do the values reflect the relevant scarcities?
    - Are there close substitutes for the ecological services being valued (i.e., other means of providing the service)?
    - Does the valuation technique adequately reflect the uniqueness of the ecosystem service or the availability of substitutes?
    - Will the values capture important nonlinearities or possible threshold effects?
- What are the data needs?
  - Are original values to be generated, or are estimates of value generated from previous studies being used ("benefit transfer")?
    - If benefit transfer is to be used, how transferable are the available estimates to the ecosystem services of interest?
  - If original estimates are to be generated, what is the appropriate sample to be used in gathering data?
    - What is the likely effect of the sample choice on the valuation estimates?
    - Have the quality of the data been evaluated adequately?
- How is aggregation handled?
  - Do benefits/values extend over time?
    - Is discounting used to aggregate over time?
    - If so, what discount rate is used?
    - What are the implications for intergenerational resource allocation using alternative decision rules?
  - How are individual values aggregated across individuals?
  - How are values aggregated across services?

- If estimates derived by different methods are combined, is there the potential for double counting? What steps have been taken to avoid double counting?

Uncertainty

- What are the primary sources of scientific uncertainty affecting the valuation estimates?
  - What are the possible scenarios or outcomes?
  - Can probabilities be estimated and with what degree of confidence?
- What methods (such as sensitivity analysis and Monte Carlo simulation) will be used to address uncertainty?
  - Can the results of the valuation exercise be used to calculate not only point estimates but also estimates of the range of values?
  - Do the value estimates capture risk aversion?
- If benefits or values extend over time, are there important irreversibilities?
  - Is it likely that significant learning will occur?
  - Is the value of being able to respond to new information (flexibility) adequately reflected in the valuation estimates?

OVERARCHING RECOMMENDATIONS

The committee recognizes that there are policy contexts in which decisions regarding ecosystem protection, preservation, or restoration will not consider the trade-offs implied by these decisions. For example, decisions may be based on rights-based decision rules, either explicitly or implicitly, where the protection of certain rights is the primary policy goal. In such contexts, valuation of ecosystem services will not play an essential role. However, when policymakers are concerned about trade-offs, then the valuation of services provided by ecosystems can inform the policy debate and lead to improved decision-making. Based on the information provided in this report, the committee has identified a number of overarching recommendations regarding the valuation of ecosystem services in such contexts. These recommendations are based on and in some cases build upon the more specific recommendations presented in the body and summaries of the six previous chapters. Two types of overarching recommendations are included: (1) recommendations for conducting ecosystem valuation and (2) research needs, which imply recommendations regarding future research funding.

Overarching Recommendations for Conducting Ecosystem Valuation

- Where possible, policymakers should seek to value ecological impacts using economic valuation approaches as a means of evaluating the trade-offs involved in environmental policy choices. If the benefits and costs of an environmental policy are evaluated, it is imperative that the benefits and costs associated with changes in ecosystem services be included as well. Without this, ecosystem impacts may not be adequately acknowledged and included (i.e., they will be implicitly given a value of zero). This does not imply that economic values are the only source of value or that decisions should be based solely on a comparison of benefits and costs; other forms of value and other considerations will undoubtedly be important as well. Rather, it implies that an assessment of benefits and costs should be part of the information available to policymakers in choosing among alternatives.
- To provide meaningful input to decision-makers, it is imperative that the valuation exercise be framed properly. In particular, it should seek to value the *changes* in ecosystem services attributable to the policy change, rather than the value of an entire ecosystem.
- A valuation exercise should recognize and delineate explicitly the sources of value from the ecosystem and identify which sources are and which are not captured in the economic approach to valuation. It should acknowledge the implications of excluding sources of value that are not captured by this approach.
- For policy evaluation, it is necessary to go beyond a listing and qualitative description of the affected ecological services. Where possible, ecological impacts should be quantified. Care should be taken to ensure that the quantification reflects the complexities, nonlinearities, and dynamic nature of the ecosystem.
- Economists and ecologists should work together from the beginning to ensure that the ecological and economic models can be appropriately linked (i.e., the output from ecological modeling is in a form that can be used as an input into economic analysis). This requires that ecosystem impacts be expressed in terms of changes in the ecosystem goods and services that people value.
- The valuation exercise should seek to value those goods and services that are most important for supporting the particular policy decision. In addition, the valuation exercise should identify the subset of services for which the economic approach to valuation can be applied with relative confidence, as well as those services or sources of value that are important but for which impacts are less easily quantified and valued. For these, it is imperative to identify the sources of uncertainty relating to the understanding of the relevant ecology, the relevant economics, or the integration of the two.

- Economic valuation of ecosystem changes should be based on the comprehensive definition embodied in the total economic value (TEV; see Chapters 2 and 4) framework. Both use and nonuse values should be included.
- The scope of the valuation exercise should consider all relevant impacts and stakeholders (although in some cases considering only a subset may be sufficient). The geographic and temporal scale of the analysis should be consistent with the scale of the impacts.
- Extrapolations across space (from one ecosystem to another), time (from present impacts to future impacts), or scale (from small changes to large changes) should be scrutinized carefully to avoid extrapolation errors.

Overarching Research Needs

Although much is known about the services provided by aquatic ecosystems and methods for valuing changes in these services exist, the committee believes that there are still major gaps in knowledge that limit our ability to incorporate adequately the value of ecosystem services into policy evaluations. Drawing from the preceding major conclusions and overarching recommendations provided above, the committee has identified the following research needs. The committee believes that funding to address these needs is necessary if progress toward improving the use of ecosystem valuation in policy decisions is to be made, and it recommends that such funding be a high priority.

- Improved documentation of the potential of various aquatic ecosystems to provide goods and services and the effect of changes in ecosystem structure and functions on this provision
- Increased understanding of the effect of changes in human actions on ecosystem structure and functions
- Increased interdisciplinary training and collaborative interaction among economists and ecologists
- Development of a more explicit and detailed mapping between ecosystem services as typically conceived by ecologists and the services that people value (and hence to which economic valuation approaches or methods can be applied)
- Development of case studies that show how these links can be established and templates that can be used more generally
- Expansion of the range of ecosystem services that are valued using economic valuation techniques
- Improvements in study designs and validity tests for stated-preference methods, particularly when used to estimate nonuse values
- Development of "cutting-edge" valuation methods, such as dynamic production function approaches and general equilibrium modeling of integrated ecological-economic systems

- Improved understanding of the spatial and temporal thresholds for various ecosystems, and development of methods to assess and incorporate into valuation the uncertainties arising from the complex dynamic and nonlinear behavior of many ecosystems
- Improvements in the methods for assessing and incorporating uncertainty and irreversibility into valuation studies

Appendix A

Summary of Related NRC Reports

Report	Summary of Content Relevant to Committee's Charge
Restoration of Aquatic Ecosystems: Science, Technology, and Public Policy (1992)	Outlines a national strategy for restoring the nation's aquatic ecosystems. The report discusses aquatic ecosystem functions in a larger ecological landscape greatly influenced by other components of the hydrologic cycle, including adjacent terrestrial systems. Because existing environmental decisions are often fragmented, the report suggests that analysis of aquatic ecosystems should be integrated into the larger ecological landscape, especially in the issue of restoration. It recommends that an aquatic ecosystem restoration strategy be developed for the nation, which includes innovation in financing and use of land and water markets
Assigning Economic Value to Natural Resources (1994)	Explores the major issues and controversies associated with incorporating natural resources and the environment into economic accounts. It also responds to the many discussions on how to make U.S. economic indicators, such as gross national product (GNP), reflect the state of the environment more accurately. The first section of the report, based largely on the results of a three-day workshop of experts in the field, discusses the possibilities and pitfalls in so-called "green" accounting. This is followed by a selection of nine individually authored papers on scientific aspects of related issues
Wetlands: Characteristics and Boundaries (1995)	Establishes a reference definition of wetlands, providing a standard by which regulatory definitions and actions can be assessed, and recommends changes in current U.S. regulatory practices to strengthen objectivity and scientific validity. The report includes a section on functional assessment of wetlands that discusses requirements and existing and future methods of wetlands functional assessments. It recommends analysis of these functions with emphasis on interactions between wetlands and their surroundings and on various classes of wetlands in a specific region
Valuing Ground Water: Economic Concepts and Approaches (1997)	Examines approaches for assessing the economic value of groundwater and the costs of contaminating or depleting this resource. It also suggests a framework for policymakers and managers to use in evaluating trade-offs when there are competing uses for groundwater. The report also discusses a number of approaches to value services of nonmarket goods—in this case, groundwater, which is a unique resource and has no close substitute

Global Environmental Change: Research Pathways for the Next Decade (1999a)	Provides guidance on formulating a framework for future U.S. research on global environmental change. The report recommends improving decisions on global change, more specifically, how to improve the estimation of nonmarket values of environmental resources and their incorporation into national accounts. It also provides suggestions for how to bring formal analyses together with judgments and to better respond to decision-making needs
Nature's Numbers (1999b)	Recommends how to incorporate environmental and other nonmarket measures into the nation's income and product accounts. The report explores alternative approaches to environmental accounting, including those used internationally, and addresses issues such as how to measure the stocks of natural resources and how to value nonmarket activities and assets. Specific applications to subsoil minerals, forests, and clean air illustrate how the general principles can be applied
Ecological Indicators for the Nation (2000a)	Provides a framework for selecting indicators that define ecological conditions and processes, along with recommendations on several specific indicators for gauging the integrity of the nation's ecosystems. Specifically, the report lists five indicators for ecological functioning: (1) *production capacity* as a measure of the energy-capturing capacity of the terrestrial ecosystems; (2) *net primary production*, a measure of the amount of energy and carbon that has been brought into the ecosystem; (3) *carbon storage*, the amount sequestered or released by ecosystems; (4) *stream oxygen*, an indicator of the ecological functioning of flowing-water ecosystems; and (5) *trophic status of lakes*, an indicator for aquatic productivity. In addition to these five indicators, soil condition, land use, and their relationship to ecosystem functioning are also discussed
Watershed Management for Potable Water Supply: Assessing the New York City Strategy (2000b)	Evaluates the New York City Watershed Memorandum of Agreement (MOA), a comprehensive watershed management plan that allows the city to avoid filtration of its large upstate surface water supply. Many of the report's recommendations are broadly applicable to surface water supplies across the country, including those concerning target buffer zones, stormwater management, water quality monitoring, and effluent trading. One of its recommendations is for New York City to lead efforts in quantifying the contributions of watershed management to overall reduction of risk from watersheds from waterborne pathogens

Report	Summary of Content Relevant to Committee's Charge
Assessing the TMDL Approach to Water Quality Management (2001a)	Reviews the scientific basis underlying the development and implementation of the U.S. Environmental Protection Agency's total maximum daily load (TMDL) program for water pollution reduction. The report includes a section on decision uncertainty that discusses a broad-based approach to address water resource problems in order to arrive at a more integrative diagnosis of the cause of degradation
Compensating for Wetland Losses Under the Clean Water Act (2001b)	Evaluates mitigation practices as a means to restore or maintain the quality of the nation's wetlands in the context of the Clean Water Act. The report discusses the array of approaches to and issues associated with wetlands functional assessment in relation to the national goals of "no net loss of wetlands"
Envisioning the Agenda for Water Resources Research in the Twenty-First Century (2001c)	Discusses the future of the nation's water resources and appropriate research needed to promote sustainable management of these resources. The report recommends developing new methods for estimating the value of nonmarketed attributes of water resources
Riparian Areas: Functions and Strategies for Management (2002)	Examines the structures and functioning of riparian areas, including impacts of human activities on riparian areas, the legal status, and the potential for management and restoration of these areas. The report discusses the environmental services of riparian areas; that is, fundamental ecological processes that they provide in the presence or absence of humans. It concludes that few federal statutes refer expressly to riparian values and as a consequence, generally do not require or ensure protection of these areas. Further, it recommends that Congress enact legislation that recognizes the values of riparian areas and directs federal land management and regulatory agencies to give greater priority to their protection

REFERENCES

NRC (National Research Council). 1992. Restoration of Aquatic Ecosystems: Science, Technology, and Public Policy. Washington, D.C.: National Academy Press.

NRC. 1994. Assigning Economic Value to Natural Resources: Washington, D.C.: National Academy Press.

NRC. 1995. Wetlands: Characteristics and Boundaries. Washington, D.C.: National Academy Press.

NRC, 1997. Valuing Ground Water: Economic Concepts and Approaches. Washington, D.C.: National Academy Press.

NRC. 1999a. Global Environmental Change: Research Pathways for the Next Decade. Washington, D.C.: National Academy Press.

NRC. 1999b. Nature's Numbers. Washington, D.C.: National Academy Press.

NRC. 2000a. Ecological Indicators for the Nation. Washington, D.C.: National Academy Press.

NRC. 2000b. Watershed Management for Potable Water Supply: Assessing the New York City Strategy. Washington, D.C.: National Academy Press.

NRC. 2001a. Assessing the TMDL Approach to Water Quality Management. Washington, D.C.: National Academy Press.

NRC. 2001b. Compensating for Wetland Losses Under the Clean Water Act. Washington, D.C.: National Academy Press.

NRC. 2001c. Envisioning the Agenda for Water Resources Research in the Twenty-First Century. Washington, D.C.: National Academy Press.

NRC. 2002. Riparian Areas: Functions and Strategies for Management. Washington, D.C.: National Academy Press.

Appendix B
Household Production Function Models

This appendix discusses in more detail the modeling of household production methods of valuing aquatic ecosystem services discussed in Chapter 4.

Household production function (HPF) approaches involve some form of modeling of household behavior, based on the assumption of either a substitute or a complementary relationship between the environmental good or service and one or more marketed commodities consumed by the household. Examples of these models include allocation of time models for recreation or other activities involving household labor allocation, averting behavior models that account for the health and welfare impacts of pollution, and hedonic price models that account for the impacts of environmental quality on choice of housing.

The underlying assumption in most HPF models is that a household allocates some of its available labor time, and possibly its income, for an activity that is affected in some way by "environmental quality" (i.e., the state of the environment or the goods and services it provides). The household therefore combines its labor, environmental quality, and other goods to "produce" a good or service, but only for its own consumption and welfare (i.e., household utility). By determining how changes in environmental quality influence this household production function and thus the welfare of the household, it is possible to value these changes.

TRAVEL-COST MODELS

Assume a representative household that allocates some of its labor time l for an "environmentally" based activity from which the household derives utility. In this example, assume that this activity is recreational fishing from a mountain lake. The household could be located near the mountains, or it could be traveling from other regions or even different countries to fish in this location.

To capture the effects that this fishing activity has on the household's welfare, one assumes that the household maximizes a utility function U, representing its welfare level and consisting of

$$U = U\left(x, l^u, z\right), \tag{1}$$

where x represents all market-purchased consumption goods, l^u is the time the household spends on leisure, and z is the number of visits the household makes to the mountain lake for fishing. The utility function is assumed to have the normal properties of being concave with respect to its individual arguments.

The number of visits by the household is its internal "production function" for recreational fishing at the mountain lake. These visits may depend on the total time l that the household spends traveling to and fishing at the site, the various goods and services v (e.g., mode of travel, expenditures during traveling, and lodging, fishing gear) that the household uses in these activities, and the overall environmental quality of the lake q that makes it particularly suitable for fishing. Thus, the household's "production" of the number of fishing visits z to the mountain lake is

$$z = z(l, v; q). \tag{2}$$

Production of z is concave with respect to l and v and will shift with changes in environmental quality of the lake q.

Finally, one assumes that the household has an income based on wage earnings and uses that income to purchase all of its expenditures, including money spent on traveling to and from the lake. Given market prices p^x and p^v for commodities x and v, respectively, and representing the market wage rate earned by the household as w, the household's budget constraint is expressed as

$$p^x x + p^v v = w(L - l^u - l) + M, \tag{3}$$

with L being the total labor time available to the household and M representing any nonlabor income of the household (e.g., property rents, interest income, dividends). Equation (3) indicates that the total expenditures of the household must equal its total income.

By assuming that the household maximizes its utility from Equation (1) subject to Equations (2) and (3), one can derive the optimal demands for the time and purchased inputs, l^* and v^*, respectively, that the household spends on recreational fishing. These input demands will depend on the prices faced by the household p^x, p^v, and w, its nonlabor income level M; and the environmental quality of the lake q. By substituting l^* and v^* into Equation (2), the household's demand for the optimal number of visits z^* to the lake for recreational fishing can be expressed as

$$z^* = z(p^x, p^v, w, M; q). \tag{4}$$

Since the number of visits for recreational fishing is observable for all households that engage in this activity, the demand function in Equation (4) can be estimated empirically across households. Moreover, it is a common practice in many travel-cost models to determine whether households would vary their number of visits if any fees for recreational fishing f also changed. As a result, the aggregate recreational visit function in Equation (4) estimated across all households would represent the willingness to pay, or demand, of these households for recreational fishing visits to the lake in response to changes in the fee rate f. Changes in environmental quality of the lake would therefore cause this demand curve to "shift," and the welfare consequences, or value, of this change in environmental quality would be measured by changes in consumer surplus from this shift in the demand for fishing visits.

AVERTING BEHAVIOR MODEL

Instead of z being a desirable commodity such as recreational visits, it could alternatively be "bad," such as the incidence of waterborne disease from use of a microbially polluted aquatic system as a source of domestic water supply. This implies that $\partial U / \partial z < 0$ in the utility function from Equation (1). The household may not be able to allocate its labor time to affect the incidence of the disease, but it may be able to allocate expenditures $p^v v$ that would mitigate the adverse effects of z or reduce its occurrence. For example, these could be purchases of marketed goods (e.g., bottled water, water filters, medical treatment) or payment for access to public services (e.g., improved sewage treatment or water supply). In addition, any improvements in water quality q may also mitigate the incidence of disease. As a result, Equation (2) is now modified to

$$z = z(v; q), \tag{5}$$

where $\partial z/\partial v < 0$ and $\partial z/\partial q < 0$. By assuming that the household's allocation of its labor time is not relevant to this simplified problem, the budget constraint in Equation (3) is now

$$p^x x + p^v v = M, \tag{6}$$

where M is total household income, including any labor income. Maximizing the utility function of Equation (1) with respect to Equations (5) and (6) yields the optimal demand for any mitigating good or service purchased v^*, as a function of prices p^x and p^v; household income M; and water quality q. By substituting latter demand for v^* into the disease incidence function of Equation (5), totally differentiating, and rearranging, one can obtain an estimable reduced form relationship between disease incidence z^* and levels of water quality q.

HEDONIC PRICE MODELS

Another possibility is that z is a desirable characteristic of certain residential property (e.g., "good" neighborhood, beautiful scenery or views, beachfront), which is in turn influenced by the services of an aquatic ecosystem (e.g., pristine environment, unpolluted water, good beaches, protected coastline). As a consequence, the market equilibrium for this residential property, and in turn its price P, will be affected by the desirable characteristic and, thus, the ecological services and environmental quality q that influences this characteristic

$$P = f(z(q)),\ \frac{\partial f}{\partial z} > 0,\ \frac{\partial z}{\partial q} > 0. \tag{7}$$

For a household purchasing this property, the budget constraint is likely to be

$$p^x x + P = M, \tag{8}$$

where M is again total household income and P is the property purchase. Substituting Equation (8) and $z(q)$ into the utility function of Equation (1) for x and z, respectively; totally differentiating with respect to P and q; and rearranging yield the following condition for optimal choice of any ecological service q that affects the value of the residential property:

$$p^x \frac{\partial U / \partial z \; \partial z / \partial q}{\partial U / \partial x} = \frac{dP}{dq}. \tag{9}$$

That is, the marginal willingness to pay for an improvement in environmental quality q must equal its marginal implicit price in terms of the impact of q on property values. Estimation of the hedonic price function in Equation (7) will allow this implicit price to be calculated.

Appendix C
Production Function Models

This appendix provides technical details on the modeling of production function approaches to valuing aquatic ecosystems discussed in Chapter 4.

The general production function (PF) approach of valuing the support and protection that environmental goods and services provide economic activity consists of the following two-step procedure (Barbier, 1994):

1. The physical effects of changes in a biological resource or ecological service on an economic activity are determined.
2. The impact of these environmental changes is valued in terms of the corresponding change in marketed output of the relevant activity. In other words, the biological resource or ecological service is treated as an "input" to the economic activity, and like any other input, its value can be equated with its impact on the productivity of any marketed output.

More formally, if h is the marketed output of an economic activity, then it can be considered a function of a range of inputs:

$$h = h(E_i \ldots E_k, S). \qquad (1)$$

For example, the ecological service of particular interest could be the role of coastal wetlands, such as marshlands or mangroves, in supporting offshore fisheries through serving as both a spawning ground and a nursery for fry. The area of coastal wetlands S may therefore have a direct influence on the marketed fish catch h, which is independent from the standard inputs of a commercial fishery $E_i \ldots E_k$.

There are generally two approaches currently in the literature for valuing the welfare contribution of changes in the ecological service S, which are referred to as *static* and *dynamic* approaches (Barbier, 2000). In static approaches, the welfare contribution of changes in the environmental input is determined through producer and consumer surplus measures of any corresponding changes in the one-period market equilibrium for the output h. In dynamic approaches, the ecological service is considered to affect an intertemporal, or "bio-economic," production relationship. For example, a coastal wetland that serves as breeding and nursery habitat for fisheries could be modeled as part of the growth function of the fish stock, and any welfare impacts of a change in this habitat support function can be determined in terms of changes in the long-run equilibrium conditions of the fishery or in the harvesting path to this equilibrium.

STATIC MODELS

To illustrate a static model, the wetland habitat-fishery linkage analysis pioneered by Ellis and Fisher (1987) and Freeman (1991) is used below. Assume that in Equation (1) there is only one conventional input or that all inputs can be aggregated into one unit (e.g., fishing "effort," denoted as E). The commercial fishery will seek to minimize the total costs of fishing C:

$$C = wE \,, \tag{2}$$

where w is the unit cost of effort.

The fishery will choose the total level of effort E that will minimize costs in Equation (2) subject to the harvesting relationship in Equation (1). This will lead to an optimal effort level E^*, which is a function of the harvest h per unit cost w and the area of coastal wetlands that support the fishery S (i.e., $E^* = E[h, w, S]$). Substituting this relationship into Equation (2) yields the optimal cost function of the fishery:

$$C^* = C(h, w, S), \ \frac{\partial C}{\partial h} > 0, \ \frac{\partial C}{\partial S} < 0 \,. \tag{3}$$

The change in costs as harvest changes is the standard marginal cost, or supply, curve of the fishery. It has the normal upward-sloping properties for any marketed supply; that is, the fishery faces increasing marginal costs as it supplies more harvested output to the market. However, as shown in Figure 4-1, an increase in wetland area leads to a downward shift of the supply curve. As a result, the marginal cost of supplying a given level of harvest will fall. More wetland habitat increases the abundance of fish and therefore lowers the cost of catch. Also illustrated in Figure 4-1 is that a new market equilibrium and price P of fish will occur, where price equals the new marginal cost (i.e., $P = \partial C / \partial h$). The welfare gains from an increase in the habitat-fishery ecological service that occurs as an increase in S can be measured by the increase in consumer and producer surplus in the market for fish.

Unfortunately, many fisheries are not managed optimally so that all fishermen can agree to maximize joint profits, or equivalently minimize joint profits. Most fisheries have the characteristics of *open access*. That is, any profits in the fishery will attract new entrants until all the profits disappear. Thus, in an open-access fishery, the market equilibrium for catch occurs where the total revenue of the fishery just equals cost (i.e., $Ph = C$). Combining the latter equilibrium condition with Equation (3) yields an average cost relationship:

$$P = \frac{C}{h} = c = c(h, w, S), \frac{\partial c}{\partial h} > 0, \ \frac{\partial c}{\partial S} < 0, \tag{4}$$

where c is the average cost of the fishery. The average costs of supplying more fish to the market are also increasing, and as shown in Figure 4-2, an increase in the wetland habitat will also lower these average costs. However, welfare gains from an increase in this ecological service are now measured by the change in consumer surplus only. Since there are no profits in an open-access fishery, there is no producer surplus gain from the improved ecological service.

DYNAMIC MODELS

A dynamic approach adapts bioeconomic fishery models to account for the role of a coastal habitat in terms of supporting the fishery, usually by assuming that the effect of changes in habitat area is on the carrying capacity of the fish stock and thus indirectly on production. Defining X_t as the stock of fish measured in biomass units, any net change in growth of this stock over time can be represented as

$$X_{t+1} - X_t = F(X_t, S_t) - h(X_t, E_t), \ \frac{\partial^2 F}{\partial X^2} > 0, \ \frac{\partial F}{\partial S} > 0. \tag{5}$$

Thus, net expansion in the fish stock occurs as a result of biological growth in the current period $F(X_t, S_t)$, net of any harvesting $h(X_t, E_t)$, which is a function of the stock as well as fishing effort E_t . The influence of wetland habitat area S_t as a breeding ground and nursery habitat on growth of the fish stock is assumed to be positive, $\partial F / \partial S > 0$, because an increase in mangrove area will mean more carrying capacity for the fishery and thus greater biological growth.

To simplify this analysis, it will be restricted to the open-access case. The standard assumption for an open-access fishery is that the effort in the next period will adjust in response to real profits made in the current period (Clark, 1976). Letting *p(h)* represent landed fish price per unit harvested, w the unit cost of effort, and $\Phi > 0$ the adjustment coefficient, the fishing effort adjustment equation is

$$E_{t+1} - E_t = \phi[p(h)h(X_t, E_t) - wE_t], \ \frac{\partial p(h)}{\partial h} < 0. \tag{6}$$

In the long run, the fishery is assumed to be in equilibrium, and both the fish stock and the effort are constant: that is, $X_{t+1} = X_t = X^A$ and $E_{t+1} = E_t = E^A$. In Equation (5), this implies that any harvesting $h(X^A, E^A)$ just offsets biological growth $F(X^A, S)$. Also, in Equation (6), all of the profits in the fishery are dissipated in the long run, that is, $p(h^A)h^A = wE^A$. The latter expression can be rearranged to solve for the steady-state fish stock X^A in terms of the equilibrium price p^A, effort E^A, and cost w (i.e., $X^A = X[p^A, E^A, w]$). Substituting for X^A in the equilibrium condition for Equation (5) yields the long-run inverse supply curve of the fishery:

$$h^A = F(X^A, S) = h(p^A, S, w), \quad \frac{\partial h}{\partial S} > 0. \tag{7}$$

For an open-access fishery, this equilibrium supply curve is backward-bending (Clark, 1976). However, since coastal wetland habitat is an argument in the growth function of the fishery, the effect of an increase in wetland area will be to shift the long-run supply curve of the fishery downward and thus raise harvest levels. This effect is shown in Figure 4-3, in the case of a loss of wetland area. Welfare losses can be measured by the fall in consumer surplus, which will be greater if the demand curve is more inelastic.

REFERENCES

Barbier, E.B. 1994. Valuing environmental functions: Tropical wetlands. Land Economics 70(2):155-173.

Clark, C. 1976. Mathematical Bioeconomics. New York: John Wiley and Sons.

Ellis, G.M., and A.C. Fisher. 1987. Valuing the environment as input. Journal of Environmental Management 25:149-156.

Freeman, A.M., III. 1991. Valuing environmental resources under alternative management regimes. Ecological Economics 3:247-256.

Appendix D
Committee and Staff Biographical Information

Geoffrey M. Heal, *Chair*, is Paul Garrett Professor of Public Policy and Business Responsibility and professor of finance and economics at Columbia University's Graduate School of Business, and Professor of International and Public Affairs in the School of International and Public Affairs. He has also served as senior vice dean and academic director of the Columbia Business School's M.B.A Program. Previously, he was a professor of economics at the University of Sussex (U.K.). His current research focuses on economics of natural resources and the environment, economic theory and mathematical economics, and resource allocation under uncertainty. Dr. Heal is a member of the Pew Oceans Commission, a director of the Union of Concerned Scientists, and a fellow of the Econometric Society. Dr. Heal received a B.A. in physics and economics from Churchill College in Cambridge, U.K., and a Ph.D. in economics from Cambridge University.

Edward B. Barbier is the John S. Bugas Professor of Economics at the University of Wyoming. Before joining the faculty of the University of Wyoming, he served in the Environment Department, University of York, U.K. and directed the London Environmental Economics Center of the International Institute for Environment and Development and University College, London. Dr. Barbier's current research includes natural resources and economic development, economic valuation and use of wetlands, land degradation issues, trade and the environment, and biodiversity loss. He earned a B.A. in economics and political science from Yale University; an M.Sc. in economics from the London School of Economics and Political Science, U.K.; and a Ph.D. in economics from the University of London.

Kevin J. Boyle is Distinguished Maine Professor of Environmental Economics at the University of Maine. Dr. Boyle's research interests are in understanding the public's preferences for environmental and ecological resources and responses to environmental laws and regulation. In particular, his work focuses on estimation of economic values for environmental resources that are not expressed through the market. Dr. Boyle has served as associate editor of the Journal of Environmental Economics and Management and of Marine Resource Economics. He has a B.A. in economics from the University of Maine, an M.S. in agricultural and resource economics from Oregon State University, and a Ph.D. in agricultural economics from the University of Wisconsin.

Alan P. Covich is a professor and director of the Institute of Ecology at the University of Georgia. He was previously a professor in the Department of Fishery and Wildlife Biology at Colorado State University and in the Department of Zoology at the University of Oklahoma. Dr. Covich's research focuses on ecosystem functioning in temperate and tropical streams, including assembly of food webs, predator-prey dynamics and chemical communication, and cross-site comparisons of drought impacts on drainage networks. For the past 16 years, he has conducted research in the Luquillo Experimental Forest Long Term Ecological Research (LTER) site in Puerto Rico. Dr. Covich is a past president of the North American Benthological Society and the American Institute of Biological Sciences. He has an A.B. from Washington University and an M.S. and Ph.D. in biology from Yale University.

Steven P. Gloss is an ecologist with the U.S. Geological Survey's Southwest Biological Science Center and is based in the school of natural resources at the University of Arizona in Tucson. Dr. Gloss was previously the program manager for biological sciences at the Grand Canyon Monitoring and Research Center in Flagstaff, Arizona and a professor of zoology and physiology at the University of Wyoming. He is a former member of the Water Science and Technology Board (WSTB), served on the National Research Council (NRC) Committee on Grand Canyon Monitoring and Research, and chaired the NRC Committee on the Missouri River Ecosystem Science. Dr. Gloss' research interests include water resources policy and management, aquatic ecology, fisheries science, and conservation of native fishes. He received a B.S. in biology from Mount Union College, an M.S. in biology from South Dakota State University, and a Ph.D. in biology from the University of New Mexico.

Carlton H. Hershner, Jr., is an associate professor of marine science at the College of William and Mary and directs the Center for Coastal Resources Management at the Virginia Institute of Marine Science. His primary research interests are in tidal and nontidal wetlands ecology, landscape ecology, and resource management and policy issues. Dr. Hershner also conducts research in resource inventory procedures, habitat restoration protocols, resource management "expert system" development, and science policy interactions. He recently served as a member of the NRC Panel on Adaptive Management for Resource Stewardship. Dr. Hershner has a B.S. in biology from Bucknell University and a Ph.D. in marine science from the University of Virginia.

John P. Hoehn is a professor of environmental and natural resource economics at Michigan State University. His primary research interests include methods for valuing environmental change, economic analysis of policies and incentives for ecosystem preservation, water quality demands, and natural resource damage assessment. Dr. Hoehn received an A.B. in anthropology from the University of California, Berkeley, and an M.S. and Ph.D. in agricultural economics from the University of Kentucky.

Stephen Polasky is the Fesler-Lampert Professor of Ecological/Environmental Economics at the University of Minnesota. He has served as a senior staff economist for the President's Council of Economic Advisers and previously held faculty positions in agricultural and resource economics and economics at Oregon State University and Boston College, respectively. His research interests include biodiversity conservation and endangered species policy, integrating ecological and economic analysis, common property resources, and environmental regulation. Dr. Polasky previously served on the NRC Committee to Review the Florida Keys Carrying Capacity Study. He is currently a member of the EPA's Science Advisory Board Environmental Economics Advisory Committee. He received a B.A. from Williams College and a Ph.D. in economics from the University of Michigan.

Catherine M. Pringle is a professor at the Institute of Ecology of the University of Georgia. Her research areas are aquatic ecology, tropical ecology, conservation biology, nutrient cycling, and effects of environmental problems on the ecology of aquatic ecosystems. Her main research sites are at La Selva Biological Station in Costa Rica, the Luquillo LTER site in Puerto Rico, and the Coweeta LTER site in North Carolina. She is past president of the North American Benthological Society and chair of the Ecological Society of America's Sustainable Biosphere Initiative Advisory Committee. Dr. Pringle received her B.S. in botany and her Ph.D. in aquatic biology from the University of Michigan.

Kathleen Segerson is a professor and head of the Department of Economics at the University of Connecticut. Dr. Segerson previously held a faculty position in agricultural economics at the University of Wisconsin. Her fields of research include environmental and natural resource economics, the economic implications of environmental management techniques, and the use of economic incentives in resource policy. Dr. Segerson previously served on the NRC Committee on Causes and Management of Coastal Eutrophication and is currently a member of the EPA's Science Advisory Board Environmental Economics Advisory Committee. She received a B.A. from Dartmouth College and a Ph.D. in agricultural economics from Cornell University.

Kristin Shrader-Frechette is the O'Neill Professor of Philosophy and concurrent professor of biological sciences at the University of Notre Dame. Dr. Shader-Frechette previously held professorships at the University of Florida and the University of California, St. Barbara. Her research focuses primarily on environmental ethics and policy, quantitative risk assessment, philosophy of science, and normative ethics. She was an associate editor of *BioScience* until 2002 and is currently editor-in-chief of the Oxford University Press monograph series on Environmental Ethics and Science Policy. She is past president of the Risk Assessment and Policy Association and the International Society for Environmental Ethics. She has served on the Board on Environmental Studies and Toxicology and several NRC committees. Dr. Shader-Frechette received a B.A. in mathematics from Edgecliff College of Xavier University and a Ph.D. in philosophy from the University of Notre Dame. She has completed post-docs in biology, in hydrogeology, and economics.

STAFF

Mark C. Gibson is a senior program officer at the NRC's Water Science and Technology Board (WSTB) and was responsible for the completion of this report. Since joining the NRC in 1998, he has served as study director for six committees, including the Committee on Drinking Water Contaminants that released three reports, the Committee to Improve the U.S. Geological Survey National Water Quality Assessment Program, and the Committee on Indicators for Waterborne Pathogens. He is currently directing the Committee on Water Quality Improvement for the Pittsburgh Region. Mr. Gibson received his B.S. in biology from Virginia Polytechnic Institute and State University and his M.S. in environmental science and policy in biology from George Mason University.

Ellen A. de Guzman is a research associate at the WSTB. She has worked on many NRC studies, including the Committee on Privatization of Water Services in the United States, Committee to Improve the U.S. Geological Survey National Water Quality Assessment Program, and the Committee on Drinking Water Contaminants. She co-edits the WSTB Newsletter and annual report and manages the WSTB web site. She received her B.A. from the University of the Philippines.